Tolley's
Managing Stress in the Workplace

Tolley's
Managing Stress in the Workplace

Carole Spiers MIHE, MISMA

Occupational Stress Consultant
Carole Spiers Group
International Corporate Wellbeing Consultants

Members of the LexisNexis Group worldwide

United Kingdom	LexisNexis UK, a Division of Reed Elsevier (UK) Ltd, 2 Addiscombe Road, Croydon CR9 5AF
Argentina	LexisNexis Argentina, BUENOS AIRES
Australia	LexisNexis Butterworths, CHATSWOOD, New South Wales
Austria	LexisNexis Verlag ARD Orac GmbH & Co KG, VIENNA
Canada	LexisNexis Butterworths, MARKHAM, Ontario
Chile	LexisNexis Chile Ltda, SANTIAGO DE CHILE
Czech Republic	Nakladatelství Orac sro, PRAGUE
France	Editions du Juris-Classeur SA, PARIS
Germany	LexisNexis Deutschland GmbH, FRANKFURT and MUNSTER
Hong Kong	LexisNexis Butterworths, HONG KONG
Hungary	HVG-Orac, BUDAPEST
India	LexisNexis Butterworths, NEW DELHI
Ireland	Butterworths (Ireland) Ltd, DUBLIN
Italy	Giuffrè Editore, MILAN
Malaysia	Malayan Law Journal Sdn Bhd, KUALA LUMPUR
New Zealand	LexisNexis Butterworths, WELLINGTON
Poland	Wydawnictwo Prawnicze LexisNexis, WARSAW
Singapore	LexisNexis Butterworths, SINGAPORE
South Africa	LexisNexis Butterworths, Durban
Switzerland	Stämpfli Verlag AG, BERNE
USA	LexisNexis, DAYTON, Ohio

Reprinted 2003

A CIP Catalogue record for this book is available from the British Library.

ISBN 0 754 51269X

Typeset by Columns Design Ltd, Reading, England
Printed and bound in Great Britain by Antony Rowe Ltd, Chippenham, Wilts

Visit LexisNexis UK at www.lexisnexis.co.uk

Acknowledgements

The author would like to express her grateful thanks for the professional input and guidance from Walter Brennan, Cathy deLacy, Andrea Eichhorn, David Franey, Jill Haywood, Michael Halpern, Gerry Jackson, Sayeed Khan, Simon Lawton-Smith, Sylvia Mauger, Roger Mead, Charlotte Rayner, Nerina Ramlakhan, Suzanna Rose, Roger Simmons, Jim Stocks, Lisa Thomas, Steve Walter, the Engineering Employers Federation (EEF), the Health & Safety Executive (HSE), and Quest International. Thanks also goes to Philip Kenton for proof reading, and Jonathan Pearce for the invaluable research carried out and editing – and Bill Gates for the track changes facility in Microsoft Word.

There are many who have great gifts and knowledge but those who share their knowledge have the greatest gift of all.

Foreword

Half a million people in Britain believe that they are experiencing work-related stress at a level that is making them ill (2001/02 survey of self-reported work-related illness (HSE) (SW101/2)). One in five individuals think their job is very or extremely stressful ('Stress and Health at Work Study' (HSE)), and in 2001/02, 265,000 people were first aware that they had a stress-related illness (SW101/2).

The Health & Safety Executive's ('HSE') guidance to managers, 'Tackling work-related stress', highlights three broad reasons as to why employers should take action to reduce these numbers, ie ethical, legal and economic.

The publication of this book is therefore most timely and will provide employers with further valuable help in dealing with stress. The HSE is working with partners to develop standards of good management practice which will provide a yardstick against which employers can gauge their performance in tackling a range of key stressors – and this book fits well alongside that work.

Elizabeth Gyngell

Preface

Stress is an inherent, albeit unwelcome, feature of the human condition. Not surprisingly, therefore, there are already many books on the subject – so why another one, and what makes this book different from others? Although some management volumes may simply end up gathering dust, this one is intended to be a practical and user-friendly handbook that I hope will sit on your shelf as a reference manual, to be dipped into whenever required.

Excessive work-related pressure can cause dysfunction in both employees and the organisations for which they work. This book is intended to offer specific guidance, not only to human resources and occupational health professionals, but to all those whose roles may involve the need to help manage stress within their organisations. Many of us are told to 'get on with the job' but are not necessarily given the correct tools with which to do so. Hopefully, this book will provide a guide to practical solutions regarding stress-related issues that can occur within the workplace.

A major aim of this book is to explain the nature of stress, its causes and effects. In addition, it looks at the problems that stress brings to organisations and individuals, as well as providing a variety of options for interventions at both a personal and organisational level. The volume of recent employment legislation can make it difficult for employers to ensure that they are complying with the law. The book therefore looks in detail at the legal liabilities of employers and the proactive steps they can take to ensure that they are demonstrating a 'duty of care'.

Stress can of course impact on all of us, eg company executives, office and manual workers, professionals, doctors and dentists, hospital staff, housewives and mothers, analysts, teachers, even whiz kids (and ordinary kids). In fact, just about everyone you can think of will have to deal with undue pressure at some point in their lives. What we need to know, therefore, is how to correctly manage this. We need this knowledge because while pressure can be a potent force in propelling us forward, if that pressure turns into stress, it can lead to acute or chronic illness, and in extreme circumstances, early retirement or possibly even premature death.

When stress is adverse and prolonged, it can have far-reaching consequences, not only for the individual concerned but also for the organisation in which s/he works. Engineers are trained to recognise and identify the early signs of weakness or undue wear in machines, buildings or bridges, long before these result in failure, overload or collapse. This is termed preventive maintenance and is exactly the approach that we need to adopt with ourselves, the people with whom we work, and those for whom we are responsible for within our organisations.

Stress itself is not a disease but it can be a contributory factor in the development of serious ill health and possibly permanent disability in those adversely affected by it. Chronic, prolonged pressure can open cracks in our natural defences, which once opened may be difficult to repair. As with a hairline crack in a cylinder block, we may be able to continue – but only slowly and for a finite time before the crack inevitably develops into a fracture and we grind to a complete halt.

Given that stress is so serious, what steps can be taken to identify its presence, pre-empt its occurrence and mitigate its effects? One of the problems is that stress is insidious. It creeps up on us slowly and we often tend to attribute its effects to some other factor in our lives. In addition, there is a pernicious view that stress is harmless and that we thrive on it. In fact, prolonged stress invariably strips away our protective shields and renders us susceptible to harm and disease. That is why we need to recognise the warning signs and act on them accordingly.

Work-related stress

Notwithstanding the contention to the contrary, work-related stress can be managed and, as we will discuss, this is the responsibility of both employer and employee alike. It is therefore important that both parties are aware of what the term 'stress' actually means, and even more importantly, what action they can *jointly* take to identify its causes and to reduce its effects.

In addition to the clear financial argument, there is also a legal requirement for organisations to formulate a stress policy to ensure that effective stress management systems are in place. If stress is detected, its causes should be identified as soon as possible and the underlying problems addressed. A systematic assessment of the causes of stress within the organisation and the mitigation of any harmful effects can give a dramatic boost to morale, staff relations and productivity levels.

It has been shown over the years that organisations that develop and implement people management policies are more successful in the longer term. Unfortunately, the lack of concern still shown by some companies for employee welfare is often a significant factor in decreased profitability and productivity. Equally, there are still many companies that are reluctant subscribers to the concept of stress prevention, and only see their responsibility as that of minimising the risk of litigation and the consequent exposure to claims for compensation.

'An organisation's most valuable asset is its workforce' is a frequently quoted truism that is all too often ignored. Organisations need to be more aware of the real costs of stress-related sickness and absenteeism and the resultant loss of highly trained personnel. It is clearly far better to retain existing employees who will bring loyalty and commitment to an organisation's future, than to accept high staff turnover, additional recruitment and the retraining costs that this will incur.

Information technology and e-business may give new, more efficient ways of working, but old-fashioned, proven concepts of employee welfare, loyalty, recognition, support and customer value are all still essential to long-term business success. Fortunately, employers are now coming to terms with new legislation and directives (much of which emanate from the European Parliament). Employees are concurrently also becoming more cognisant of their rights, leading to a greater need for stress awareness policies and proactive management initiatives to be put in place.

In the UK, we are fortunate in being a multicultural society. However, with this comes a greater need for sensitivity to the needs of individual employees. People of various ethnic origins bring a wealth of difference and diversity, and this needs to be appreciated and managed and the benefits of these differences properly valued.

Organisational interventions

Interventions such as stress management training and counselling support need to be extended to all employees – from the shop floor right through to senior management. Such measures also need to be endorsed at board level and become part of the culture and philosophy of the organisation.

Managers need to be seen to 'walk and talk the job' in order to appreciate working conditions and the work environment at first hand. Employees frequently tell me that they do not feel valued and that their work goes unrecognised. Management training should emphasise the importance of treating staff with respect and of acknowledging good work rather than always complaining about errors. It is not acceptable for managers to say 'I have an open door policy' if, in the event, they are rarely available when employees need to see them.

Managers need to listen to their employees, as it is important for them to feel and know that their presence and opinions matter. It should be incumbent upon managers to be trained to listen in order that they can motivate staff and develop team spirit. Listening to colleagues, airing and sharing concerns and developing forward-thinking strategies will all help to bring about a proactive, cohesive organisational infrastructure.

Stress management is an integral part of human resource management and needs to be included within the portfolio of skills of every manager with a responsibility for staff. However, if proven principles of 'man management' are not adhered to, and that includes treating all members of an organisation with respect and dignity, the result will be a disillusioned and de-motivated workforce, with all the negative consequences that this inevitably brings.

On a personal note

I have written this book in an attempt to share my 30+ years' experience of working in the field of stress management and industrial counselling in the corporate, public and voluntary sectors. There have been many changes during that time – many for the better and some for the worse. I have seen many disillusioned employees, some of whom have coped effectively, as well as those who have not. A number of these people have not known to whom to turn in times of trouble, because when they did approach their management for assistance, there was no one available and they often regretted having spoken out in the first place.

All of the case studies in this book are taken either from my own personal experience or from those of my colleagues (names have been changed to maintain confidentiality). The target readership includes employers whose responsibility is to identify and manage workplace stress, and those employees who need further knowledge and understanding of this important subject.

This book has been written with the forebearance of my children, and open-ended support from my professional colleagues and friends (in particular my good friends and colleagues Andrea Eichhorn and Michael Halpern) – who have given their untiring support throughout this project, and I gratefully acknowledge their help and expertise. Their willingness to give their valuable time and knowledge has reinforced my long-held views of the value and importance of team spirit and support.

I believe that understanding the strengths of one's colleagues and valuing their capabilities and experience, enables the benefits of these to be utilised to the full. These abilities, together with a willingness to assist one another when required, are at the very core of building successful relationships in the workplace – which at the end of the day, is intended to be the essence of this book.

Carole Spiers
Carole Spiers Group
International Corporate Wellbeing Consultants

About the Author

Carole Spiers MIHE, MISMA, is the Director and founder of the Carole Spiers Group – International Occupational Stress and Corporate Wellbeing Consultants. She is an international presenter and Occupational Stress Counsellor and consultant with over twenty years' experience in the field of stress management. Formerly the chair of ISMA^{UK} (International Stress Management Association) from 2002–2003, Carole is now one of their Vice-Presidents as well as being Deputy CEO of the Andrea Adams Trust (the UK charity dedicated to tackling bullying in the workplace). In her professional capacity, Carole is frequently called upon by the BBC and other media outlets to provide informed comment on stress-related issues, and regularly makes keynote presentations to seminar audiences. Carole lives in London and has two daughters and one son.

Contents

1 The Nature of Stress

Introduction

1.1

> 'The optimist sees opportunity in every danger, the pessimist sees danger in every opportunity'. *Winston Churchill (1874–1965)*

The human stress reaction which occurs in response to the impact of excessive pressure, is a concept which has only become accepted in recent years as a significant factor to be considered in the organisation and management of the workplace.

Over the last few years there has been considerable research into work-related stress and it has now become possible, with confidence, to help employers to understand stress and its associated health problems.

Understanding is complicated by the fact that stress affects individual people in different ways and what may be stressful for one person, may leave another either completely or partially unaffected.

It is also recognised that there is often a stigma attached to any stress-related condition, whereby an employee who is unable to cope adequately with his or her workload may be perceived as being weak. The resultant position being that 'stress' is an area that still tends to be the subject of some misunderstanding and controversy.

In this chapter, we will look at the theory of stress and its accepted definition; why some people are more susceptible to it than others, and also dispute and disprove some of the myths that abound – most notably that 'stress is good for you'!

The mechanics of stress

1.2 From the time of the hunter-scavenger, about two million years ago, stress has helped us to survive danger. As soon as we perceive an impending threat to our safety, our possessions or our space, our brain then sends messages to our nervous system to prepare to either stand and fight, or run to escape from the danger, ie 'fight or flight'.

When Neolithic Man went out to forage for food and came across a sabre-toothed tiger with the same intent, he had to make an instant

[1]decision to either fight with the beast (and possibly lose life or limb) or to make a quick escape. In the split second that the body's fight/ flight response was activated, the sympathetic nervous system would come into play and instantly kick-off the 'fight or flight' process.

These days, although we not meet many tigers, we do meet a wide variety of threats in many other forms and when we are suddenly faced with a threat of danger, our innate survival reaction 'kicks-in' automatically as follows (see also Figure 1):

- The cardiovascular and respiratory systems go into overdrive as a result of messages from the endocrine, immune and central nervous systems.
- Blood circulation and oxygenation increase and the adrenal gland, (which secretes the hormones adrenaline and noradrenaline), can automatically give vastly increased strength during the emergency (eg as in a train crash when a survivor finds the strength to pull a victim out from under the rubble).
- Cortisol, which helps the body adapt to stress, is also released via this gland, which in turn is stimulated by the pituitary gland in the hypothalamus that links the endocrine and nervous systems.

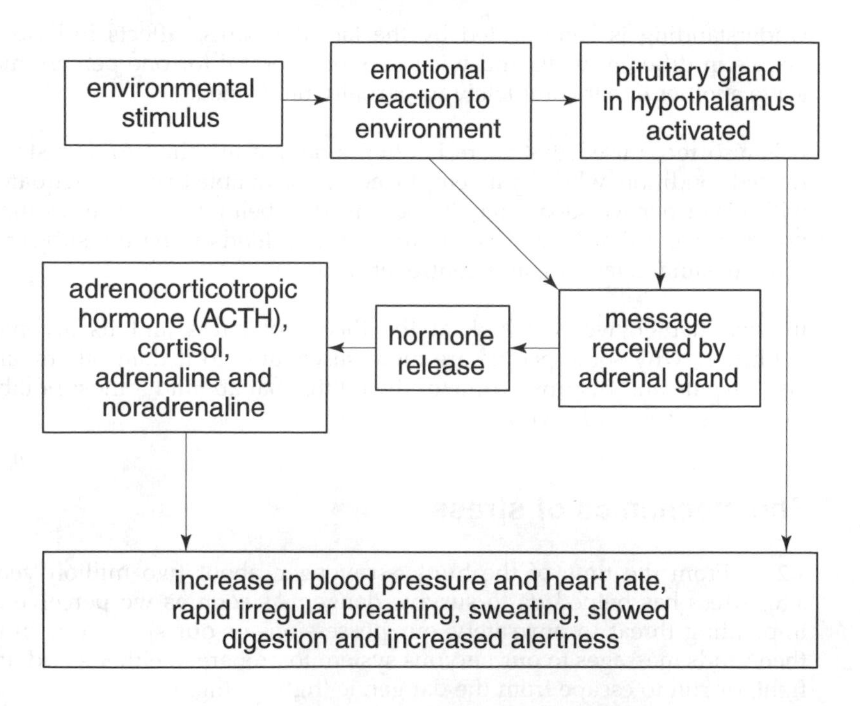

Figure 1: Physiological changes upon the perception of danger

- Glucose and fatty acids are increased in order to provide energy for the muscular activity that will be required to deal with the perceived threat.

- All these reactions occur automatically and very swiftly at the first signs of danger and in accordance with our innate, adaptive mechanism that has evolved to create physiological changes that increase our chances of survival.

Case study

1.3

Mechanics of stress

Philip worked in an advertising agency creating promotional copy for various account executives who would make formal presentations to clients, at meetings arranged at his firm's offices. One day, just a few hours prior to a scheduled presentation with a major client, the wife of the particular Account Executive called in to say her husband had been involved in a car accident. The Sales Director asked Philip to stand in to stage the presentation, as there was no one else available that afternoon.

Philip, who was not used to meeting clients, and certainly not major ones, had no option but to prepare the presentation. There was little problem with the slides that needed just two more hours' work for the finishing touches, but he had no experience of how to speak about the product and the proposed promotion. Having been informed that the client would be attending at 2 pm with his two sales managers, and that they would already be on the plane from Frankfurt, he started to feel extremely nervous.

The more he thought about it the more nervous he became, until half an hour before the appointed time, he actually felt sick. Drinking a glass of water, after returning from the washroom, he could feel his collar becoming damp with perspiration. As the minutes went by towards 2 pm, the sweat started to pour down his face and he could feel his heart thumping.

Knowing there was no alternative, he tried to focus on what he had to do, but his mind refused to clear and he was reduced to wiping his head every few seconds to take away the sweat. No matter what

(cont'd)

he tried to concentrate on, his body simply ignored his wishes, as his autonomic nervous system continued to pump stress hormones within him.

It was less obvious to anyone else, who merely took the impression that he seemed rushed and maybe had too much work to do and insufficient time to finish it. Philip, himself, however, felt as if he was in the hold of a ship rolling about on top of the ocean in a force-eight gale, with no power. His shirt by now was soaking wet as his body was in the grip of a vicious circle of perceived threat, acute anxiety, stress reaction, hormonal response leading to further fear and the resultant physical manifestations.

The client's party arrived on time at 2 pm. Philip welcomed them into the conference room, and hesitatingly started the presentation. After a few minutes, he forgot his apprehension and became completely engrossed in describing the slides and extolling the virtues of the product. Without any conscious action by Philip, the sweating ceased, his heart rate returned to normal and his face to its usual colour.

The presentation went well and the client expressed delight with the proposed campaign. When they had left, Philip tried to understand what exactly it was that had brought him to such a nervous and stressed condition. He now felt confident and relaxed. Only his damp collar and slightly matted hair, reminded him of his previously anxious state, just one and a half hours before.

Today, we are likely to be 'threatened' by late trains, long hours, overwork, harassment, bullying or the thousand and one other causes of stress in the 21st century workplace. However, the threats and prolonged pressures we face are often long-term rather than the instant ones that our forebears were likely to meet – and herein lies the difference.

We currently live in a post-industrial period of transition, the so-called 'information age', which has introduced new technology systems and innovative management philosophies. These have brought about very rapid changes in the workplace and increased expectations of those who work there. Even the speed with which we are now expected to communicate via emails, faxes, etc, leaves us little time for reflection, before we are obliged to move swiftly on to the next task.

In addition to this, we are often required to work against a background of corporate mergers, short-term contracts, downsizing and possible redundancy (see CHAPTER 4), which can pose very real threats to our day-to-day lives. These developments clearly increase the incidence of stress, while on a personal level, should we be unable to cope with our work, any prolonged stress may influence the development of a wide range of medical complaints and diseases (see CHAPTER 9).

The origin of stress as a concept

1.4 If we leave aside the physiological reactions, what does the term 'stress' actually mean? Could it be that we have now found a 'label' that accurately describes one of the more significant side effects inherent in 21st century methods of working and living?

The word 'stress' itself possibly derives from the Latin '*strictus*' meaning tight or narrow or more likely from the Middle English '*stresse*' meaning hardship or distress.

The originator of the biological concept of stress was the Nobel Prize winner Hans Selye (1907–82), a Canadian physician and medical educator born in Austria – and a pioneer in the field of stress research. Selye noted that a person who is subjected to prolonged stress goes through three phases: alarm reaction, stage of resistance, and finally exhaustion. He termed these responses the General Adaption Syndrome (GAS).

- *Alarm reaction*: This is the fight or flight response when the body's resources are mobilised and include the various neurological and physiological responses that occur when confronted with a stressor, ie anything that causes excessive pressure to an individual, whether it be from external sources, inter-personal relationships or internal tensions.
- *Resistance*: If we continue to experience stress, the body enters the second stage, during which it is more able to cope with the original stress but its resistance to any other stress is lowered. If the threat is brief, there are usually sufficient reserves available to adapt.
- *Exhaustion*: After prolonged resistance, energy reserves are depleted and breakdown occurs. We do not have the energy to continue with the adaptation to the stress and the body fails to return to normal. Depending on the individual and the stressor, continued stress can lead to 'burnout' (breakdown), serious disease, organ failure or even death.

Overall, Selye's view of stress was that it was largely physiological. He described the demands placed upon the human body at such times as 'stressors'.

Selye's definition of stress was that it was 'the non-specific response of the body to any demand made upon it' and he referred to 'stress' as 'the rate of wear and tear in the organism'.

There are many critics of Selye's work, and his theories do not fit into modern definitions of stress as his model ignores both the psychological impact of stress upon individuals and their own innate ability to cope, ie to recognise stress and take steps to change their situation.

In the attempt to develop a model of stress that reflects the diversity of its causes and effects, three different but overlapping models, two from America and one from Scandinavia, are now paramount in the understanding of the nature of work-related stress.

- Physiological model

  This looks at the effects of a range of potentially damaging stimuli on the human body, when present over a prolonged period.

- Interaction with environment model

  This looks at an individual's interaction with their working environment and there are two approaches to this model that are most relevant:

  - Person environment fit theory (J P R French, R D Caplan, and R Van Harrison (1982)).
  - Job demands/ decision latitude theory (Karasek (1979)).

- Transactional model (Lazarus (1966))

  This model considers the psychological state involving both cognition and emotion.

The individual nature of stress

1.5 Whilst a modicum of pressure is necessary to ensure that we are kept motivated, excessive pressure can turn into stress that adversely effects our everyday life, our health, our performance and our personal relationships, as can be seen in Figure 2.

Furthermore, prolonged or continuous stress can demoralise and weaken us, reducing our capacity to change the situation.

When we are exposed to stressors, either in a social or working environment, we may react emotionally by becoming anxious or depressed, demotivated, restless or possibly angry. Such exposure also usually affects our behaviour. We may start to drink heavily, over-eat, become withdrawn, smoke to excess or take 'recreational' drugs in an effort to seek release from tension. Some may find themselves driving dangerously too fast or being quick-tempered and irritable with their family, friends or colleagues.

Symptomatically, an individual may not recognise their own change of behaviour and is often likely to deny it when it is brought to their attention. This is because it is usually not possible to see ourselves objectively, particularly when we are under excessive pressure.

Stress-related behaviour patterns, when sustained over a prolonged period, may well cause illness and in severe instances, premature death or

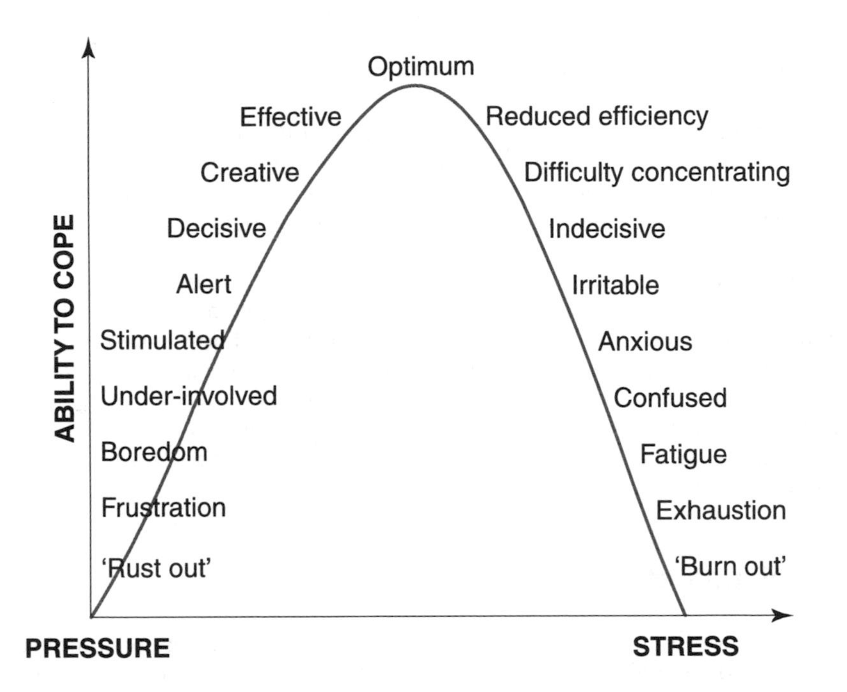

Figure 2: Nature of stress

suicide (as discussed in CHAPTER 7). Other effects may be physiological eg heated arguments (particularly when prolonged), can cause an increase in blood pressure and/ or heart rate; gastric problems can lead to ulcers and muscle tension can cause headaches, stiff neck, back pain and disturbed sleep. All these effects are clearly detrimental to health and well being and can cause dysfunction and possibly serious illness.

We are all at risk, although our individual susceptibility depends on our:

- life experiences and conditioning;
- state of health and beliefs;
- personality type; and
- inherited genetic influences.

Other factors include our:

- age and gender;
- religion, culture and race;
- income and level of education;

- family, social partnership and parental status;

– all of which can determine predisposition.

As we have seen earlier, stress manifests itself differently to each person. What is accepted as a motivating pressure for one person, may manifest itself as stress to another, and what may appear to us to be stressful on one day, may actually be seen as a positive pressure the next.

Our vulnerability to stress can also be influenced by life events that may put us under emotional strain. However, some people are more resilient than others as they have better coping resources and know when to seek support and/ or guidance. In the main, people often need to adapt their behaviour and learn coping skills in order to manage their stress levels. Relaxation, rest, exercise and a good diet, all help to build natural resistance to stress and to boost our immune systems by lowering our reactions to stressful events.

Definitions of stress

1.6 It should already be apparent that stress is an extremely complex issue. Unfortunately, however, definitions on the subject of stress are many and even the experts do not always agree.

Hans Selye defined it as:

> 'The lowest common denominator in the organism's reactions to almost every conceivable kind of exposure, challenge and demand' (in other words the reaction to all kinds of stressors).

More recently, in 2001, the Health & Safety Executive wrote:

> 'Stress is the adverse reaction people have to excessive pressures or other types of demand placed on them. It arises when they perceive that they are unable to cope with those demands. It is not a disease, but if stress is intense and goes on for some time, it can lead to mental or physical ill health, eg depression, nervous breakdown or heart disease'.

Other definitions reflect the nature and complexity of the subject (that is frequently the subject of ill-informed press comment).

The National Institute of Occupational Safety and Health (1999), stated that:

> 'Job stress can be defined as the harmful physical and emotional responses that occur when the requirements of the job do not match the capabilities, resources or needs of the worker. Job stress can lead to poor health and even injury'.

The European Commission in their 'Guidance on work-related stress' (1999) defined it as:

> 'Emotional, cognitive, behavioural and physiological reaction to aversive and noxious aspects of work content, work organisation and work environments. It is a state characterised by high levels of arousal and distress and often by feelings of not coping'.

Some real life examples

1.7 In an attempt to summarise the above, particularly in relation to work-related stress, the consensus appears to be that it is essentially the result of an apparent mismatch between individual capacity and the work required to be carried out.

The following examples are intended to demonstrate the types of situations that can result in stress and how they typically manifest themselves. Many other examples will be explored in detail throughout this book.

- We may have the inclination and belief that we have the resources to carry out a particular task but in fact, the task may in reality exceed our ability to cope with it. This mismatch in capacity may be due to adverse environmental factors such as excessive noise, heat or cold, that can weaken our resolve or hinder our performance. In addition, resources that we thought we could rely on, may not in the event, actually be available to us.

- Too high a self-expectation or the excessive demands that others may make upon us, can engender a need to prove ourselves. Our subsequent inability to cope or to perform at the required standard, can easily cause us stress.

- Having inadequate, internal resources with which to meet a challenge or too little control over a situation, may also cause stress. We may be asked to complete a particular job but have insufficient skills or equipment to do so, thereby making the situation potentially stressful. Once given the appropriate support however, it is possible to view the situation or problem as a challenge that can be evaluated and met, thereby keeping us out of the 'stress zone'.

- We have all experienced occasions when the anticipation of an unwelcome or embarrassing event has caused us great anxiety but in reality the expected problem either did not materialise or the magnitude was less than expected. It is often the case that we overestimate the problem and under-estimate our coping resources.

- Stress can also be caused by the lack of opportunity to utilise one's own ability effectively. For example, someone may seek work to pay off a mortgage or loan and has taken a job that gives them no other satisfaction other than the monthly salary gained. However, the experience of working daily with little or no job satisfaction may well induce feelings of frustration and anxiety that will eventually become stressful.

Case study

1.8

Stressful or not stressful?

John and Bill are two Sales Managers who have each made appointments to visit one of their respective clients. Both encounter a major traffic jam on the motorway and, as they have mobile phones, they each let their clients know that they are going to be late. From there on, however, their reactions are totally different. John does all he can with the resources that are within his control. He telephones the client, listens to music on the radio and accepts that there is little he can do to get to his destination any quicker.

Bill on the other hand, grasps the steering wheel ever tighter, continuously switching lanes in a vain attempt to move forward by leap-frogging the car in front – all the time watching anxiously as the minutes tick by on the clock.

In the end, Bill arrives late and inevitably in a stressed condition. John arrives late but in a controlled state of mind. Both were frustrated at being late but it is Bill who experiences high levels of stress, whilst John does not.

As this example shows, it should be appreciated that innate resistance and reactions to stress vary from person to person and within the person themselves, either from day to day or perhaps from month to month.

As another example of these differing reactions demonstrates, one person may see a new method of working or a company expansion as an opportunity for promotion, whereas another may see it as a threat to their security.

Conversely, an individual who initially sees this type of change as a threat, may well over time, come to appreciate its advantages and may ultimately be even more enthusiastic about the change than another person who initially welcomed it.

The previous definitions and examples illustrate just some of the many and varied facets of stress, its root causes and its effects.

These root causes are referred to as 'stressors', ie anything that causes excessive pressure to an individual, be it from external sources, interpersonal relationships or internal tensions.

We will be looking in detail at these stressors, both those affecting the organisation and those impacting on the employee in CHAPTER 7.

This chapter will now examine the theory of stress, starting with the importance of control.

Locus of control

1.9 There is no universal agreement about the theory of 'control' and its measurement. In *Psychological Monographs* (1966) Vol 80, Julian Rotter, an American clinical psychologist, maintains that the 'locus of control' is the development by an individual of an expectation in respect of whether events that happen to them are caused by internal or external factors, ie by their own mental or emotional processes, by those of others or by fortune or chance. An individual may say, for example 'I will never get this job because the system is against me'. Locus of control is therefore a generalised concept that refers to a person's internal beliefs regarding whether or not they can rely sufficiently on their own resources.

Rotter's contention is that those with a strong internal locus of control believe that success and failure lie ultimately with or within themselves and are consequently less likely to experience stress. Those who believe in an external locus of control subscribe to the theory that life events are controlled by chance or luck or by some other powerful influence. They therefore attribute little or no impact on their lives from their own or internal efforts. Rotter consequently classified all people as either 'internals' or 'externals' depending on where they see the responsibility for their own behaviour, ie the locus of control.

How much or how far we are in control of our response reactions is a matter of much theory and discussion. Dr Richard S Lazarus, Emeritus Professor of Psychology at the University of California, in *Stress, Appraisal and Coping* (1984) wrote that a person's conviction about the level of (his) control is central to the degree to which he feels threatened by events and his ability to influence his inherent coping mechanism.

Lazarus also wrote about the importance of how the individual appraises both the threat and their resources to deal with it. This reinforces the concept of the individual's perception to classify the threat as a problem or a challenge.

Case study

1.10

Perception

Donald visits the dentist – an event he perceives as involving near unbearable pain and discomfort. On his way to the surgery, he experiences increasing muscle tension and nervous feelings in his stomach through apprehension. During the journey, he earnestly wishes for some outside intervention that would prevent his arrival. Donald has been careful not to plan anything for the rest of the afternoon in order to allow himself time to recover from the expected trauma.

Another patient, Richard, also perceives his visit to the dentist as one that will be unpleasant and uncomfortable but unlike Donald, he reminds himself that he has been through it many times before and has always survived. He realises that any pain he may experience will be very short-lived and hardly to be considered 'the end of the world'. He also recognises the importance of looking after his teeth and is therefore prepared to accept the necessity of regular check-ups. He approaches the surgery feeling reasonably relaxed and has planned to return to work in the afternoon.

Modifying expectations

1.11 It can be a very demoralising experience to realise that, under certain circumstances, we seem to be incapable of meeting the demands of everyday living and/ or working. However by modifying our expectations, setting more reasonable goals and re-evaluating our lifestyle, we do have the opportunity to reduce the effect of pressures upon us and to regain some or all of our lost control. If we are able to do that successfully, then we may have learned to live with stress.

To 'cohabit' with life's stressors instead of constantly fighting them is an art, which when mastered, results in a feeling of equilibrium and of being at one with the world in which we live.

It would appear that by setting priorities, modifying our goals, delegating responsibilities and having an accurate appraisal of the stressors that we encounter, we can significantly reduce the pressures upon us, thereby maintaining a sense of control and of well-being.

The role of personality

1.12 We all cope with pressure differently and of course, our personality plays an essential part in this.

- *Type 'A' personalities*

  Typical type 'A' personalities are likely to be hard working, setting high goals for both themselves and others. They will have a great sense of urgency, often setting unnecessary deadlines to drive themselves forward. They will find it difficult to relax and are likely to have a feeling of guilt if they are taking 'time out' for themselves. Evidence of these traits includes high levels of energy, treating everything as time-sensitive, often trying to multi-task unnecessarily, completing sentences for other people and sometimes exhibiting explosive mannerisms such as table banging or shouting. They also find difficulty in delegating, show frequent irritability and suffer fools not gladly.

  Type 'A' individuals exhibit the behaviour styles and associated coping mechanisms that cause them to have a high predisposition to suffer stress-related problems.

- *Type 'B' personalities*

  Type 'B' characteristics are most likely to be able to better cope with stress.

  Typical type 'B' behaviours exhibit almost the reverse traits to type 'A' personalities. Such individuals will be more in tune with themselves – in that they feel no need to impress others with their achievements in order to gain personal satisfaction. They will be much more able to relax and for example, if they engage in sporting activities, they are likely to be in less competitive arenas and regard taking part to be equally as important as winning. They are also much less likely to suffer from anticipatory emotions, such as anxiety.

Meyer Friedman, Professor of Cardiology at San Francisco Medical School, argues that the risk of heart disease is higher in individuals whose personality conforms to his description of type 'A' behaviour, ie characterised by aggressiveness, competitiveness, impatience and a sometime desperate need for advancement and achievement. He maintains that the risk is lower in individuals whose personality conforms to his description of type 'B' behaviour, ie characterised by a relaxed, deliberate, patient and non-competitive approach to life, with a minimum need for advancement and achievement.

Type 'A' and type 'B' behaviour are also descriptive of personality types that identify those individuals who are likely to be susceptible to stress. The original research by Meyer Friedman and Ray Rosenman (1974) found that typically, many patients with coronary heart disease exhibited remarkably similar personal characteristics and that by testing for these attributes, susceptibility to coronary heart disease could be predicted.

It is important to note however, that the majority of us are somewhere on the continuum between 'A' and 'B', rather than at the extremities!

The concept of the 'stress-hardy' personality

1.13 Dr Suzanne Kobasa developed the concept of 'stress hardiness' (or resistance to stress) at City University in New York after carrying out a long-term research study on the impact of stress on senior executives at a leading US communications company, during the time of its enforced break-up.

Employees at the company were either losing their jobs or being reassigned and, over an eight-year period, Kobasa found that there were two different patterns in the way these executives responded to the consequent stress of the events. Staff in one group showed increasing medical and psychological symptoms and visited their doctors more frequently. While in contrast, the second group showed no difference in symptoms during this stressful period as compared to before its onset. Surprisingly, they seemed healthier and more robust. Essentially they rose to meet the challenge and so Dr Kobasa referred to this second group as having a 'stress-hardy' personality.

Kobasa postulated the theory that people possessing a 'hardy' personality trait are very much less likely to suffer stress-related illness after prolonged exposure to adverse circumstances. She described the ability to control circumstance as an important component of 'hardiness', a personality trait that has the effect of being able to ameliorate or mitigate the stress of major life events such as bereavement, marriage, divorce, or a change of job or moving house.

Psychologists have been attempting to isolate the components of the stress-hardy personality ever since. The belief is that the approach to life employed naturally by stress-hardy individuals, incorporates mental and behavioural skills that can be taught to others. Over time, the regular use of these skills can become effective, healthy habits that can replace those that are less functional. According to Dr Kobasa, those who seem to cope with their job stress (ie have 'hardiness' to it) exhibit three specific characteristics:

- *Commitment*

  People who have a strong commitment are fully involved in what they are doing and are unlikely to be easily demotivated. They have a curiosity about whatever is happening to them as opposed to feeling alienated from people or the environment.

- *Control*

  Individuals who feel that they can influence events and surroundings, or that their input is important, have a strong sense of self-achievement and effectiveness. These equate to an internal locus of control, as opposed to feelings of impotence or being a victim of circumstances.

- *Challenge*

People who take on life as a challenge, welcome new situations for the opportunity to learn, grow and develop on a personal level rather than looking at any change as a possible threat.

The common element in all these approaches is that their purpose is to build a foundation of well-being, rather than to treat symptoms. These individual approaches are based upon the belief that real change in automatic reactions (behavioural, mental and physiological) can be taught and can become lifetime, healthy habits.

Case study

1.14

Commitment and challenge

Daniel Libeskind, a Polish American, was born the year after the end of World War II, in Lodz, Poland. His parents were a Jewish couple who had managed to escape from the horrors of the Holocaust by entering the Soviet Union, but over 80 of their relatives lost their lives. The family eventually emigrated to Israel and then to New York, where Daniel studied music and architecture, as well as studying at the University of Essex in the UK.

Daniel subsequently took teaching jobs in Milan, and also at Yale and Harvard universities, but his decision to enter a competition for the design of a new Jewish museum in Berlin changed the course of his life. In July 1989 he received news that he had won this prestigious competition and he chose to move to Berlin to realise the building of his conceptual scheme. The success of his design won him international acclaim, even though he was previously completely unknown as an international architect.

In 2003, Libeskind won the competition to design a scheme on the site of the World Trade Centre in New York – one of the most significant projects of the new 21st century. Initially, he was not even included on the list of those invited to submit designs. Instead he was approached to be one of the judges. Not satisfied with that, Libeskind asked to be allowed to submit his own design.

(cont'd)

> Although basically shy, following the submission of his design he has spared no effort in appearing on television in various countries to lobby support for his concept, which includes a 1,776 ft skyscraper. He also decided to move from Berlin back to New York to exert maximum pressure on the authorities concerned to ensure that his project design is actually built. Perseverance and courage are the hallmarks of this small, fast-talking architect, who only 14 years previously was an unknown academic. Commitment and challenge were, and are, factors integral to his success.

Cognitive behavioural therapy (CBT)

1.15 These three elements of (1) thoughts, (2) behaviour and (3) physiology are also the foundation of cognitive therapy – an approach developed by Aaron Beck in the 1960's, which subsequently developed into cognitive behavioural therapy ('CBT'). The essential concept of CBT is that if you change the way you think and behave, then you will also change your negative feelings. Cognitions (thoughts or thinking style), physiology, emotions and behaviour are all inter-linked.

If cognition is negative, it will affect the other three areas. For example, if someone thinks that they will perform poorly when making a presentation, telling themselves that they will stutter, make mistakes and forget what to say (cognitions), then they may feel very nervous and experience trembling, nausea and fear (physiology and emotion) and consequently perform poorly (behaviour) as shown in Figure 3.

Conversely, someone who prepares well and visualises themselves speaking with confidence (cognitions and behaviour), may well feel a little nervous (emotion) but will counteract the stressed feelings with reminders of their good preparation (cognitions) and is likely to perform effectively (behaviour).

People who can learn to adopt a more rational and positive approach to life will consequently find its inevitable stresses much easier to deal with, as negative thinking causes much stress. A great deal of scientific research has underpinned CBT, which has become a well-respected and trusted psychological approach and one for which there appears to be considerable evidence proving its effectiveness in many areas of stress.

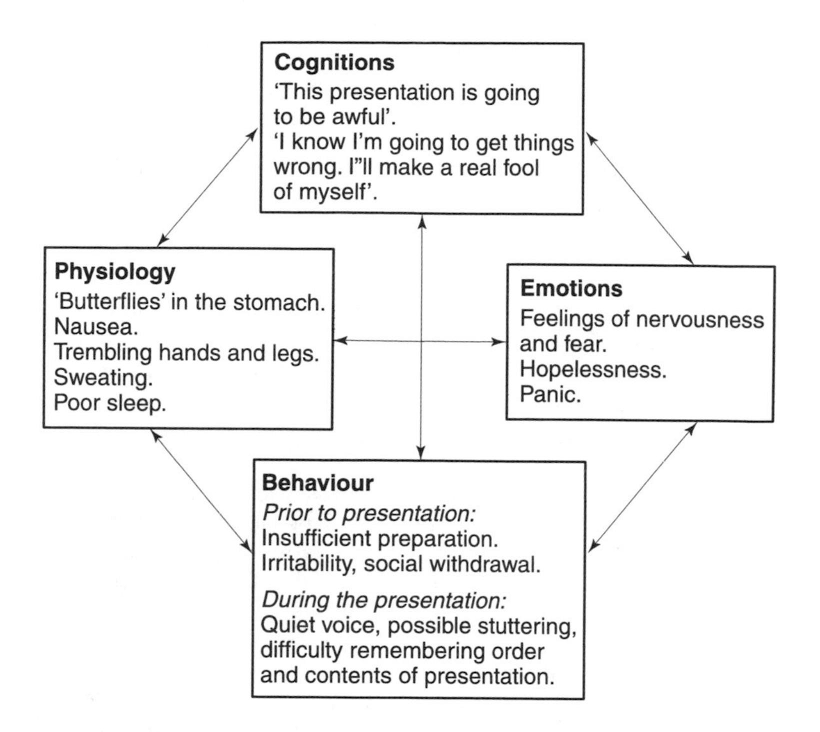

Figure 3: Cognitive behavioural model

Case study

1.16

Cognitive behavioural therapy (CBT)

Marilyn was an industrial chemist, but two years previously after a long-running disagreement with a colleague in her department, she began to suffer from 'social anxiety'. Whenever she found herself in the company of more than one person, she would become tongue-tied and experienced symptoms of panic that were sometimes so bad that they caused in her an urgent desire to run from the room.

Only feeling relaxed when analysing compounds in the laboratory, she learned to avoid, most days, mixing with other staff in the dining room, preferring instead to eat in her office. She often found ways not to attend meetings by taking time-off for sickness. These

(cont'd)

methods of avoidance required being economical with the truth and that eventually led to Marilyn suffering more anxiety and consequent depression.

Clearly, this acquired problem circumscribed her life and potentially jeopardised her promotional career path.

Eventually, her GP had sent her to a cognitive behavioural therapist, with whom she consulted regularly for two months in an attempt to find a solution to her phobic behaviour. The weekly sessions would each last for 50 minutes, and she soon gained an understanding of her physical, emotional and behavioural reaction to the stress she experienced, by the process of examining her thought processes when meeting others, either at work or outside.

Working together with her therapist, Marilyn started to deliberately put herself in positions whereby she would be obliged to talk to other colleagues at work, for instance, by attending continuing professional development (CPD) lectures in the conference room instead of through the use of online facilities.

Gradually, she gained sufficient self-confidence to walk into meetings and maintain eye-contact without undue concern. In time, she became more open again in her verbal dealings with others and experienced a palpable reduction in the occasions where she felt an inability to speak her mind.

Whenever she felt the old feelings starting to rise and the start of panic beginning, she would take some deep breaths and use the techniques she had learned during her CBT sessions. Slowly but perceptively, the adverse feelings ebbed away and her confidence returned. 'Just like turning off a switch, opening a door and allowing the sunshine in', she thought to herself.

Her department head was surprised, but pleased, at her new-found confidence, and after her annual appraisal recommended her for promotion.

Myths and misconceptions

1.17 There are many myths and misconceptions about stress and people may possibly cause significant damage to themselves by basing their lives on these inaccuracies. Three of the most commonly quoted include the following.

'Stress is good for you'

1.18 It is often mistakenly thought that stress is good for people, when in fact long-term stress is, invariably, harmful. A certain amount of pressure can indeed motivate and can therefore be useful but stress is never so. A probable explanation of the myth that people perform well under stress is that in fact they perform well under pressure that is controlled, ie when that pressure is effectively managed.

Pressure is useful when our body and mind are finely tuned in a way that enables them to achieve optimum results and performance (see Figure 2 at 1.5 above). A feeling of nervousness before giving a presentation will often result in increased mental acuity and responsiveness, which will stimulate the audience. The relevant factor in this context is pressure that is within our control. However, if one arrives late, inadequately prepared and the laptop or projector fails to operate properly, then the presentation would indeed inevitably be stressful.

However, the word 'stress' itself, is often applied incorrectly. Many people will use it when they have a temporary work overload, whereas, in fact, stress only occurs when a person perceives (over a prolonged period) that they have insufficient personal resources to cope with a given situation.

We can think of stress as a light switch that our body turns on automatically under specific circumstances. The foundational basis of stress management is the need to learn how to turn the switch off. This is a learning curve that needs to be taught as we have to train ourselves how to manage our body's natural response to perceptions of danger.

Case study

1.19

> **'Stress is good for you'**
>
> David T was a junior barrister in London. Married, with one daughter, he had originally trained as a solicitor, but found that he was more motivated by the idea of sharpening his wits against opposing counsel than by sitting in a law office.
>
> *(cont'd)*

Six months after starting work in chambers in London, the frequency and size of the briefs that David was receiving were slowly increasing. Even so, he still became nervous when standing before a judge for the first time, particularly in a strange courtroom. David had mentioned these unfortunate feelings to the clerk in his chambers, who had seen many faces come and go over the years. The clerk's advice, that 'stress is good for you – it keeps you on your toes', was invariably his reply.

Soon after, during an opening address in the High Court, while defending a client in an important case, David completely forgot his client's name. As he frantically searched his mind, he felt a wave of embarrassment surge over him as those present in court stared curiously at him. Eventually he remembered, but the incident ruined the thrust of his address.

Subsequently, every time David stood up in court, he automatically remembered the image of himself standing, silent and embarrassed. This negative self-image became an ever increasing problem, that threatened his career, and led him to question the wisdom of changing from the comparative safety of the solicitor's office.

At one stage, the steady flow of cases appeared to be drying up. He was even informed by a close friend of rumours that 'David has lost his bottle', and that 'if he doesn't buck his ideas up, he'll very soon be for the chop'. There was no doubt that, for David, the stress of facing a courtroom was certainly not 'good for him'.

As a last resort, David sought help from his GP, who recommended a course of cognitive behavioural therapy with a local psychologist. After a few weeks, David gradually started to regain the confidence he had previously lost.

That was three years ago. Today he is a much sought after barrister in the Queen's Bench Division, but has not forgotten his former problems, and realises very well that undue stress, far from being a benefit, can have serious consequences. He was fortunate in being able to counter it with professional help – before it was all too late.

'Suffering from stress is a sign of weakness'

1.20 Many people think that if they admit to experiencing stress, it is a sign of failure, weakness or ineptitude. An individual working in an organisation where there are imminent redundancies, for example, may well seek to cover up any sign of stress in the belief that they may be regarded as unable to cope with his/ her job and might therefore be regarded as expendable.

Many employees are wary of any mention of stress being noted on their work record in case it might prejudice their chances of promotion and so will not be inclined to discuss the problem with colleagues. This is why it is so important that the workplace culture embraces the notion that to be stressed occasionally is a normal human condition and that to admit to it, initially to oneself, is the first step in modifying the situation or meeting the challenge.

Case study

1.21

Stress not a sign of weakness

Colin Dale had joined a publishing house straight from university, where he had obtained a MA in English literature, nearly fifteen years previously. During that time he had worked his way through the ranks – from sub-editor, through sales, to a position on the board and finally to Chief Executive Officer.

The previous two years in this executive position had been rewarding, but not without problems – some of them, serious. The auditors had found a 'black hole' in the accounts, of about £2.5 million. The subsequent investigation by the fraud squad had uncovered a scheme to defraud the company and a senior department head was on bail awaiting trial.

The fallout from this disclosure was that the company's stock market value had fallen by nearly 25%, making it a prime take-over target. Colin Dale had been well aware of this and had been quoted as saying that the company could absorb the losses. Notwithstanding this, a hostile takeover bid had been made by a competitor and Colin had been forced to call in their financial advisors to discuss a scheme in order to mount a defence.

During the three months following when the problem came to light, Colin had been suffering increasingly from stress that manifested itself in severe headaches that incapacitated him for the rest of the day. This left the remaining board members to steer the ship without the captain, thereby making successful headway, extremely difficult. Colin knew full well that everything that he had worked for over the past 15 years was now at risk and that he himself might lose everything.

(cont'd)

His wife, Janet, was supportive, but she was secretly resentful that the apparent negligence of her husband had jeopardised her life and that of their children. Colin, however, instinctively felt now very insecure, but was determined to put on a brave face lest his fellow directors, and his family, perceived him as being weak. He repeatedly said 'everything will be alright it's not a problem – trust me!'.

In the final analysis, this attitude fooled no one, and he was ignominiously dismissed from the board in favour of a representative from the bank, who held a debenture over the company's assets. Colin was devastated. His health started to deteriorate further to a point where he was hospitalised and medicated with anti-depressants.

He did not know how to deflect the stress, to handle the upheaval to his life or to deal with the adverse effects on his health. After consultation with a counsellor, he was advised to rest for two weeks, take a holiday and then to try to re-order his life and amend his lifestyle.

In the end, Colin came to the decision that he was unable to maintain any position in which there was inherent pressure and he decided to put all his energies into opening a small garden centre on the outskirts of his village. This venture was extremely successful from a financial point of view, and the benefit to his health was incalculable.

'Stressors affect everybody equally'

1.22 An employer or manager should appreciate that not all members of their team will react in the same way to any given problem and that which one person perceives merely as pressure, another may perceive as stress.

Managers and supervisors need to be aware of the symptoms of stress and have the skills and expertise to diffuse or mitigate any issues before they become potentially serious or disruptive.

The facility to be able to talk over difficult situations can often help those employees who are under excessive pressure and managers can often provide the first line of support in encouraging staff to take steps to combat the problem. This could be through an in-house referral, eg human resources and occupational health departments or to an external counselling service, eg employee assistance programme or to another outside agency.

Case study

1.23

'Stessors affect everybody equally'

Gillian and Joy were sisters, and were also the managing team responsible for interior design at a large and international firm of fabric manufacturers – whose customers were primarily in the hotel furnishing sector. Their new boss, Helen, had arrived a month ago from the New York branch, and during her new 'reign' had already imposed strict, new sales targets. Every department felt the impact of this new directive and, what had been previously a relaxed working environment, suddenly changed overnight.

Notwithstanding the fact that they were sisters, Gillian had a remarkably different attitude to that of her elder sibling. Whilst Joy was happy to work at her own pace, Gillian was actually motivated by pressure. In the context of their jobs, this manifested itself by one sister reacting with some apprehension and signs of stress to the new targets, whilst the other sister positively welcomed the challenge.

However, after three weeks the changes became more marked. Their new boss, Helen, called them both into her office to tell them that the interior design department was being axed, and that all future design work for customers would be handled by an outside agency. She had not anticipated that these new arrangements would cause any problem and expected both sisters to accept the offer of new positions within the company.

The sisters were taken by surprise at this sudden and unexpected development and Gillian decided to leave and join a competitor firm, in the city. In fact, she was secretly excited by the new and challenging opportunity. Joy, on the other hand, felt the loss of her position very hard and she felt it to be a reflection on her ability which made her worried about her future, especially after all the years that she had worked in the firm.

In the event, she was offered a sideways move into customer liaison, but felt that as the company appeared to have had lost all interest in her, she had, as a result, lost all interest in the company, and gave in her notice. After a month or so, she decided to go back to college and take a post-graduate degree in teaching, which she did successfully. She now teaches Italian and French to students in London and is happier than she has ever been.

What makes events stressful?

1.24 Negative events are more likely to be stressful than are positive events, although not exclusively so. Uncontrollable or unpredictable events are more stressful than those that are not, and ambiguous situations are often perceived as more stressful than those that are clear-cut.

Overworked people are invariably more stressed than those with fewer tasks to perform and often have difficulty in balancing their home and work lives. They frequently cannot set their priorities correctly and as a result, may only require something comparatively inconsequential to upset their emotional balance.

The key aim of everyone should be to endeavour to maintain an appropriate work-life balance and that means, amongst other factors, making informed choices, setting priorities and employing effective time management techniques.

Life events

1.25 Our experiences of having previously dealt with situations and people, whether adaptively or maladaptively, will make a difference to how we react and cope with our future life events.

The loss of a close relation or life partner is acknowledged as being a highly stressful event, whilst marriage and divorce also rate high on the scale – albeit that marriage is ostensibly, a happy event.

The social readjustment rating scale (SRRS) developed by Holmes and Rahe (1967) lists life events from the most to the least stressful. The scale, as shown in Table 1.1, assigns relative values to a variety of life events with a baseline of 100 points for the death of a spouse and a mid-line baseline of 50 points for marriage. The numerical values indicate the relative impact of stressful events and show the wide range of stressors that affect our lives.

It should be noted however that the scale does not differentiate between happy and sad events, both are simply assigned a relative number of 'stress points'. Moreover it should also be borne in mind that there are critics of this work who make the point that it is the perception of the event that is important rather than the event itself.

Table 1.1: The social readjustment rating scale (SRRS) developed by Holmes and Rahe (1967)

Event	*Scale of Impact*
Death of spouse	100
Divorce	75
Marital separation	65
Jail term	63
Death of a close family member	63
Personal injury or illness	53
Marriage	50
Dismissal from work	47
Marital reconciliation	45
Retirement	45
Change in health of family member	44
Pregnancy	40
Sex difficulties	39
Gain of a new family member	39
Business readjustment	39
Change in financial state	38
Death of close friend	37
Change to different line of work	36
Major mortgage	36
Foreclosure of mortgage or loan	31
Change in responsibilities at work	30
Son or daughter leaving home	29
Trouble with in-laws	29
Outstanding personal achievement	29
Partner begins or stops work	28
Begin or end school	26
Change in living conditions	26
Revision of personal habits	25
Trouble with boss	24
Change in work hours or conditions	23
Change in residence	20
Change in social activities	19
Small mortgage or loan	18
Change in sleeping/ eating habits	17
Change in family get-togethers	16
Holidays	15
Christmas	13
Minor violations of the law	12

It is instructive to appreciate the common thread that links these events. In the vast majority of cases, it is the emotional impact of change that (fortunately) is usually short-lived and self-limiting. However, when we have a severe emotional reaction to an event or circumstance that is prolonged, then this can cause psycho-physiological damage.

Unfortunately, such an occurrence is by no means exceptional in today's often frenetic and pressurised lifestyle. In addition, economic, political and social stressors must also be taken into account. These may include such factors as crime, drugs, increased violence, natural disasters such as flooding, uncertainty, social isolation, media intrusion, together with the sheer pace of life in the 21st century.

The importance of a support network

1.26 In light of the aforementioned, it is essential to be aware that coping with stress can be made demonstrably easier with support from colleagues, family or friends. Strong and caring relationships are of primary benefit in helping any individual learn to cope with pressure and to counteract stress.

These networks can be reinforced by external sources of assistance, ie professional counselling or possibly a strong religious affiliation.

Key learning points

1.27 Stress has many causes and affects people in a variety of ways. In the workplace, stress can have devastating effects, not just on the individual but on the organisation as a whole. Where an employee is suffering any combination of the emotional, physical and behavioural symptoms which stress induces, performance will clearly suffer and this in turn, may well affect others within the organisation.

Although stressful situations are sometimes unavoidable, it is very often possible both for management to foresee and pre-empt their occurrence and also for employees to learn to effectively cope with the consequent pressure. A proactive management culture can avoid the worst effects of stress by means of risk assessment, improved communication, ongoing performance reviews, education and training (as will be explained in later chapters).

- Stress is the reaction to an inappropriate amount of pressure or responsibility when the individual being subjected to these feels inadequate and unable to cope.
- Pressure provides the stimulation and challenge that we use to achieve job satisfaction and self-esteem. An optimum amount of pressure assists performance but if it develops into stress, then performance will be affected.

- Excessive pressure is not 'good for us'.
- To suffer from stress is not 'a sign of weakness'.
- The ability to control circumstance, ie to have some input over events is generally accepted to be an important contributor to a sense of well-being in the individual. This understanding has led to the development of management interventions which assist individuals to cope with pressure when it turns into stress.
- Stress manifests itself in cognitive, psychological, physiological and behavioural changes.
- Every individual responds to stress to a different degree and in their own way.
- It is possible to learn to manage stress successfully, rather than allowing it to overwhelm us to an extent when our physical and mental health is damaged.
- Avoiding the effects of excessive stress requires us to identify those stressors which affect us personally and to learn how to control them, either by ourselves or with professional help.
- What is accepted as a motivating pressure for one person, may manifest itself as stress to another and what may appear to us as stressful one day, may actually be seen as a positive pressure the next.
- Stress is like a light switch. Our mind turns it on automatically but we need to learn how to turn it off.

2 Current Legislation and the Employer

Introduction

2.1

> 'All human beings are born free and equal in dignity and rights'.
> *Universal Declaration of Human Rights 1948, Article 1*

In CHAPTER 1 we looked at the nature of stress and some of the issues involved in arriving at a precise definition of the term. Given the difficulty of achieving this, it will come as little or no surprise that the legal background to stress is no less complicated!

There is a wealth of UK and European legislation, either directly or indirectly related to stress and this chapter will enumerate the main points of the current legislation and the responsibilities of employers that arise from it. This chapter will also examine examples of some of the legal precedents as far as stress-related litigation is concerned, the types of claims for which employers have been brought to court for, and the compensation awarded as a result of such claims.

UK legislation, regulations and codes of practice

2.2 There are three basic reasons why employers need to take action to identify and deal with work-related stress: legal, economic and ethical. Employers therefore need to ensure that:

- Work, or the work environment, does not cause damage to the health or safety of any employee or to any member of the public.
- There is no failure to take appropriate action; the consequences of which can result in litigation costs and potentially substantial compensation awards against the organisation for breach of duty of care.
- Individuals who have been ill, as a result of work-related stress are supported in rehabilitation to work.

In this chapter we will focus, in particular, on the legal position: how the law sees work-related stress; its consequences and what provisions it has made in requiring employers to take proper precautions regarding the health and safety of their employees.

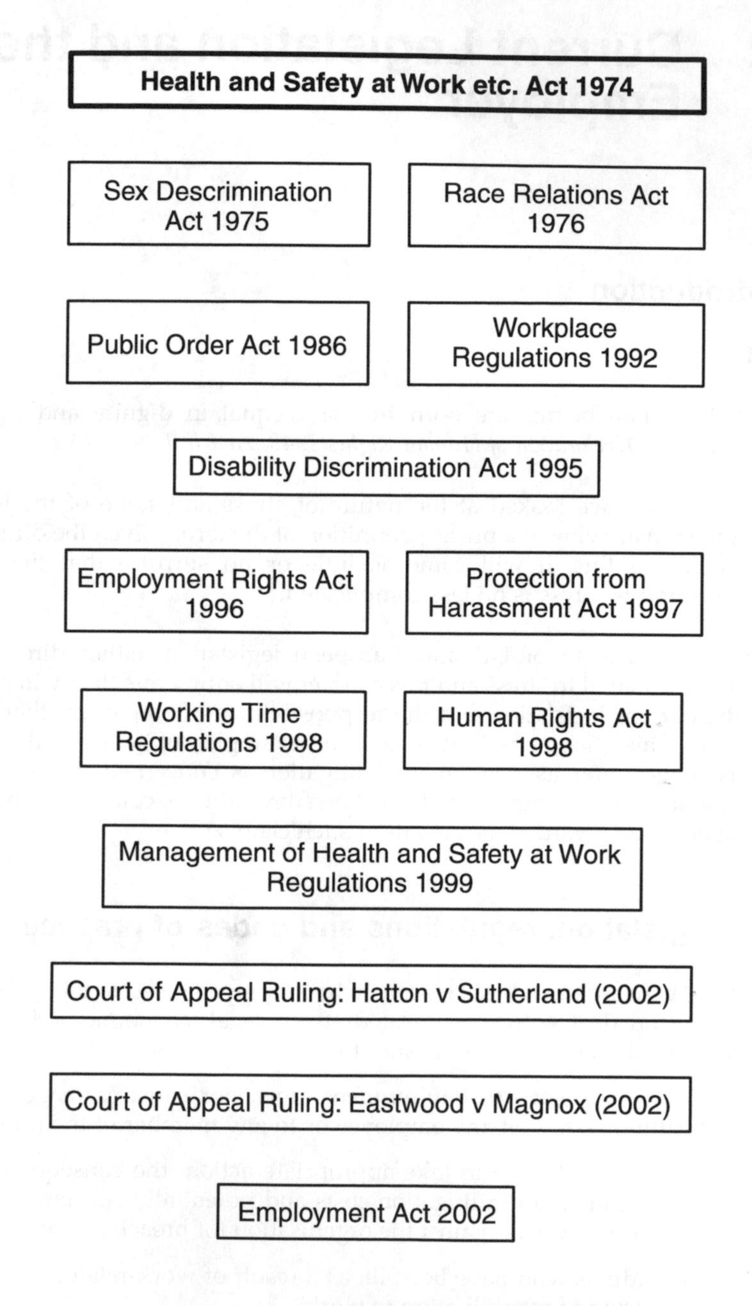
Health and Safety at Work etc. Act 1974
Sex Descrimination Act 1975
Race Relations Act 1976
Public Order Act 1986
Workplace Regulations 1992
Disability Discrimination Act 1995
Employment Rights Act 1996
Protection from Harassment Act 1997
Working Time Regulations 1998
Human Rights Act 1998
Management of Health and Safety at Work Regulations 1999
Court of Appeal Ruling: Hatton v Sutherland (2002)
Court of Appeal Ruling: Eastwood v Magnox (2002)
Employment Act 2002

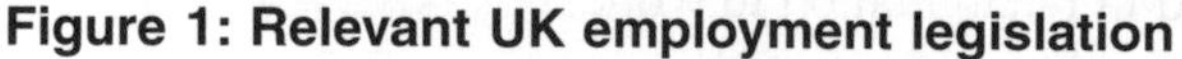

Figure 1: Relevant UK employment legislation

The Health and Safety at Work etc Act 1974

2.3 In the United Kingdom, the *Health and Safety at Work etc Act* came into force in 1974. This is an umbrella piece of legislation which in *section 2* covers the broad principles of the employer's 'duty of care' – to ensure, as far as is reasonably possible, the health (including mental health), safety and welfare of all employees whilst at work and to create safe and healthy working systems. This general duty of care includes pre-emptive action to prevent and control workplace stress.

The 1974 Act requires employers to provide a safe working environment and equipment, in addition to safe methods of work, together with information and training. Inadequate or insufficient provision of any of these can increase the incidence of workplace stress and the risk of stress-related illness.

Where work procedures and systems give rise to risk of stress, the employer has a legal duty to eliminate or control that risk as far as is reasonably practicable. This means that a balance has to be found between the costs, (including time), of making the work system safe and the likelihood of that work practice causing injury or sickness.

Employers must identify and eliminate any foreseeable risk unless the cost of doing so grossly outweighs that risk (ie a cost benefit analysis needs to be undertaken, which should take into account the benefits of reduced absence and sickness, improved morale, productivity gains, etc).

Finally, if a member of staff (or a staff association) brings problems in the workplace to the attention of management, the employer is obliged to investigate and to introduce measures to eliminate or control these issues, where they exist.

UK courts now impose a stringent duty on employers in relation to workplace stress, recognising that it is one of the most important causes of employee sickness and absence.

Employers must provide a safe workplace that is free of potentially dangerous equipment or machinery and, in addition, must also ensure that fellow workers are reasonably competent. In the event of the employer failing to conform to these requirements, then the law could find that a duty of care has been breached and that reasonable care has not been taken to protect the employee from foreseeable injury. Any criminal negligence would be dealt with by the criminal court and compensation might also be claimed through the civil courts.

It must be borne in mind that it is an implied term of every contract of employment that the employer will take appropriate steps to safeguard the health and safety of their employees. This is a contractual obligation.

Management of Health and Safety at Work Regulations 1999

2.4 Work-related stress is also covered by the provisions of the *Management of Health and Safety at Work Regulations 1999 (SI 1999/3242)* (including the Approved Code of Practice).

- *Regulation 3* provides for an assessment to be undertaken by the employer of all risks and potential risks to the health and safety of employees. This is to ascertain the effectiveness of pre-emptive and control measures regarding known hazards including stress. Where such measures exist, comprehensive and relevant information regarding those risks must be provided to the employees.
- *Regulation 3(3)* requires that the assessment be reviewed where there is reason to suspect that it is no longer valid or there has been a significant change within the organisation.
- *Regulation 3(6)* requires that any organisation having more than five employees must record the significant findings and the identification of any particular group found to be especially at risk. Assessment findings should be communicated to all staff.
- *Regulation 4* requires that where control measures do not exist, then they must be formulated and implemented as soon as practicable after the risk assessment is completed.
- *Regulation 6* requires that every employer shall ensure that his employees are provided with such health surveillance as is appropriate having regard to the risks to their health and safety which are identified by the assessment.
- *Regulation 10* requires that every employer shall provide his employees with comprehensible and relevant information regarding the risks to their health and safety as identified by the assessment, together with the preventative and protective measures to be implemented.
- *Regulation 13* requires the employer to take into proper account the individual employee's capabilities when assessing risks and to provide suitable training as necessary.
- *Schedule 1* to *Regulation 4* specifies the general principles of prevention set out in Article 6(2) of Council Directive 89/39/EEC, as follows:
  - Avoiding risks.
  - Evaluating the risks which cannot be avoided.
  - Combating the risks at source.
  - Adapting the work to the individual, especially as regards the design of workplaces, the choice of work equipment and the

choice of working and production methods, with a view, in particular, to alleviating monotonous work and work at a pre-determined work-rate and reducing any adverse effect on health.

- Adapting to technical progress.
- Replacing the dangerous with the non-dangerous or the less dangerous.
- Developing a coherent overall prevention policy that covers technology, organisation of work, working conditions, social relationships and the influence of factors relating to the working environment.
- Giving collective protective measures priority over individual protective measures.
- Giving appropriate instructions to employees.

UK and European law

2.5 A significant proportion of UK law now originates in, or is derived directly from, membership of the EU together with its treaties and directives.

European Union treaties are legally binding and are directly applicable to all EU Member States and are therefore enforceable in UK courts. European Union directives, on the other hand, are required to be implemented by individual states, usually by means of independent legislation or other binding regulation within their respective jurisdictions.

In all cases, EU law takes precedence over UK (or other independent state) law and is the overriding authority – the final court of the EU being the European Court of Justice.

HSE guidelines and the Tokyo Declaration

2.6 The current UK legislation regarding the prevention of work-related stress owes its impetus and rationale to the General Principles of Prevention originally set out in Article 6(2) of Council Directive 89/391/EEC, and the principle of Article 152 of the Treaty of Amsterdam regarding human health protection.

Directive 89/391/EEC was eventually followed by the European Commission's Green Paper on 'Partnership for a new organisation of work' in April 1997 and the 'Reports on the integration of health protection requirements in community policies' (1998).

Between 1995 and 1998, the Health & Safety Executive ('HSE') issued guidance to help employers manage work-related stress. Its publication, 'Stress at work: A guide for employers' stated that:

> 'Ill health resulting from stress caused at work has to be treated the same as ill health due to other physical causes present in the workplace. This means that employers do have a legal duty to take reasonable care to ensure that health is not placed at risk through excessive and sustained levels of stress arising from the way work is organised, the way people deal with each other at their work or from the day-to-day demands placed on their workforce. Employers should bear stress in mind when assessing possible health hazards in their workplaces, keeping an eye out for developing problems and being prepared to act if harm to health seems likely. In other words, stress should be treated like any other occupational health hazard'.

It goes on to say,

> 'Employers are not under a legal duty to prevent ill health due to stress arising from circumstances outside work, such as personal or domestic problems. But it may be in their own, as well as their employees' interests to deal sympathetically with staff whose domestic circumstances or state of health make it difficult for them to cope for the time being with pressures of work'.

It was as a result of the increasing awareness of stress-related illness in the industrialised nations that in 1998 the Tokyo Declaration, compiled by delegations from the US, EU and Japan, advocated various ways to prevent work-related stress. Eventually, these recommendations were reflected in Resolution No. A4–0050 passed by the European Parliament on 29 February 1999. In it, it resolves, *inter alia*, that:

- work must be adapted to peoples abilities and needs and not vice-versa and notes that by preventing a disparity arising between the demands of work and the capacities of the workers, it is possible to retain employees until retirement age; and considers that new technologies should be used in order to achieve these aims
- the European Commission be urged to investigate the new problem areas which are not covered by current legislation, ie stress, burnout, violence and the threat of violence by customers, and harassment in the workplace;
- it be noted that musculoskeletal diseases and psycho-social factors constitute the greatest modern threat to workers health;
- attention is drawn to the problems resulting from a lack of autonomy at the workplace, monotonous and repetitive work, as well as work with a narrow variety of content – all of which are typical of womens work in particular; and calls for attention to be

paid to the importance of ergonomics in the improvement of health and safety conditions in the workplace;

- attention is also drawn to the health and safety at work of groups which, at present, now largely fall outside the scope of legislative protection, such as home workers and the self-employed; and recommends the principle of safety management whereby the management of the risks of the working environment and development of the safety and welfare of workers are regarded as part of the normal activity of the workplace and that this should be undertaken in co-operation both with the management and the workforce.

In accordance with the philosophy of the above framework directive, the primary task of every employer should be to try to eliminate or control stressors, ie the root causes of work-related stress.

The Human Rights Act 1998

2.7 The *Human Rights Act 1998*, (which incorporated the European Convention on Human Rights into English law), affects every area of public and private rights, including relations between individuals and public authorities and certain employers. It gives effect to the European Convention for the Protection of Human Rights and Fundamental Freedoms, which must be respected by all public authorities.

The Convention incorporates highly developed case law and these European-defined rights affect the behaviour not only of public authorities and the courts, but also a broad range of company activities including employment disputes and contracts. The 1998 Act has had a significant impact on disputes surrounding disability, mental health and personal injury. All these areas fall within the domain of responsibility of most managers, who should be aware of the main provisions of the Convention.

Other stress-related legislation

2.8 The particular issue of stress at work has been highlighted by both the CBI and the TUC as one of the key employment law issues of the 21st century. The initial focus in the 1990s was on the liability for illness due to work-related stress. However, subsequent legislation, including the *Disability Discrimination Act 1995*, has ensured that tribunals look more closely at employees who claim to have been treated unfairly as a result of depression or other stress-related symptoms. As a result of this, it is increasingly likely that employees will seek compensation as a result of work-related stress and will claim for this through the courts.

In addition to the above, there are other areas of health and safety law that may apply in certain circumstances. These are as follows.

The *Employment Rights Act 1996* provides for an employee to bring a complaint of unfair dismissal to an Employment Tribunal by way of a 'constructive dismissal'. A constructive dismissal arises where the employee has resigned but the resignation is construed as a dismissal by the tribunal because the employer has acted so badly that the employee has effectively been forced to give up his or her job. Workplace bullying or victimisation, which leads to a resignation, can give rise to a claim of constructive dismissal.

The *Sex Discrimination Act 1975 and Race Relations Act 1976* apply in cases where the stress is due to sexual or racial discrimination or harassment. The employer can be found liable for the actions of its employees (even if they are not aware of the behaviour in question), unless the employer has taken all reasonable steps to train employees to prevent discrimination arising.

The *Working Time Regulations 1998 (SI 1998/1833)* implement the provisions of the European Working Time Directive, which are:

- 48 hours to be the maximum of the average hours that a worker can be required to work by the employer in a week;
- 8 hours' work in 24 being the limit that nightshift workers can be required to work by the employer;
- A right for nightshift workers to free health assessments;
- A right to 11 hours' rest a day and a day off each week;
- A right to a rest break where the working day exceeds 6 hours;
- A right to four weeks' paid leave per year.

There are also certain provisions regarding workers under 18 years of age.

The *Workplace (Health, Safety and Welfare) Regulations 1992 (SI 1992/3004)* cover workplace conditions such as lighting, temperature, ventilation, space, cleanliness and rest-room facilities for washing. *Regulation 25* of the 1992 Regulations requires suitable facilities to be provided by the employer for pregnant or breast-feeding female staff to rest and also for the provision of rest facilities for non-smokers, ie free of tobacco smoke.

Other legislation and regulations that are relevant to the issue of work-related stress are:

- *Public Order Act of 1986;*
- *Manual Handling Operations Regulations 1992 (SI 1992/2793);*
- *Provision and Use of Work Equipment Regulations 1998 (SI 1998/2306);*
- *Personal Protective Equipment at Work Regulations 1992 (SI 1992/2966);*

- *Health and Safety (Display Screen Equipment) Regulations 1992 (SI 1992/2792);*
- *Protection from Harassment Act 1997;*
- *Public Interest Disclosure Act 1998;*
- *Disability Discrimination Act 1995;*
- *Employment Act 2002.*

It can be seen from the above legislation that the Government is intent on confronting and dealing with the problem of work-related/ occupational stress as one of the most serious health and safety issues facing British employers and employees.

Failure to comply with the provisions of certain current legislation may be a criminal offence carrying penalties as laid down by statute and, in addition, there may well be evidence to support a civil litigation claim for damages based upon negligence.

Current initiatives

'Securing Health Together'

2.9 Work-related stress is one of two health-related programmes (the other being musculoskeletal disorders) in eight priority programmes that the Health & Safety Executive ('HSE') has chosen to meet the targets for in its occupational health strategy 'Securing Health Together'. This strategy forms a key part of the Health & Safety Commission's 'Revitalising Health and Safety' initiative.

'Securing Health Together', is a long-term occupational health strategy for England, Scotland and Wales, (which follows on from the Revitalising Health and Safety initiative). It is a combined initiative of the HSE in association with the departments of health, social security, education, the Scottish Executive, and the National Assembly of Wales, and sets out an occupational health strategy for Great Britain.

As part of this initiative, government departments, in co-operation with employers and their organisations, employees, trade unions, health professionals and voluntary groups, have set several challenging targets to be achieved by 2010. These are as follows:

- education in incidence of work-related ill health;
- reduction in ill health to members of the public, caused by work activity;
- 30% reduction in the number of working days lost due to work-related ill-health problems;

- anyone not currently in employment due to ill health or disability is, where necessary and appropriate, made aware of and offered opportunities to prepare for and find work.

To achieve these targets, the strategy is to take forward five key programmes of work. Five Programme Action Groups are identifying areas for action under each programme, setting targets, identifying key partners and initiating action.

Under these five programmes (listed below), the Action Groups are undertaking the following work in respect of work-related stress:

- *Securing compliance through improvements in the law relating to occupational health*

  ie developing appropriate standards.

- *Striving for excellence through continuous improvement in occupational health*

  ie to try and establish centres of best practice that can share their knowledge with local enterprises.

- *Obtaining essential knowledge on occupational health*

  ie commissioning research to provide the knowledge needed to underpin the standards.

- *Ensuring that all interested parties have the necessary competences and skills*

  ie ensuring enforcement officers are being equipped with the skills they need to accurately identify work-related stress and take appropriate action.

- *Ensuring that the appropriate support mechanisms are in place to deliver information, advice and other support on occupational health*

  ie the HSE has already started a publicity drive which includes the 'Tackling work-related stress' series of booklets.

New legislation and anticipated future law and initiatives

2.10 In addition to these current provisions, a variety of other legislation and initiatives with implications in respect of work-related stress have come into force or are expected within the next few years.

These include:

- *Flexible working*

  This came into force in April 2003 and will allow employees (with at least six months service), the right to request flexible working. It is the duty of the employer to consider the request seriously.

- *Adoption leave*

  The right to paid adoption leave came into force in April 2003.

- *Maternity and paternity leave*

  The rights to maternity and paternity paid leave came into force in April 2003.

- *Equal Treatment Directive*

  This Directive comes into force on 2 December 2003. It implements the principle of equal treatment by employers on the grounds of sexual orientation, age, disability, religion or belief. Employers have until December 2006 to implement certain provisions relating to age and disability discrimination. The Directive is enforced under the European Council Directive 2000/78/EC.

- *Management standards*

  The HSE is planning to develop management standards. These standards will provide a clear yardstick against which to measure an employers management performance in preventing stress. The first pilot phase of the standards will happen in 2003, with the final phase in 2005.

- *Age discrimination*

  Legislation on age discrimination is expected by 2 December 2006.

Stress-related litigation

General issues related to negligence

2.11 In a substantial number of cases, specific health and safety legislation will not necessarily be the most useful route of choice for addressing problems of work-related stress. This is because it is usually only an effective option when its provisions can be applied to work organisation and management systems.

It may, in many instances, be preferable to rely on other legislation that relates to harassment, sexual discrimination and employment law, etc, for a satisfactory result to be achieved.

When considering the duty of care under health and safety legislation, consideration must be taken of the fact that in certain instances, control of the exposure to work-related stressors may be outside the power of the employer to influence. However, in such cases, employers might be able to take steps to minimise exposure and/ or implement measures to mitigate any detrimental effects to employees. For example, there may be instances where an employee is working on an external site (of a customer) and is subject to, for instance, excessive noise or extremes of temperature. In this case the employer would need to exert influence on the site-owner, ie the customer, to remedy the problem or to take action to mitigate the effect(s).

A common stressor is work overload and this can occur through a variety of factors. Should the overload be the result of unrealistic work targets in terms of production or quality, then this would come under the provisions of the health and safety legislation.

However, defining an unrealistic workload can be far from easy, and individual cases would have to be considered in the light of particular circumstances and evidence. Effective measures should be implemented to control this hazard, which is often widespread.

It may be necessary to seek expert opinion on the predictability of the causes of ill-health that result from particular work practices or work environments, which may cause stress to specific individuals, under certain circumstances.

The costs of interventions or remedial measures will vary dependant on the particular circumstances. There may be a need, for instance, in the case of high production targets, to balance the associated risks involved with the costs of employing more staff.

The provision of training is an example where the costs of such interventions may be self-financing. This can be achieved by employing a workforce that is better equipped to handle the demands of its workload and who will eventually contribute towards both higher productivity and a significant reduction in absence due to sickness.

Consultation with employee representatives is vital in securing co-operation in the implementation of preventative measures against work-related stress.

Employers will leave themselves open to allegations of negligence if they:

- Knowingly re-expose workers to environmental or situational factors that have previously been implicated in causing stress-related illness.
- Designate duties or assign work schedules to employees who are not sufficiently capable or who are insufficiently trained to carry out the work or duties required.
- Fail to address known or previously reported issues of harassment, bullying or workplace violence and stress.
- Refuse to take measures to reduce the impact of organisational or management imposed change on employees where it is known to have directly or indirectly caused stress-related problems of ill-health or absence.

Examples of stress-related litigation

2.12 'There is no logical reason why psychiatric damage should be excluded from the scope of an employer's duty of care.' These were the words used by the judge in his summing up of the *Walker* case below and in speaking them, the judge established a principle of considerable significance, ie that a duty of care to prevent stress is an important issue for every employer.

- *Walker v Northumberland County Council [1995] 1 All ER 737*

  John Walker had been employed by Northumberland County Council from 1970 until 1987. He was employed as a Social Services Officer and was responsible for teams of field-workers whose work included dealing with problem families and child care. He suffered a stress-related depressive illness following a prolonged period in which his work pressures increased significantly and in November 1986, he suffered a mental breakdown. Although Mr Walker had brought his employer's attention to the problems faced, they failed to provide him with additional staff or resources. He was absent from work for 4 months after which period he was pronounced fit to return to work, provided that he was given more support and a lesser workload. His employer failed to comply with these requirements, and an increased load of highly stressful work rapidly induced a second breakdown, which rendered him permanently unfit for carrying out social work.

  Mr Walker claimed damages for the effect of both breakdowns. The court held that the first illness was not reasonably foreseeable, but that he was entitled to compensation in respect of the second, as the employer then knew that he was at risk but failed to take any adequate steps to prevent recurrence of his illness.

  Both parties set out to appeal against the decision: the employer appealed on the basis that it was not liable for either breakdown, while the employee sought to establish that the employer was liable for both. The case was due to be heard by the Court of Appeal in June 1996. However, the parties decided not to appeal, but rather to agree damages for the employee.

  The cost to Northumberland County Council of this one case was more than £400,000. Damages of £175,000 were paid to the employee. The costs of the two-week trial alone were £150,000. In addition the council paid the costs of sick leave and an ill-health pension. Total costs were in the region of £500,000 for the damage caused to one man. The council also suffered the loss of a well-trained resource, additional pressure on those remaining, together with damage to its image and unwanted publicity.

- *Lancaster v Birmingham City Council (1999) (Unreported)*

  Beverley Lancaster, the claimant, was a draftswoman with the Birmingham City Council when she suffered a nervous breakdown after she was moved to a front-line post in housing, without proper training. She was awarded £67,400 when the council admitted liability for personal injury caused by stress. This was the first case in which an employer admitted liability for causing personal injury due to stress.

- *Noonan v Liverpool City Council (1999) (Unreported)*

  A home help supervisor was paid £84,000 in damages after quitting her job claiming she was suffering from stress after being bullied for five years by a colleague.

- *Benson v Wirrall Metropolitan Council (1999) (Unreported)*

  The Association of Teachers and Lecturers gained an award of £47,900 in an out of court settlement, which is believed to be the first compensation case for stress in the classroom.

- *Suicide due to bullying behaviour (1999)*

  A verdict of suicide was made by a Surrey coroner after stating that an employee's job at Schroeders Bank had subjected him to pressure or bullying. 'It would seem that Mr Mason was subject to bullying and pressure that made life very difficult for him', the coroner said.

- *McLeod v Test Valley Borough Council (2000) (Unreported)*

  A former senior housing officer was paid £200,000 in damages for suffering from a mental breakdown after being bullied by his line manager.

- *Ingram v Worcester County Council (2000) (Unreported)*

  A warden, at campsites for gypsies, was awarded a record £203,000 in damages for suffering work-related stress. The council was accused of failing to support the wardens' work, leading to a loss of authority on the sites. Mr Ingram subsequently became the victim of violent and abusive behaviour.

- *North v Lloyds Bank plc (2000) (Unreported)*

  Leslie North, a former bank manager, won an out of court settlement of £100,000 compensation for stress from his former employers. He had fought a seven-year battle for compensation after he was reduced to tears by the hostile attitude of his boss. He was diagnosed as suffering from post-traumatic stress after being asked to meet sales targets as well as carrying out day-to-day branch duties.

- *Howell v Newport County Council (2000) (Unreported)*

  Janice Howell reached a settlement of £254,362 after Newport County Council admitted liability for the intolerable working conditions at Maindee Junior School in Newport. She suffered a breakdown after trying to cope unassisted, with pupils who were either disturbed or suffered from learning/ behavioural difficulties.

- *Prison Officer v Home Office (2001) (Unreported)*

  In what was seen as the first successful jail-based claim against the Home Office, a prison officer was awarded more than £100,000 damages following a claim for work-related stress. He claimed he was unable to work under the mounting pressure of his job, where he regularly had to deal with incidents of self-harm, violence, bullying and verbal abuse. His solicitor said 'This was a breakdown with lasting consequences for a dedicated officer'.

- *Conway v Worcester County Council (2001) (Unreported)*

  A former residential social worker with Worcester County Council won £140,000 compensation after developing a stress-related illness through work. Ms Conway was put in sole charge of a home for people with learning difficulties, after the manager resigned. She received no additional training and was working up to 80 hours a week. She became depressed as a result. The council admitted liability and her settlement was based on the injury she suffered, claims for loss of earnings, loss of pension, medical treatment and retraining costs.

- *B v Somerset County Council (2001)*

  Mr Barber, a former head of mathematics, was awarded £100,000 damages and costs by Exeter county court against Somerset County Council who ran the East Bridgewater School in Somerset. Mr Barber developed symptoms of depression brought on by his unremitting workload of up to 70 hours a week, and being subject to bullying by a head teacher.

In all the afore mentioned cases above, the employer was in breach of the provisions of current health and safety legislation. This was invariably related to the required duty of care laid down in the *Health and Safety at Work etc Act 1974* (and/ or other relevant legislation).

Court of Appeal ruling in Hatton v Sutherland

2.13 (See APPENDIX at 2.17 for a full report on the case.)

In February 2002, the Court of Appeal ruled in four appeal cases involving stress at work (*Hatton v Sutherland and three other appeals [2002] ECWA Civ 76*). Three claimants who had been successful in the lower courts and had obtained awards of almost £200,000 between them for work-related stress, lost the money on appeal in a landmark decision that established an important precedent. The Court of Appeal ruled that signs of stress in a worker must be plain enough for any reasonable employer to realise that some action must be taken before the employer can be sued for negligence.

Three appeals by employers were allowed and a fourth dismissed. The four cases were all appeals from different county courts where the judges had awarded damages for negligence after hearing how claimants had to stop work because of stress-related psychiatric illness. (The fourth appeal was dismissed on the basis that the signs of stress were clear whilst in the three other cases they failed to meet this criterion.)

The Court of Appeal has now given very clear guidelines on the standards expected of employers. In their decision, the judges said:

1. 'Employers are generally entitled to take what they are told by employees at face value and do not have to make searching enquiries. There are no occupations which should be regarded as dangerous to mental health'.
2. 'Employers are usually entitled to assume that the employee can withstand the normal pressures of the job unless they know of a particular problem or vulnerability'.
3. 'To trigger a duty on the part of the employer to take action, indications of impending harm to health arising from stress at work must be plain enough to show that action should be taken'.
4. 'If the only reasonable and effective step would be to sack or demote the employee, the employer is not in breach of duty in allowing a willing worker to continue in the job'.
5. 'Any employer who offers a confidential counselling advice service with access to treatment is unlikely to be found in breach of duty'.

It is worth considering point 4 (above) of the judges' decision in light of a later judgment made by the Court of Appeal in another case. In *Wayne Coxall v Goodyear GB Ltd [2003] 1 WLR 536*, the Court of Appeal considered whether an employer had a duty to dismiss an employee in circumstances where that employment continued to expose the employee to a risk of injury. Although this concerned an occupational asthma claim, it is interesting because the Court of Appeal upheld the lower court's decision to distinguish *Hatton* and found that there was a positive duty to dismiss the employee in the circumstances. The two decisions do no sit comfortably together.

Another important case that should be considered is *Young v The Post Office [2002] EWCA Civ 661*, where an employer was successfully sued by a claimant following a second breakdown. The Court of Appeal said that it would be unusual, but not impossible, for a finding of contributory negligence in a claim for damages as a result of stress.

The guidelines in the *Hatton* case have provided a good starting point from which the courts can consider claims in regard to work-related stress; although how this area of the law will eventually develop, is yet to be seen, (as of the date of writing) and the outcome of the appeal in the case of *Barber v Somerset CC* (one of the appeals heard in *Hatton v Sutherland*) which is currently progressing its way in the House of Lords – a judgement from which is anticipated by the end of 2003.

The NUT commented, 'Unions will certainly make sure that employers know they must assess the risks of stressful occupations'. The Institute of Directors remarked that 'Employers are not responsible for non-work pressures on employees, and if employees are over-worked they have a responsibility to say so'.

The impact of the Court of Appeal judgments

2.14 The decision by the Court of Appeal in February 2002 has indicated that the focus of employment thinking should be on the foreseeability of psychiatric illness as a result of work-related stress.

Employers will need to assess much more carefully the risks facing employees and employees will need to be more diligent in advising employers of undue pressure. Employees will also need to take more responsibility for their own welfare, and in cases where the employer is not informed regarding a particular risk, they are unlikely to be found negligent. The employer is however required by law to carry out regular risk assessments.

A very important point to arise from the ruling is that the provision of counselling and advice services for employees will demonstrate the employer's regard for health and safety and therefore make them less likely to be in danger of breach of duty of care.

Some factors relating to foreseeable injury are:

- The nature of the work to be carried out.
- Level of workload.
- Undue or excessive emotional or intellectual demands.
- Indications of stress from others engaged in similar work.
- Frequency and length of absence due to sickness.
- Obvious indications of employee ill-health.
- Particular vulnerabilities of a particular employee.
- History of problems relating to the work situation.

Factors that are likely to be examined in the case of injury are:

- The extent of the risk.
- The seriousness of the injury.

- Steps which reasonably should have been taken.
- The estimated costs of such preventative action.
- The likely effectiveness of such action.

The Court of Appeal sets out 16 key principles to be applied to cases:

(i) The normal principles of employer liability are key factors.

(ii) Whether the injury/ sickness was reasonably foreseeable and was attributable to workplace conditions.

(iii) Foreseeability depends upon the employer's current knowledge of the employee. The employer is entitled to assume that normal pressures of work will not cause injury.

(iv) No job or occupation is intrinsically more stressful.

(v) Control measures should ascertain the nature and extent of work required; whether the demands are reasonable and whether there is any undue level of absence attributable to sickness; if there are any signs of impending harm to the worker and whether there are any specific vulnerabilities.

(vi) The employer is entitled to take information at face value and need not make special enquiries.

(vii) Any indication of possible injury or harm must be plain to see.

(viii) Breach of duty of care can only be established where the employer fails to take reasonable steps in all the circumstances having regard to the potential risk and the costs of prevention.

(ix) Relevant factors are the size and scope of the business, its resources and the interests of the other employees.

(x) Employer's actions required must be substantiated.

(xi) Counselling, support and training services offered by the employer will render him unlikely to be found in breach of duty of care.

(xii) The employer is not in breach in allowing the employee to continue where the only alternative is dismissal.

(xiii) Establishing a breach of duty entails identifying steps that could have been taken.

(xiv) A plaintiff must establish that breach of duty exacerbated or contributed to the harm or injury suffered.

(xv) Where the injury results from more than one cause, the employer is liable only for that which was under his control.

(xvi) The level of damages or compensation will be influenced by any pre-existing conditions.

Court of Appeal ruling in Eastwood v Magnox Electric plc

2.15 The number of recent large compensation awards could also be significantly reduced following the Court of Appeal ruling in the case of *Eastwood v Magnox Electric plc [2002] EWCA Civ 463*, if upheld. (But see also *Robert McCabe v Cornwall CC and Mounts Bay School (18 December 2002)* (*Unreported*) where the Court of Appeal refused to accept that wherever a dismissal eventuates, a common law claim is excluded.)

The Appeal Court ruled in the *Eastwood* case that a claim for damages as compensation for stress arising from unfair dismissal should be decided at an Employment Tribunal rather than a county court. This would have the effect of limiting the maximum award for unfair dismissal due to stress to £52,700 – rather than an unlimited award potentially running into hundreds of thousands of pounds that could be awarded by a county court.

The ruling will apply to the majority of stress claims because most are made by claimants who have already left an organisation because of the alleged effects of workplace stress.

Key learning points

2.16 As stated at the beginning of this chapter, if there is a belief that an organisation may be in breach of current legislation on work-related stress, or could be facing litigation, then all reasonable steps should be taken to ensure that tackling stress is a priority for the business. In the light of this, the following key points should be considered.

- Every employer has a general duty of care to protect its employees from foreseeable injury, and that includes pre-emptive action to prevent stress-related injury.
- A primary task of every employer is to try to eliminate or control stressors (ie the root causes of work-related stress), and employers are obliged to carry out risk assessments not only for physical risks but also for psychological risks.
- European Union regulations are legally binding and enforceable in UK courts.
- EU law takes precedence over UK law and is the overriding authority.
- Breach of duty of care regulations could lead to a charge of criminal negligence and claims for compensation might also be made through the civil courts.

(cont'd)

- Employers are usually entitled to assume that the employee can withstand the normal pressures of the job unless they know of some particular problem or vulnerability.
- Any employer who offers a confidential counselling advice service with access to treatment is unlikely to be found in breach of duty of care (although at the time of this publication going to press this has yet to be tested in the courts).
- Employers without systems in place to tackle employee stress are breaking the law. The Health & Safety Commission has stated that 'You do not have the option of not managing stress. The law requires you to assess the risk to the mental health of employees and take steps to reduce the risks as far as is reasonably possible.'

APPENDIX
Hatton v Sutherland; Barber v Somerset County Council; Jones v Sandwell Metropolitan Borough Council; Bishop v Baker Refractories Ltd, Court of Appeal: Brooke, Hale and Kay LJJ (5 February 2002)

2.17 Claims for psychiatric injury arising from stress at work were to be considered in accordance with the ordinary principles of employer's liability. The test was the same whatever the employment: there were no occupations which should be regarded as intrinsically dangerous to mental health.

The Court of Appeal so held in a reserved judgment of 5 February 2002 in:

(1) allowing an appeal by the defendant, Terence Sutherland, (the Chairman of the Governors of St Thomas Becket RC High School), from the decision of Judge Trigger in the Liverpool county court on 7 August 2000 whereby he awarded the claimant teacher, Penelope Hatton, damages of £90,765.83;

(2) allowing an appeal by the defendant, Somerset County Council, from the decision of Judge Roach in the Exeter county court on 8 March 2001 whereby he awarded the claimant teacher, Leon Alan Barber, damages of £101,041.59;

(3) dismissing an appeal by the defendant, Sandwell Metropolitan Borough Council, from the decision of Judge Nicholl in the Birmingham county court on 31 October 2000, whereby he awarded the claimant, Olwen Jones, an administrative assistant at a local authority training centre, damages of £157,541; and

(4) allowing an appeal by the defendant, Baker Refractories Ltd, from the decision of Judge Kent-Jones in the Leeds county court

on 26 January 2001, whereby he awarded the claimant, Melvyn Edward Bishop, a raw materials operative in the defendant's factory, general damages of £7,000. In each case the claim was against the claimant's employer for damages for psychiatric illness which the claimant alleged had been caused by stress at work.

HALE LJ, giving the judgment of the court, said that liability in negligence depended upon three inter-related requirements: the existence of a duty to take care; a failure to take the care which could reasonably be expected in the circumstances; and damages suffered as a result of that failure. Claims for psychiatric injury fell into four different categories:

(1) tortious claims by primary victims (usually those within the foreseeable scope of physical injury);

(2) tortious claims by secondary victims (those outside that zone who suffered as a result of harm to others);

(3) contractual claims by primary victims (where the harm was the reasonably foreseeable product of specific breaches of a contractual duty of care towards a victim whose identity was known in advance); and

(4) contractual claims by secondary victims (where the harm was suffered as a result of harm to others, in the same way as secondary victims in tort, but there was also a contractual relationship with the defendant).

The instant claims fell within category (3). From the discussion on the law, the following practical propositions emerged:

(1) There were no special control mechanisms applying to claims for psychiatric, or physical, illness or injury arising from the stress of doing work the employee was required to do. The ordinary principle of employer's liability applied.

(2) The threshold question was whether this kind of harm to this particular employee was reasonably foreseeable. That had two components: (a) an injury to health, as distinct from emotional stress, which was (b) attributable to stress at work, as distinct from other factors.

(3) Foreseeability depended upon what the employer knew, or ought reasonably to have known, about the individual employee. Because of the nature of mental disorder, it was harder to foresee than physical injury, but might be easier to foresee in a known individual than in the population at large. An employer was usually entitled to assume that the employee could withstand the normal pressures of the job unless he knew of some particular problem or vulnerability.

(4) The test was the same whatever the employment: there were no occupations which should be regarded as intrinsically dangerous to mental health.

(5) The factors likely to be relevant in answering the threshold question included (a) the nature and extent of the work done by the employee and (b) signs from the employee of impending harm to health.

(6) The employer was generally entitled to take what he was told by his employee at face value, unless he had good reason to think to the contrary. He did not generally have to make searching enquiries of the employee or seek permission to make further enquiries of his medical advisers.

(7) To trigger a duty to take steps, the indications of impending harm to health arising from stress at work had to be plain enough for any reasonable employer to realise that he should do something about it.

(8) The employer was only in breach of duty if he failed to take the steps which were reasonable in the circumstances, bearing in mind the magnitude of the risk of harm occurring, the gravity of the harm which might occur, the costs and practicability of preventing it, and the justifications for running the risk.

(9) The size and scope of the employer's operation, its resources and the demands it faced were relevant in deciding what was reasonable; these included the interests of other employees and the need to treat them fairly, for example in any redistribution of duties.

(10) An employer could only reasonably be expected to take steps which were likely to do some good; the court would be likely to need expert evidence on that.

(11) An employer who offered a confidential advice service, with referral to appropriate counselling or treatment services, was unlikely to be found in breach of duty.

(12) If the only reasonable and effective step would have been to dismiss or demote the employee, the employer would not be in breach of duty in allowing a willing employee to continue in his job.

(13) In all cases, therefore, it was necessary to identify the steps which the employer both could and should have taken before finding him in breach of his duty of care.

(14) The claimant had to show that breach of duty had caused or materially contributed to the harm suffered. It was not enough to show that occupational stress had caused the harm.

(15) Where the harm suffered had more than one cause, the employer should only pay for that proportion of the harm suffered which was attributable to his wrongdoing unless the harm was truly indivisible. It was for the defendant to raise the question of apportionment.

(16) The assessment of damages would take account of any pre-existing disorder or vulnerability and the chance that the claimant would have succumbed to a stress related disorder in any event.

With acknowledgement to:

The Incorporated Council of Law Reporting for England & Wales (ICLR), Megarry House, 119 Chancery Lane, London WC2A 1PP (tel: 020 7242 6471; fax: 020 7831 5247).

Reported by: Carolyn Toulmin, Barrister.

For the full reports of the case as appearing in the ICLR's Industrial Cases Reports, see [2002] ICR 613 and [2002] ICR 1026, HL(E).

3 The Health and Safety Framework

Introduction

3.1

> 'Perfect freedom is reserved for the man who lives by his own work, and in that work does what he wants to do'. *R G Collingwood*

Employers have a legal responsibility to identify, remove or reduce the effects of hazards at work and these include both physical and psychological hazards, such as stress.

Workplace stress is an important health and safety issue and employers ideally need to develop policies, implement procedures and offer guidance, so that they can fully comply with their legal duties.

As discussed in CHAPTER 2, in 2002 the Health & Safety Commission developed a strategy for dealing with work-related stress, based on its long-term occupational health strategy, 'Securing Health Together'. This work is intended to lead to a number of initiatives including the publishing of standards, in 2003, for measuring employers' performance in preventing work-related stress. The standards will facilitate the enforcement of stress-related offences and are intended to form the basis of an Approved Code of Practice regarding stress.

The Health & Safety Executive ('HSE') is encouraging firms to become part of 'good practice' networks – a move that is based on recent research, which suggests that many firms currently manage only certain aspects of stress satisfactorily and need assistance and/ or training in dealing correctly and comprehensively with the problem. These networks allow firms to share that which they have achieved, whilst accessing support for areas which they find challenging.

The HSE is training enforcement officers, who will be equipped to deal with organisational stress as part of their case load. In the interim, employers have been given guidance in the form of 'Tackling work-related stress: A manager's guide to improving and maintaining employee health and well-being' (2001), which describes how to approach work-related stress.

Health and Safety Policy Statement

3.2 Having a Health and Safety Policy Statement is a legal requirement under *section 2(3)* of the *Health and Safety at Work etc Act 1974* and many employers find it a useful method of underlining the commitment that they, as employers, have for the welfare of their employees. It should form the foundation of a safety culture in which both employers and employees work together for their mutual health and safety.

Every employer with five or more employees must have a written Health and Safety Policy Statement that sets out how to correctly manage health and safety issues.

This statement should contain information specific to the organisation. It should also outline who within the organisation has responsibility for health and safety and what are their responsibilities, together with an indication of the remedial action that would be taken in the case of identified, workplace hazards. In this context, employers have a duty to consult with their employees when formulating health and safety policies and procedures.

Stress policies

3.3 Ideally, stress policies should be included as part of the organisation's Health and Safety Policy Statement. Work-related stress, however, is a complex subject and many employers find that having a separate stress policy document is a preferred course of action. Whatever form the policy takes, it is normal practice to write guidelines for managers and staff detailing the recommended procedures and interventions necessary for dealing with both the causes and the effects of stress within the organisation.

A stand alone stress policy should be structured in the same way as an organisation's main Health and Safety Policy and follow the HSE guidelines.

It would be normal practice to start with a clear statement of the organisations position in regard to stress, for example:

> 'This policy indicates our determination to tackle the effects of workplace stress on our employees. We recognise that work related stress is harmful and we are committed to take all reasonable and practical steps to prevent stress and where that is not possible to minimise its effects. We will ensure that our staff, at all levels, are aware of the steps that they can take to minimise its effects on themselves and their colleagues.'

It will then continue in line with the following outline (see list below), ensuring that it is complete in providing details of the responsibilities of both the management and individual employees and provides details of the methods that will be used to assess the risks and to provide support to employees who may be under pressure or suffering from stress.

It would be considered good practice to treat this in the same way as any other safety documentation and for employers to consult with the recognised health and safety committee or trade union during it's preparation.

A typical stress policy will follow the following outline, and like any other safety document it needs to be tailored to the specific organisation taking into account the specific risks of the industry and the needs of the employees.

- A clear definition of stress that will usually be:
  - The HSE definition (which is the one usually accepted and understood by most employers).
- An explanation of the management hierarchy and the communication channels within the organisation.

  This should clearly identify:
  - Who is individually responsible for taking action if and when a particular stress factor/ hazard is identified.
  - Who is responsible for considering the stress impact of any change in working practices.
  - The correct management procedure to be initiated when an employee either reports work-related stress or has taken sickness absence believed to have been brought about as a result of it.
  - The correct procedure for an employee to take when reporting instances of work-related stress.
  - The responsibility of the employer towards members of staff who may be suffering from stress due to external factors unrelated to the workplace.
  - The steps that individual employees should take in relation to their own well-being in order to protect themselves against stress and to minimise the effects as far as possible.
- The management systems used to monitor health and safety policy including:
  - HR data such as sickness records and staff turnover figures.
  - Compliance with procedures and recommendations.
  - Other audit methods including stress surveys.

- Details of safety training relating to the management of stress at work:
  - This may include both stress management training for managers and stress awareness training for all employees.
- Occupational health facilities, including:
  - Occupational health support, employee assistance programmes, and in-house counselling support services where applicable.
- Reporting procedures:
  - How data will be recorded and published.
  - Method of recording stress-related incidents.
- Consultation arrangements:
  - How the trade union and staff safety representatives will be kept informed on stress-related issues.

A stress policy that stands alone from the Health and Safety Policy Statement, has the advantage of:

- promoting an understanding towards those people who are potentially at risk or already suffering from stress;
- indicating that management views stress as a specific health and safety issue that can affect both employees and the organisation as a whole;
- committing management to identifying and assessing the risk of stress and undertaking actions to remove or minimise the stressors involved.

Management guidelines

3.4 Alongside a stress policy, it is considered 'best practice' to write guidelines outlining the organisation's commitment to the prevention, reduction and elimination of stress in the workplace. This guidance would:

- Normally contain the procedures to follow when dealing with individual cases of work-related stress.
- Detail reporting procedures for staff who feel they are under pressure.
- Recommend interventions that could be taken to identify and deal with the causes of work-related stress.
- Detail 'back to work' procedures including return to work interviews for those who have experienced work-related sickness absence.

These procedures may well be linked to or mirror other procedures existing within an organisation including: absence management, discipline, bullying, harassment, job capability and staff development.

Risk assessment

3.5 Employers have an absolute duty under the *Management of Health and Safety at Work Regulations 1999 (SI 1999/3242)* to carry out 'adequate' or 'suitable and sufficient' risk assessments of hazards that might cause harm to employees. [*SI 1999/3242, Reg 3*].

This is an all-encompassing duty, which is related to both the physical and psychological health of employees. The Regulations also apply to any contractors or self-employed persons who are working within the organisation.

For certain categories of work, organisations are also obliged to assess any potential risks to members of the public who might be affected by their activities, although this is unlikely to be necessary when assessing stress-related risks.

A risk assessment for stress is quite simply an investigation into what might cause employees harm at their place of work. Its aim is preventative in that it is designed to identify potential stressors and areas that may cause stress to staff or to minimise exposure to potential psychosocial hazards.

In this context, a 'hazard' is defined as anything to which the employee is exposed at work that may possibly present a risk to their health. 'Risk' is the probability, (whether great or small), that someone may be harmed by a hazard that has been identified.

Figure 1: Hazard, risk and harm

(Figure 1 is contained in 'Organisational interventions for work stress: A risk management approach' (HSE Contract Research Report 286/2000) and is reproduced with the kind permission of the HSE.)

Clearly, the first stage in any risk assessment process is to identify the potential hazards. This is then followed by an assessment of risk to ascertain whether there could be harm to health.

'The five steps to risk assessment' is a guide published by the HSE in 1999 that outlines the most common procedures for risk assessment at work.

Employers also have other risk assessment responsibilities which are covered by regulations such as the *Control of Substances Hazardous to Health Regulations 2002 (COSHH) (SI 2002/2677)*. Certain parts of these Regulations insist that the risk assessment is completed before commencement of any new work.

This 'prior' risk assessment is not mandatory in reference to stress, though in practice, organisations should always carry out a risk assessment before commencing new work or making any changes either to the work environment or work practices that could present a potential risk to employees.

Small firms, with five or less employees, have a dispensation with regard to the recording of risk assessments. They are, however, still required to assess risks and it is, therefore, a precautionary step to record their assessments.

Risk assessment: the principles

3.6 *Regulation 3* of the *Management of Health and Safety at Work Regulations 1999 (SI 1999/3242)* is not specific about the form that a risk assessment should take but it does provide some assistance to employers about the steps to take in evaluating the risks.

Most employers are aware of the principles of risk assessment in relation to the more common workplace hazards such as electrical risks or manual handling, where potential danger is apparent or identifiable. When it comes to stress, however, employers are often uncertain as to exactly how and what to assess and which tools and measures should be utilised.

As already stated, the purpose of a risk assessment is to identify workplace hazards that could potentially cause harm to employees. The first step subsequent to identifying a potential hazard is to make an informed decision as to whether or not it is significant and, if so, to then quantify the risk.

It is possible to consider stress at work, under three main categories:

- Some stress is avoidable and the risk assessment will identify the cause and enable employers to design and implement interventions which will remove the stressor(s) completely.
- There will be cases where one or more stressors are an integral or intrinsic part of the work. For example, shift working, working in a potentially hazardous environment or dealing with complaints from customers. The purpose of the risk assessment will then be to identify the stressors but also to devise ways in which their effects can be reduced. This might be by designing a more flexible shift pattern, offering staff adequate training to cope with difficult people or possibly a system whereby individuals are rotated through jobs that have a mix of both high and low pressure, at varying times.

- The third category is the area where workplace stress is more likely to have been caused, or exacerbated, by domestic or personal factors external to the workplace, eg health or domestic problems. It is this last category that is more focused on the individual and their robustness in being able to cope with their work. A stressful situation may well be temporary and the employer might well identify ways of supporting the individual on a temporary basis, possibly by bringing forward leave entitlement, time off or a role change. These courses of action may prove to be essential where an employee is in a safety critical role.

  Should there be an ongoing inability to cope with work demands, it would be in the employer's best interests to play an active role in providing (where possible and practicable) support to the individual, notwithstanding that the stress may be from an external source.

  However, nothing in this advice should be taken to recommend the shifting of the onus of personal responsibility onto an employer. Neither should an employee with domestic, stress-related problems of an intractable nature be guaranteed continuous employment regardless of his or her ability to carry out the job for which they were originally hired.

It is considered important that organisations prepare the ground with employees (or their representatives) prior to undertaking a risk assessment. It should be explained that the intention and purpose is to identify any potential areas of risk in the workplace, and then to undertake all necessary action to remove or reduce any such identifiable risks that could be the source of potential harm to anyone within the organisation.

It is recommended that organisations also seek the co-operation of employees, should they decide to undertake an audit questionnaire. It should be made clear at the outset that they will be informed of the results of the assessment when they become available.

The five steps of risk assessment

3.7 The five steps of risk assessment are:

- *Step 1*: Identify the hazards.
- *Step 2*: Decide who might be harmed and how.
- *Step 3*: Evaluate the risks and decide whether the existing precautions are adequate or whether more should be done.
- *Step 4*: Record the significant findings of the assessment.
- *Step 5*: Review the assessment and revise it at appropriate intervals – in particular, it should be revised subsequent to any major organisational change.

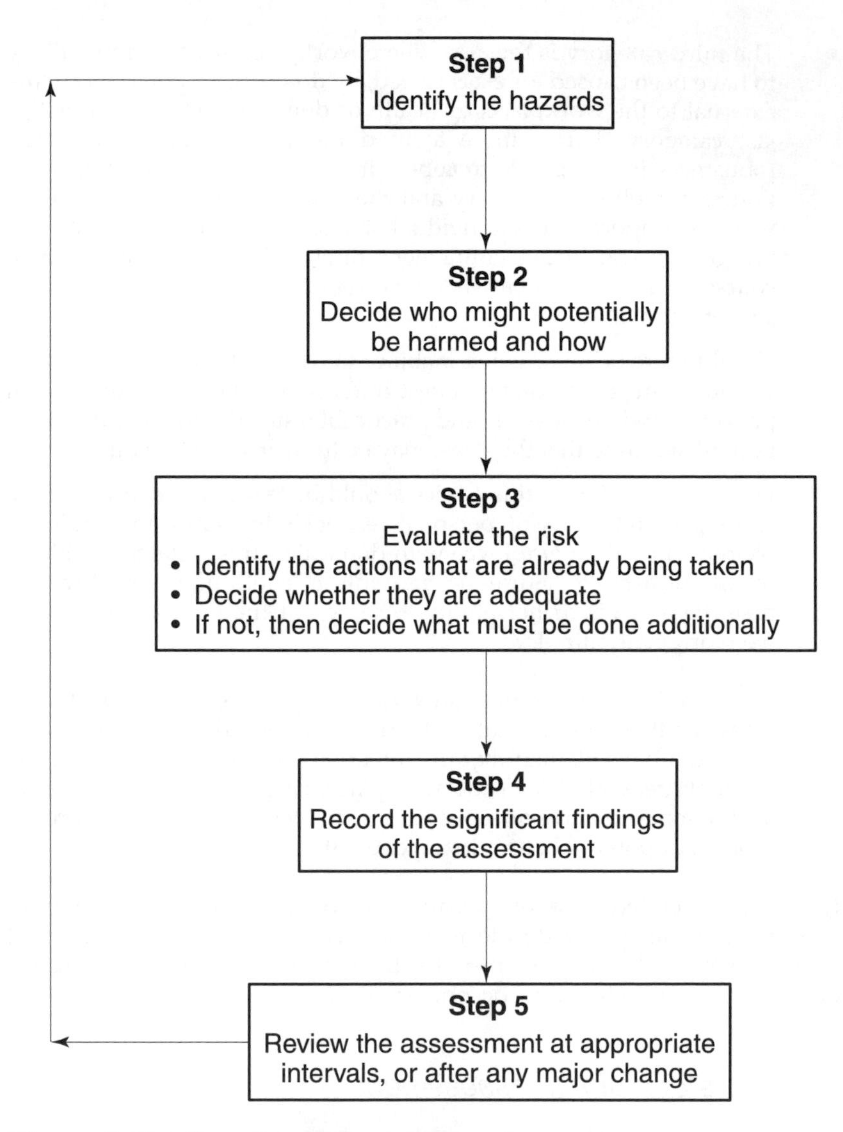

Figure 2: The five-step risk assessment process

Step 1: Identify the hazards

3.8 The risks referred to in this section are directly related to those hazards which may be generated within an organisation and that may well impact adversely on its staff (see Figure 3 below). It is important to remember that this is an assessment of the organisation, its systems and processes and *not* the individuals it employs.

It has been argued that there are specific jobs or tasks which are inherently stressful and that these should be considered when making these initial decisions. However in the Court of Appeal ruling (see *Hatton v Sutherland and Others [2002] ECWA Civ 76* at 2.13) it said that 'there were no occupations which should be regarded as intrinsically dangerous to mental health'. The invariable rule when carrying out risk assessments is to consider the whole workplace in the first instance. If this is not considered then warning signs may be overlooked in other staff categories which may have been prejudged not to be at risk.

An initial risk assessment can normally be undertaken by examining the existing data that is held by most organisations within their human resources department ('HR'), eg:

- Examine the records of stress-related absences. Are they increasing year on year? Do they appear more in certain departments than others; among particular categories of staff or possibly in specific teams? Additionally, in view of the fact that stress is often either under or misreported, it is essential to look at overall patterns of absence as well as those relating to identifiable stress.
- Review the figures on staff turnover. Are they significantly higher in certain areas without any obvious reason?
- If exit interviews are usually carried out when an employee leaves the organisation, is there any identification of stress (or excess pressure) as a common thread?
- Have there been any recorded increase in workplace accidents and, if so, has a human factor (as opposed to a systems or mechanical failure) been identified as the probable cause?
- Have there been any incidents of bullying or harassment and, if so, did they occur across the organisation or only at particular levels of authority or in specific departments?
- If there is an occupational health service and/ or an employee assistance programme, what level and degree of feedback is being received?

In addition to looking at the existing HR data, it is recommended that informal discussions with staff be held. These can be at all levels – from briefings with team leaders to seminars where management conduct question and answer sessions with staff.

It is advantageous for managers and supervisors to be visible and approachable. The particular points to look out for are:

- Are individual members of staff reasonably happy in their work?
- Is morale generally high across the organisation or are there pockets where staff morale needs bolstering?

- Are there cross-sectional concerns over specific organisational pressures or company demands?

Performance appraisals are another effective source of information enabling organisations to establish how individuals perceive the pressure they experience and whether they might be having difficulty in coping.

It should be possible, from this data, to identify stress-related risks or problems within specific areas of the organisation. However, that data may also identify illness or behaviour that requires further investigation, as compared with identifiable stress-related absence.

In cases where an assessment does not highlight any particular concerns, there is still a need to record findings and make arrangements to monitor the situation and fix a date for review. Should any major changes to working conditions and/ or procedures be implemented, then it will be necessary to subsequently re-assess the situation.

It is now usual for organisations or companies to have a formal risk assessment procedure to be used in relation to changes in working practices and this is referred to later in the chapter (see 3.10 below).

It is necessary to be aware that human resource procedures need to be sufficiently robust in order to provide an accurate picture of stress and stress-related absence. Is the culture such that employees are encouraged to talk openly about pressure and stress rather than maintaining a discreet silence to management? It should be ensured that methods are qualitative as well as quantitative in order to provide a true picture.

If the figures examined imply that there may indeed be some hazards that are stress-related, then it becomes necessary to refine the data gathering process and to carry out further investigation to identify the causes and to quantify the likelihood of any individual becoming harmed.

A further essential point in seeking to identify any work-related risks is to consider any common factors that might conceivably cause stress at work This is because it is often the case that some concerns raised by small groups (or individuals), are actually part of a wider organisational issue. The HSE has identified seven such factors that employers should consider when carrying out risk assessments for stress.

These are in the areas of:

- culture;
- demands;
- control;
- relationships;

- change;
- role;
- support.

These factors will be examined in more detail later on in the book but in the context of carrying out a risk assessment, it is essential that there is a clear understanding of these factors and how, in certain circumstances, they can become stressors.

Figure 3 (below) shows how these seven factors are linked to both individual and organisational symptoms. It is these symptoms among others which can be risk assessed. This model then continues by identifying some of the negative outcomes and costs which follow from allowing stress to continue unchecked.

Factor 1: Culture

3.9 Culture can be a major factor in determining the level of stress inherent in an organisation. It will also be a critical factor in how successfully stress is handled by management. The best way of describing the culture of an organisation is to consider the way that it depicts its personality. Is it a progressive organisation that recognises and respects the contribution of every employee or does it have an unhealthy, macho, almost type 'A' personality? (as discussed in CHAPTER 1 at 1.12). Is it a culture that is resistant to change and one that views success only as the product of an authoritarian, management style?

At the most basic level, a macho culture within an organisation will discourage employees from coming forward and speaking out to inform management that they are being placed under too much pressure, resulting in difficulty coping with their workload. Where this is the situation, the first indication that an employee is suffering from stress may well be when they become absent on sick leave. If this absence is not properly investigated, then the first intimation that it is stress-related might be when the individual leaves the company and/ or seeks compensation from their employer.

If a culture can be created where managers value and have consideration for the feelings and well-being of staff under their control, then cases of stress will be less frequent.

It is equally importantly that where there is a possibility that excessive pressure will turn into stress, the affected employee should pre-empt the potentially harmful situation by communicating the facts to his, or her, line manager. The management can then take proactive steps to reduce the risk of this pressure harming the individual employee and leading to stress-related absence.

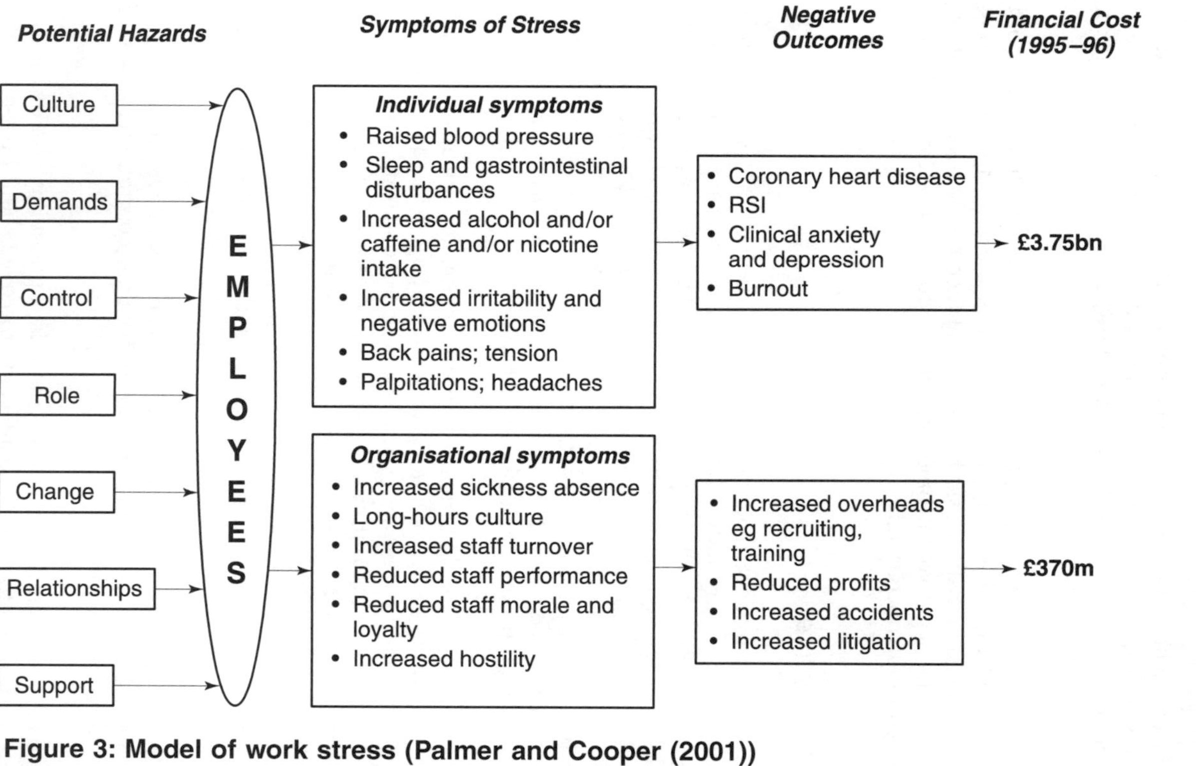

Figure 3: Model of work stress (Palmer and Cooper (2001))

(Figure 3 reproduced from S Palmer, C Cooper, and K Thomas 'Model of organisational stress for use within an occupational health education/ promotion or well-being programme – A short communication' (2001) *Health Education Journal* 60(4).)

Factor 2: Demands

3.10 The 'demands' to which the HSE refer to here, are directly related to the amount of work that employees are required to do, as well as the environmental factors, both physical and psychosocial, within which the work is carried out.

Employees may be overloaded in one of two ways:

- either by being given too large a quantity of work to complete in a specified time with the resources available, ie a 'quantitative overload'; or
- where the overload is qualitative, ie where the work is beyond the individual's ability to carry it out. This may be due in part to a lack of training, intellect or physical ability.

Whatever the reason, it may make it impossible for the employee to meet the demands that are being placed upon him (or her).

Employers have a mandatory duty under *Regulation 13(1)* of the *Management of Health and Safety at Work Regulations 1999 (SI 1999/3242)* to take into account the capabilities of employees as regards health and safety. Specifically, Paragraph 80 of the Approved Code of Practice to the Regulations states: 'When allocating work to employees, employers should ensure that the demands of the job do not exceed the employee's ability to carry out the work without risk to themselves or others'.

This implies that, where an employee is suffering from excessive pressure or stress, then their capacity and capability to carry out their workload should be examined with the aim of identifying whether additional training may help them to develop the skills needed to meet the demands of their job.

It will be recalled from CHAPTER 1 that the reverse situation may also apply. Where a person is given either an insufficient volume of work or work that is not sufficiently challenging, then that situation may induce or contribute to stress. However, it should be remembered that we all need a minimum level of pressure in order to become motivated and to obtain job satisfaction.

With regard to the demands placed upon employees by the physical environment, it is necessary to consider factors such as excessive noise, vibration or temperature; insufficient ventilation or number of air changes and/ or any other working conditions which may not be conducive to stress-free working.

For example, a risk assessment may have been carried out for excessive noise to ascertain the risks of physical damage to the hearing of employees and as a consequence, protective equipment such as ear defenders to minimise those risks may have been provided. However, in this example, the continuous wearing of ear defenders during the working day, (thereby preventing any conversation with workmates), may possibly cause isolation of the employee. This in itself could be an alternative source of stress, and reinforces the need to properly assess all working practices, whether existing or proposed.

The most extreme psychosocial demands are those where members of staff might be exposed to mental or physical abuse. Typically this situation may occur in the service industry where staff frequently face significant abusive (and sometimes violent) behaviour from members of the public, eg in transport or community-based services.

Factor 3: Control

3.11 The third factor to consider is 'control'. Employees will report higher levels of job satisfaction where they have a reasonable amount of control over their working lives. To avoid stress, the types of questions that should be asked, are: How much control can be delegated to staff? Are they able to plan their own workload? Can they work collaboratively in teams? The key is to provide reasonable support instead of continuous supervision.

Factor 4: Relationships

3.12 The HSE uses the term 'relationships' to describe the way we interact with people with whom we work. Work colleagues can either be of great help in supporting colleagues who may be feeling less robust or, conversely, work colleagues can conceivably be a cause of stress. This latter situation occurs when a superior officer or manager exhibits bullying or harassing behaviour to subordinates. These areas are covered in detail in CHAPTER 6, but it is worth mentioning that bullying and harassment may not only be a cause of stress but also a symptom of stress, as in a situation where a manager or employee are themselves under excessive pressure.

Factor 5: Change

3.13 The HSE recognises that change can be stressful for many people, often because the outcomes of change are not clearly communicated to those who may be affected by it. To address this, the HSE has focused on the way in which change is managed. Change itself can be beneficial by adding to the interest and job satisfaction that is obtained from the job. However, if the organisation is implementing a major change (or even some minor changes that will impact on only a few employees), then the psychological effects that such change(s) might possibly have on any staff directly or indirectly affected, should first be properly assessed.

Factor 6: Role

3.14 'Role', in this context, has two dimensions. The first is 'role conflict', which occurs when an employee feels that they are being asked to carry out work that is not part of their ordinarily designated duties and which they prefer not to undertake. The second, is 'role ambiguity', that comes from the employee not having a clear picture of that which is expected from him (or her).

Factor 7: Support

3.15 There are required levels of support, which if inadequate, may transform manageable problems into workplace stressors. These include situations where employees should correctly receive specific training to carry out their role but fail to receive it (although this does not necessarily mean training in aspects of stress management). Employers have a general duty under *Regulation 13* of the *Management of Health and Safety at Work Regulations 1999 (SI 1999/3242)*, which recommends that 'every member of staff receives sufficient training to undertake the core duties of their jobs'.

Support from within an organisation to assist individual employees to carry out their duties is also important. This may be at the level of a manager taking time out to talk with staff under his/ her control, providing appropriate supervision, or the provision of an in-house employee counsellor or an employee assistance programme. If no such support is provided, then the risk of stress may be increased. As already discussed in CHAPTER 2, the Court of Appeal ruled in February 2002 that 'employers who provide a confidential counselling service are unlikely to be found negligent of a duty of care'.

Measurement tools

3.16 One of the most frequently used methods of identifying risk, is the use of an employee questionnaire or stress audit. Before using this tool, either on the whole or part of the organisation, it is necessary to consider the advantages and disadvantages.

The HSE recommends that if a decision is made to use a questionnaire, then professional advice should be sought from an external source. Not only will they be able to advise on what is appropriate for the organisation but they will also be able to ensure that the responses given by the respondent employees will remain confidential. Questionnaires carried out in this way are likely to have a higher and more accurate response rate than where they are carried out completely from in-house resources.

The advantages of using a structured questionnaire from an external supplier are that a high response rate should be able to be achieved and the views of a large cross-section of the organisation obtained. The information received can also be used as a benchmark, should there be a requirement for the same questions to be repeated at a later date. The downside is that questionnaires can prove to be a time-consuming and expensive process and may not provide substantially more additional detail, in some instances, than can be obtained from the alternative methods outlined above.

It should be pointed out that the HSE carried out research entitled, 'A critical review of psychosocial hazard measures' (Contract Research Report 356/2001), that contained some criticism of existing stress questionnaires. The review looked at 26 measures that could be used for the risk assessment for stress. The findings were that although most of the measures examined were reasonably good at measuring various hazards, there seemed to be insufficient underpinning research as to whether they were measuring the right hazards.

The HSE subsequently, in its first recommendation, called for a re-appraisal to determine how these measures were actually being used and how they link into other health and safety policies and practices.

The second recommendation made was to the effect that organisations should be prepared to consider developing their own measures, which should be:

- focused on their specific job and role requirements;
- based on local knowledge and an appropriate understanding of the context;
- formulated with consideration of best practice;
- incorporated, where possible, into some form of risk management framework.

The third recommendation from the HSE was that organisations should continue to develop other ways of assessing hazards within their workplace, apart from self-reporting questionnaires. These included structured observation, task analysis and detailed job descriptions together with comprehensive examination of reports of harmful incidents. The organisation should ensure that their conclusions regarding cause and future recommendations, are made available to all staff concerned.

The foregoing discussion is not intended to imply that organisations should cease to use 'off the shelf' stress questionnaires but rather that they should consider such as merely one part of the process. Indeed, there is a wealth of experience in using these tools, which when combined with the benchmarking data that is available, makes them extremely useful as part of an assessment.

There are now continuing advances in this field with new proprietary audit methodologies being designed to better reflect new understandings of the causes of stress. Some of these have been developed to take advantage of new computer technology which may well reduce costs and provide easily accessible and flexible data. If it is necessary to use such a system to measure stress as part of the risk assessment procedures, then professional advice should be sought on the range that is available, in order to consider whether they will fulfil the specific requirements. Where they do not, then the best option is to devise a dedicated questionnaire to meet the needs of the particular organisation.

Irrespective of the form of audit measure that will be used (and notwithstanding that it is considered good practice to seek the views of a larger group as possible), it is advisable to consider whether it is actually necessary to include the whole of the workforce. If, for example, the other methods outlined at the start of this chapter, eg using HR data etc, have already identified a 'hot spot' in a particular working group or at a particular location, then it may be better to concentrate the survey on that particular group rather than on the workforce as a whole. It should be borne in mind that the assessment of risk needs to be 'adequate' or 'suitable and sufficient'.

It is also useful, on occasion, to look at individual areas or teams that carry out the same or similar function to ascertain what, if any, are the differences between them. If two groups have identical workloads but one is more prone to stress than the other, the data will help in deciding which interventions may be most effective at reducing stress within the identified group.

It is important to understand that the primary purpose of a risk assessment is not to identify whether one part of an organisation is suffering from more stress than another but whether there are stressors existing in the workplace and whether those stressors are causing stress and risk of damage to employee health (and to the organisation).

Step 2: Decide who might be harmed and how

3.17 Having identified the possible stress-related risk areas within an organisation, the next step is to consider how many of the staff might potentially be harmed by them. Given that stress is an individual response to pressure, this cannot be an exact science. Different employees will have differing reactions to the same workplace stressors. This may be because they have better coping skills, or alternatively there may be those who are concurrently also facing pressure from a home-related situation which, when combined with additional pressure at work, might be sufficient to trigger the stress response.

This part of the risk assessment should enable a recognition of the larger picture and the ability to set priorities for actions recommended. If there is a group of 300 staff involved in a change programme that is not progressing smoothly, then they may very well be a higher priority than, for example, two or three staff in the accounts department, who may be facing additional pressure at the end of the financial year. It is possible, therefore, to tailor the speed and complexity of the response, in the next stages of the risk assessment, to more accurately match the perceived threat.

Step 3: Evaluate the risk

3.18 During the first two steps of risk assessment employers will have identified the hazards and then evaluated who might be harmed and how.

The framework for this part of the risk assessment is to:

- identify the actions and responses that are already in place;
- decide whether these are adequate and if not;
- take and implement decisions for further interventions.

If it is determined that some of the risks are not adequately controlled then it may be necessary to prioritise the actions that are taken so that the most serious situation is dealt with first.

The questions to ask include:

- How many employees are at risk?
- How severe will the pressure or stress be to those individuals?
- Have previous employees in this category suffered from stress?
- Will the duration of the stress be permanent or is it related to short-term pressure, such as the financial year end or the timely completion of a manufacturing contract?

There are various ways to prioritise the risk related to stress. Some methods are one dimensional, ie taking into account one factor such as how many people are involved. However, this approach will rarely be appropriate to stress and a value judgement will need to be made based upon all of the factors which you identified. Similarly risk assessment for stress is rarely based upon technical or hard criteria and there will be an element of judgement. If necessary seek additional expert advice in order to rank the priorities.

In broad terms maximum priority should be given to any situation which may give rise to severe stress in any number of employees, however small, especially if there is a history of stress among employees in that part of the organisation and also if the pressure is likely to be prolonged.

Because risk assessment is viewed across all of the risks in the organisation, priorities may need to be weighed up for action between stress and other risks to employees. As an example would it be better to invest in stress awareness or manual handling training? Only the management team can make these judgements within the organisation and as with all health and safety actions, it is important to show that a decision has been made and the justification for the priority for action is recorded.

Regulation 4 of the *Management of Health and Safety at Work Regulations 1999 (SI 1999/3242)* requires that the work on controlling risks follows a hierarchical procedure.

In outline the procedure is as follows:

The first step must be to identify and deal with risks that could potentially cause the highest levels of damage or harm to any individual. Following that, it is necessary to look at those more widespread risks that could conceivably cause some harm to a greater number of employees.

Once it has been decided which situation is going to be tackled first, then the first priority for intervention must be:

- can the hazard be eliminated altogether?

Where this is not possible, then management should move on to the second tier of interventions which are designed to minimise, or control, the stressor and thereby to mitigate its effects.

In this instance, the following are factors that should be investigated:

- **Risk avoidance:** Is there another way of carrying out this work that would entail a lower risk of excessive pressure on staff?
- **Combating risks at source:** Can the method of working be reorganised in order to reduce exposure to the hazard? It may be, for example, that in particularly pressurised environments, there is a need to give the employee additional work breaks. An example here might be air traffic controllers whose work incorporates scheduled break patterns during each shift.
- **Adapting work to the individual:** Can staff be rotated through alternative duties to provide a relief from the pressures of one of the aspects of the task? Is the working environment right for the individual?
- **Developing a coherent overall prevention policy:** This should cover all aspects of the work including ergonomic workplace and equipment design, technology and social interactions.
- **Prioritising collective measures over individual measures:** This can be achieved by tackling the sources of stress and considering the most beneficial assistance available for the person affected. It is at this stage that consideration may be given to providing stress awareness and other coping skills training. Consider how employees who are suffering from stress, be it from work or home, can best

be supported? Are there procedures in place to offer confidential counselling and support? What action should be taken regarding the management of any employees already off work due to stress-related causes?

The format of Step 3 of the risk assessment will depend very much on the circumstances of the particular organisation and on the specific hazards that have been identified in the initial two steps.

A good starting point is to return to the (HSE's) seven risk factors to identify which of them could be identified as the possible causes of stress. Consideration should then be given to eliminating or modifying those hazards that were identified as potentially having the most serious effects. Where the result shows that there are remedial aspects that are within the scope of management, then an action plan should be implemented to improve the situation, as early as possible. (CHAPTER 10 will deal with some appropriate interventions.) It will also be necessary to make sure that the effectiveness of the interventions is measurable. For each of the seven factors, there is a requirement to ask the following types of questions.

Factor 1: Culture

3.19

- What steps have been taken to make the culture of the organisation one that is positive, staff supportive, open and honest?
- Are there good channels of information and communication open in both directions? Is the information available as complete and clear as possible?
- Are employees able to contribute to the way that the work is carried out? Is there a team-working approach to problems?
- Are facilities available to employees to report when they feel under excessive pressure or are suffering from stress, without it being career threatening? If not, why not? Is stress treated as a sign of weakness? Is it part of the organisational culture to ignore the consequences of undue pressure on staff?
- How committed is the organisation to the welfare of its employees? How committed are the employees to the organisation? Would they recommend a friend to apply for a job within the same firm?

Factor 2: Demands

3.20

- Does each member of staff have sufficient resources to tackle their required level of work?

- If they are unable to cope, do they feel able to discuss priorities easily with their line manager?
- If it is not possible to change work priorities, are there procedures in place to supply additional resources when needed, such as additional people, more equipment etc?
- How often do the team leaders involve their teams in discussions about the level and quality of work required? Do they encourage participation from all of the team in order to devise the best ways of working?
- Are there procedures in place, in and around the workplace, to protect staff from violence and aggression? Are bold, published statements prominently displayed to the effect that legal action will be taken against those who threaten and abuse employees? Has training been provided for employees so that they can defuse rather than escalate these situations?

Factor 3: Control

3.21

- Careful consideration should be given as to how much control staff can be given to manage their own workload. Is it possible to allow staff some discretion in planning for themselves the way that they work?
- Are staff suggestions about the way they carry out their duties evaluated and properly considered, or are they merely ignored or even rejected out of hand?
- Are they being over-managed?
- Do they feel that trusted to work without their every move being scrutinised?
- Is there organisational encouragement for the formation of self-managed teams? If so, are they allowed some degree of autonomy without constant checking?

Factor 4: Relationships

3.22

- Is there an existing culture whereby every individual is treated with a dignity and respect? Does this permeate to the far corners of the organisation so that everybody understands that bullying and abusive behaviours are totally unacceptable?

- Are there agreed and published procedures for dealing with bullying and harassment? If so, how effective are these policies? (See CHAPTER 6.)

Factor 5: Change

3.23

- How well does the organisation manage change?
- Are all employees kept informed about potential work or organisational changes and the reasons for them? Are they informed of the timetable and whether they themselves will be directly involved or not?
- Does the official communication channel work better and faster than the grapevine, ie do employees currently hear information directly from the management team or is it received piecemeal in the canteen?
- Is there a specific policy or risk assessment procedure for dealing with change? Are the psychosocial hazards as well as all of the other consequences of the changes that are planned, properly considered?
- Remember to re-assess a situation subsequent to the actual change taking place. Are the overall risks to staff less or more than previously? This is particularly important when the change has been a reduction in staffing levels. Is there a possibility that the remaining staff will be overloaded?

Factor 6: Role

3.24

- Does each staff member have a clear job description and a contract of employment, specifying in clear language their duties and responsibilities?
- Does each employee know exactly what is expected from them in terms of output and scheduling etc?
- Are appraisal and performance review procedures in place and effectively carried out?
- Is there an effective and comprehensive induction process for all new members of staff?

Factor 7: Support

3.25

- Is there an existing culture that supports staff when mistakes occur and does it encourage a fair investigation of problems so that appropriate remedial action can be taken?
- Are staff encouraged to seek support for stress in the early stages? Do they know how to access the support that is available?
- Do managers set good examples regarding work/ life balance? Are staff encouraged to work reasonable hours and to take their full leave entitlement?

Another method that can be used to assess the risks of stress is participative risk analysis (PRA), which is widely used in Belgium (Van Emelen J 'Bilan et perspectives de médecine de travail', *Cahiers de médecine de travail* (1996) Vol XXXII, pp 9–15). This is a two-stage facilitated process to enable stress-related risks to be identified and analysed. It is especially useful where there are groups of employees who seem to have high stress levels but without a seemingly clear stressor. It is also used to assist in bringing hidden causes into open debate in a non-threatening manner.

Methodology of a participative risk analysis

3.26

Stage 1	An external facilitator meets with a representative group of employees and introduces and explains the process. Participants write down anonymously and succinctly what they think are the main problems. The facilitator then, through a discussion process assists the group to reformulate the stressors and helps them to collectively identify the causes and consequences. Prior to Stage 2, the facilitator prepares a structural map of the problems showing their relationship, causes and outcomes.
Stage 2	At this second stage, the group meets to discuss the structural map and then evaluates the problem in terms of importance, frequency and feasibility of solutions. A formula is then applied to indicate the impact of each stressor. The group then proceeds to prepare a summary report and an action plan to indicate the priority areas for intervention and to propose suitable actions to take.

Step 4: Record the significant finding of the assessment

3.27 Once the risk data has been collected, it is then necessary to consider the overall picture and devise suitable interventions that will address the problems that have been identified.

Assuming that there are more than five employees, then details of the risk assessment must be recorded together with the significant findings and the interventions proposed. The record should show that all required checks have been made, with consideration given to who might be affected and that the employer has dealt with any significant findings.

It is also necessary to show that, as far as is practicable, the identified hazards have been removed and that any remaining risks have been minimised. The *Management of Health and Safety at Work Regulations 1999 (SI 1999/3242)* require the employer to make 'suitable and sufficient' risk assessments. It is not necessary to cover any and every possible situation, but to ensure that hazards that can be identified, have been so addressed.

These should be in sufficient detail to enable an inspector to follow the subsequent decision-making processes and provide further advice and recommendations should that be necessary. This record will become important evidence should the organisation be involved in any future proceedings for civil, or even criminal, liability.

Apart from recording the findings for record purposes, it is essential that the findings are communicated to all staff.

There is a requirement under the *Safety Representatives and Safety Committee Regulations 1977 (SI 1977/500)* to not only inform but consult with the safety representatives of recognised trades unions. All other employees not represented by trades unions must be consulted directly or via their elected representatives under the *Health and Safety (Consultation with Employees) Regulations 1996 (SI 1996/1513).*

'However successful organisations often go further than strictly required by law and actively encourage and support consultation' ('Successful health and safety management', HSE (2000) (ISBN 07176 1276 7)).

Typically effective communication would be ongoing throughout the risk assessment process. The final communication would include a recap of the project, a review of the methodology, and an outline of both the results and the recommendations. This communication should be both through the formal channels outlined above and by other channels of communications used in the organisation such as team briefings, newsletters and the intranet. Part of the reason for this extensive communication is to prepare the employees for the changes that may ensue from the implementation of any recommendations.

Step 5: Review the assessment at appropriate intervals

3.28 It is essential that a timetable of reviews is set into this process and recorded within the assessment.

The timing of reviews needs to be appropriate and this is a judgement that only management can make. When introducing stress-management interventions, it is necessary to check their effectiveness and that they are achieving the desired effect. If further changes are being planned, then the assessments should be reviewed before the changes are implemented.

The organisation should also review the situation in relation to individual employees who are known to be facing pressures in their own domestic life. People in this position may be more vulnerable to any additional pressure from work. Their minds may be distracted and consequently their decisions may be affected. It is, therefore, important to take special care if such employees are in safety critical roles.

Key learning points

3.29 Risk assessment is a straightforward and logical procedure as well as being a required legal compliance and can prove to be of positive benefit to both an organisation and its employees. By carrying out thorough risk assessments staff can be prevented from being adversely affected by stress, by planning the most suitable interventions that will effectively reduce or eliminate the problem(s).

- A written Health and Safety Policy is mandatory for all firms with five or more employees.
- Employers have a legal duty to carry out risk assessments.
- The primary purpose of risk assessment is to identify and eliminate or reduce the risks to employees of workplace hazards.
- The HSE publication 'Tackling workplace stress' (2001), includes the five step risk assessment process as applied to stress-related risks.
- There are seven factors which have been identified to be stress factors which employers should consider when carrying out a risk assessment for work-related stress.
- The results of the risk assessment and any actions taken must be recorded.

(cont'd)

- The assessment must be reviewed at appropriate intervals of time or when changes to work systems are being considered.
- A stress audit can be a useful method of identifying workplace hazards and those employees who might be at risk.
- It should be borne in mind that the act of carrying out a risk assessment is in itself only valid, provided that it is carried out correctly and effectively. Failure to perform a risk assessment or carrying one out in an invalid manner could leave an employer in breach of necessary compliance under the law and open to litigation or possibly prosecution.

4 Identifying Current Workplace Stressors

Introduction

4.1

> 'To take great pride in the effort of one's work is of benefit both to the soul and the body, for he that uses the gift of his talents during the day will surely sleep soundly with satisfaction during the night!' *Anon*

The first three chapters of this book looked at the nature and definition of stress, the legal position and how standard procedures can be used to assess the risk of stress within an organisation. In this chapter we will therefore develop this theme by focusing on the causes of undue pressure in the workplace in order to help identify typical causes of potential stress. We will also look at the scale of these stressors and where an organisation will need to focus its efforts in order to eliminate or to reduce the effects on employees.

Throughout the chapter case studies have been included to illustrate potential problems. All these case studies have been taken from real life examples, although the names have been changed to maintain confidentiality.

The first step for any organisation in beginning to combat workplace stress is to recognise, in the first place, that there might be a problem, ie that any organisation might be open to the possibility. This recognition could be as a result of personal observation and/ or as an outcome of the risk assessment process described in CHAPTER 3. Whenever, or wherever, the occurrence of stress is identified, either in an individual or in a group of employees, action must be taken to investigate the cause(s) as early as is practicable.

The focus of this chapter will be to look at some of the most common causes of workplace stress and it will also introduce a simple tool – 'stress mapping' – which can be used to assess the extent to which various stressors might be affecting any particular employee or group, thereby creating a platform from which to begin to address the problem.

Who suffers most from stress in the workplace?

4.2 In its 1992 publication *Preventing Stress at Work*, the International Labour Organisation (ILO), reported evidence that workers in a wide and growing range of occupations were at risk from stress. This list has itself grown steadily during the intervening years and the following are just some of the occupations that appear to have potentially higher levels of stress than others:

- Air traffic controllers.
- Blue collar workers.
- Bus and truck drivers.
- Civil servants.
- Construction workers.
- Fire fighters.
- Healthcare professionals.
- Journalists.
- Miners.
- Police.
- Postal workers.
- Social workers.
- Teachers.

The relationship between occupational and home-based stress

4.3 In order to consider the treatment of stress in its correct context, it is worth commenting on the relationship between occupational and home-based stress. Whilst employers do not have any direct duty of care in regard to ill-health caused by stress that results from factors external to the workplace, stress from whatever cause can nevertheless have a measurable effect on the performance of employees and so may make it difficult for them to withstand the normal pressures of work.

As an employer, it is therefore important to periodically evaluate the robustness of all employees, in relation to their work, and to offer support to any individual who appears to be suffering unduly from stress.

As Figure 1 shows, employees can be exposed to a range of stressors, both from work and from home, some of which they will be able to cope with (the arrows that are shown as bouncing off) and some of which they will not.

The most serious problems can occur when home-based and work-based stressors are concurrent, with the consequence that the combined pres-

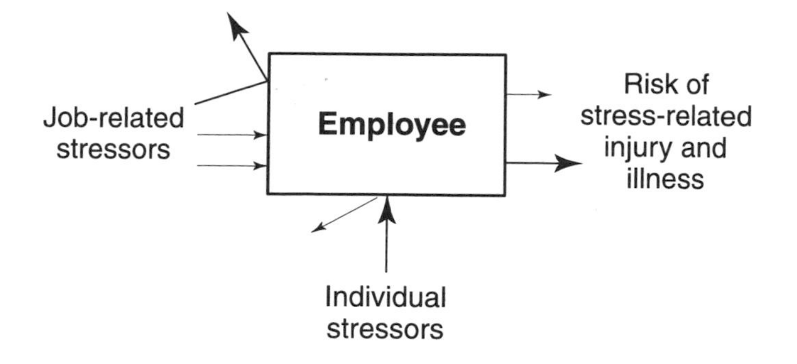

Figure 1: Impact of occupational and home-based stressors

(Figure 1 adapted from NIOSH model of work stress.)

sures faced, will adversely influence the ability to cope. These influences may well increase the risk of stress-related illness. This is the rationale for providing 'stress awareness and coping skills' training for all employees, regardless of the cause or location of the individual problem. Such training can assist people to identify signs of stress in either themselves, or in others, as well as giving them the skills they need to better counter excess pressure(s) and to perform more effectively.

Stress mapping

4.4 Stress mapping is a tool which can be used (with the help of a trained facilitator) to provide a visual representation of the perceived sources of stress between two or more persons within a group, or possibly from other (inanimate) factors such as adverse working conditions. It should provide an indication, both of the inter-relationship(s) between the various stress factors together with a measure of their relative importance. When used to determine problems between individuals, stress mapping provides a method of helping them to identify the causes of their own personal stress. When used within a team, it can help identify the apparent sources of interpersonal and environmental stress.

Whether the subjects are individuals or a team, the technique is very similar. The subject is depicted in a box in the middle of the map and around which are placed boxes for each of the factors that are perceived to be the causes or potential causes of stress. The links between these are then scored to show the perceived stress level passing in one or more directions – with '10' being the highest perceived pressure and '0' being nil pressure. For example, if an individual perceives that another person in their team is causing them stress, they would draw an arrow between their two boxes and then write a score next to each box representing the notional amount of pressure being received by each person from the other and vice versa.

In this example (Figure 2), Mike is the individual under pressure and Sue is the colleague whom he perceives as causing him stress. The arrow represents the flow of pressure in each direction. Mike feels a substantial amount of pressure from Sue and has scored this as a 10. He accepts that he is also putting some pressure on Sue but rates this much lower at 2.

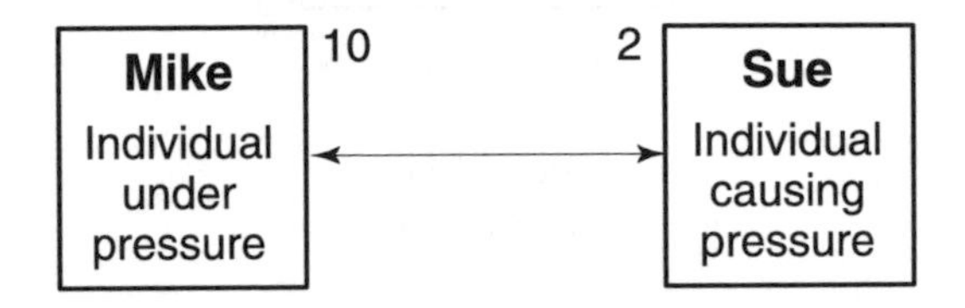

Figure 2: Two-way stress map

This approach can be extended by drawing boxes to represent various colleagues and/ or fellow workers within the work environment by indicating and scoring each of the links between particular individuals. This helps each individual to identify those areas of their working life that are either within, or not within their ability to cope. Either they, or the facilitator, can then discuss the different reactions they are experiencing to the pressures to which they perceive themselves to be exposed.

Clearly, it is the perceptions of the individual or team that matter, as the directions and levels of stress shown in the map will (necessarily) be matters of subjective opinion.

This is particularly relevant when working within a team for, as we have seen in CHAPTER 1, it is commonly found that different individuals have very differing perceptions of stress. However, it is important to note well these differences as they can often provide valuable insights into the cause(s) of individual stress. Talking to those members of a team who perceive lower stress levels can also be useful in providing input on the coping mechanisms they employ in order to reduce or minimise excess pressure.

Case study

4.5

Stress mapping

Richard is a Service Manager in a manufacturing company and is responsible for a staff of twenty four. Unfortunately, he has frequent disagreements with his departmental head. He is also usually under pressure from customers, primarily relating to deliveries and parts. In addition, he has the maintenance of the company's ageing plant with which to contend, together with plant or machinery breakdowns that tend to disrupt delivery schedules. Richard feels that the frustration engendered by these aspects of his work is the main factor in causing him to suffer disturbed sleep but he is unable to decide where his priorities should lie in dealing with the stresses of his job.

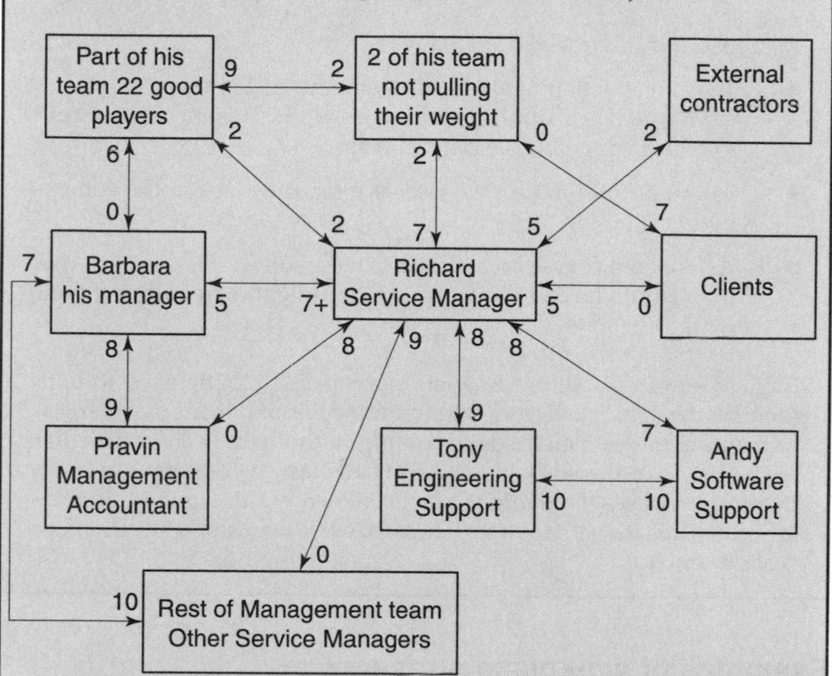

Figure 3: Stress mapping

From Richard's stress map, it was clear that he was the subject of undue pressure, only some of which was directly related to his role as a Service Manager. These indirect causes of stress included:

- Barbara, his Line Manager. (Score 7+)
- The two members of his own staff who were not pulling their weight. (Score 7)
- Pressure from the two support services (Engineering and Software Support). This was mainly caused by the lack of co-operation between the engineering areas, where the managers were blaming each other for problems not being resolved (both causing a score of 8 to Richard and causing each other a score of 10).
- Stress being caused by his manager's relationship with the rest of the management team (Score 10) and the company's accountant (Score 9).
- Richard was to an extent the 'agony uncle' and spokesperson for these other members of the management team. He was

interested to see that his direct problems with Barbara, were in effect, less than those faced by other members of her team. (Score 0)

As a result of this analysis, Richard:

- Took firm action against the members of his own team who were under-performing and, as a result, causing aggravation to both their colleagues and clients.
- Set up a multi-disciplinary service team to service the company's equipment.
- Arranged a course on stress management, for his team, so that they would be better equipped to handle the pressures of their work.

Richard's example shows how stress mapping can be used to both identify the causes of stress within an individual, team or organisation and to indicate methods of dealing with them. In this particular example, all of the stressors noted by Richard were people-related, but in other cases there might also be boxes containing environmental or organisational factors such as adverse working conditions or work overload.

Examples of workplace stressors

The organisation as a stressor

4.6 In simple terms, work-related stress can occur when there is an imbalance between:

- the *requirements* of the job and the *ability* of the employee (using the resources available to them) to complete it; and
- the level of control (if any) that employees have over the way that their work is carried out.

We now go on to look at the many ways in which such imbalances can occur.

Demand and control

4.7 As already mentioned in CHAPTER 3, demand and control are two of the seven key factors identified by the Health & Safety Executive (HSE (2001)) as being the likely causes of excessive pressure and the resultant stress. The others are: culture, relationships, change, role and support, and factors unique to the individual (we shall be dealing with these later in the chapter).

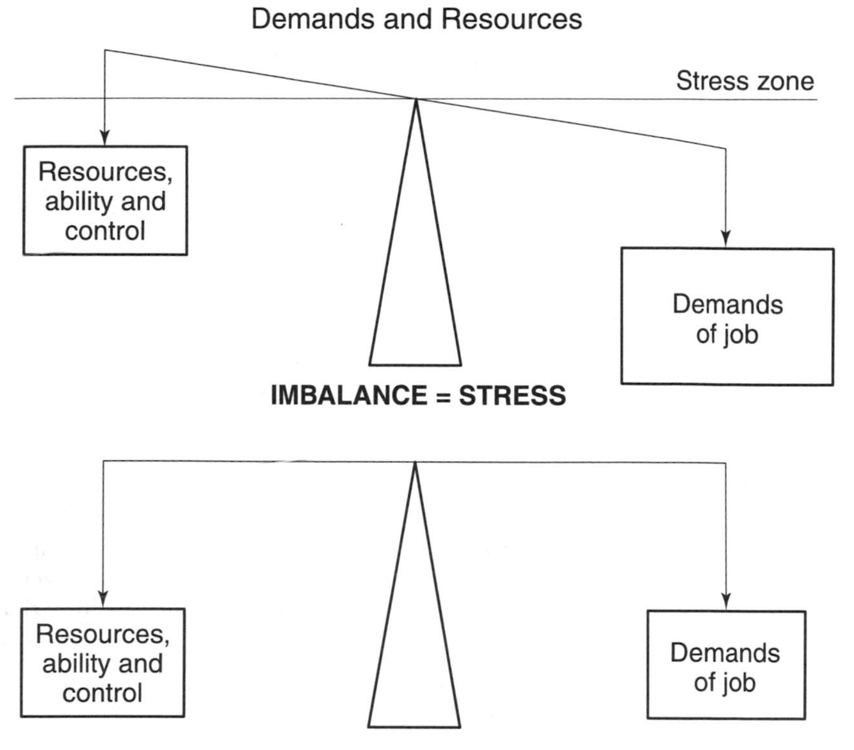

Figure 4: Balancing demand and resources

Demand can be caused by:

- Work overload

  For example, either too large, (or too difficult), a workload to complete either in the time allocated or by having too few resources such as time, staff, equipment or personal ability.

- Work underload

  For example, either an insufficiency of work or boring, repetitive tasks that provide little stimulus or challenge, often leaving the worker frustrated and irritated.

- Physical environment

  For example, factors such as excessive noise, vibration, extremes of temperature, bad ventilation, insufficient fresh air, inadequate lighting or poor hygiene.

- Psychosocial environment

  For example, issues such as the non-organisational characteristics of work structures, including, for instance, workplace violence by one or more individuals (either actual, verbal or threatened).

Control, on the other hand, is the amount of influence that an individual has over the way his or her work is carried out. As an example of the effects of this, according to 'Work-related factors and ill health: The Whitehall II study' (HSE266/2001), an insufficiency of control may indirectly increase the incidence of alcohol abuse and poor mental health. The study looked at work-related factors and ill-health within the civil service and concluded that increased participation by employees in decision-making produced greater job satisfaction and higher self-esteem.

The demand and control model (DCM)

4.8 The demand and control model as shown in Figure 5 (Johnson and Hall (1988); Karasek and Theorell (1990)) illustrates the links between demand and control and has recently been extended to also include the social support, if any, that the individual receives.

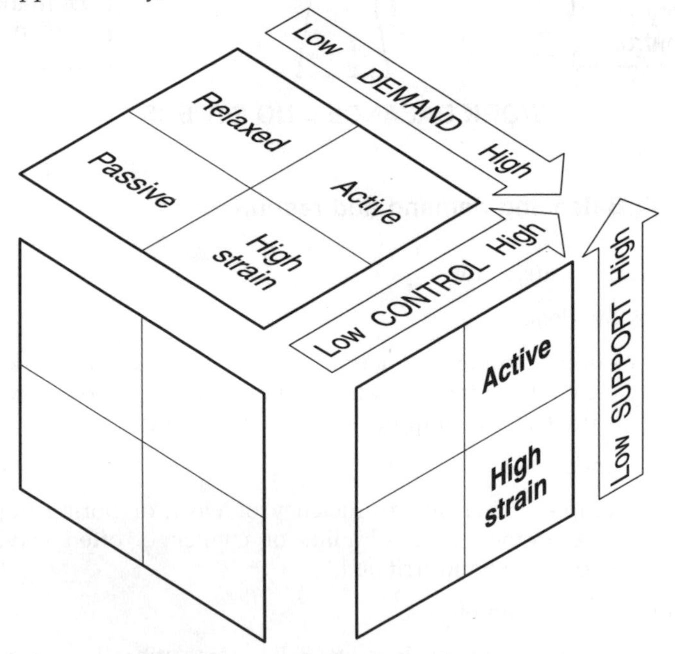

Figure 5: Demand and control model

(Figure 5 reproduced from B Gardell, and G Johansson, eds *Working life: A social science contribution to work reform* (1981) with permission of John Wiley & Sons Limited.)

According to the model, high demands should not automatically be regarded as stressors. Highly complex and demanding jobs that match the qualifications and ability of the individual may well provide high levels of job satisfaction and may even have a positive effect on health.

Taking each of the top four quadrants in turn:

- *High strain* can occur when there are high demands on the individual and they have little or no control over the way in which their work is carried out.
- *Relaxed* is when the demands are lower and the employee has a large degree of control. The employee will feel relaxed and is unlikely to feel much, if any, pressure.
- *Passive* is where both demands and control are low. This can lead to stress through work underload and the consequent lack of challenge and stimulation.
- *Active* is the area where demands are high and control lies with the employee. Here the risks of excessive pressure are lower and productivity is likely to be high.

The third dimension in the model indicates the support that the individual receives. An employee who is in the 'active' quadrant and working well under high demands can be pressurised by a lack of support and can quickly move into a 'high strain' situation. If their manager does not recognise their effort, then the employee will feel that his or her efforts are not rewarded. Productivity will drop, and the pressure and the possibility of stress will increase. Support can also include training, resources and to an extent, reward.

It should be borne in mind that high levels of support and appreciation will lessen the pressure in all of the top four quadrants. Conversely, where the manager takes no constructive interest in the work of his or her team, the situation can damage morale and productivity and lead to increased pressure. Any incidence of harassment or bullying will also cause increased pressure and stress, no matter what category the demand and control profile (see CHAPTER 6).

Case study

4.9

Demand and control

Simon, a newly appointed Account Manager in a small PR company, supervised two Account Executives, Mandy and Richard, who were assigned to his team. They were both conscientious workers and produced work and results of a similar standard. Simon and Mandy

(cont'd)

> shared a love of sport and had also a similar sense of humour, but Simon and Richard, on the other hand, did not. Over the six-month period that the three worked together as a team, Mandy's work was invariably held up as an example to the rest of the organisation, while Richard's was simply viewed as that which 'was expected'. As Richard was already contributing to the best of his ability, he quickly became disenchanted and demotivated by what he saw as favouritism and 'office politics' and decided to seek a position with the competition, where he would receive more appreciation and stand a better chance of promotion.

Work overload and underload

Work overload

4.10 In CHAPTER 3, we saw that work overload can be either quantitative or qualitative.

Quantitative overload results when the demands of the job exceed the capacity of the person to carry it out correctly and within a given timescale.

This occurs when an employee is required to do too much work within the time available. Some jobs, for example, may have unreasonable targets or demand completion dates that are beyond the ability of the individual.

Qualitative overload refers to work that is too difficult for the individual to complete.

This is the position when the job taxes either the mental or technical skills of the employee beyond his or her individual abilities.

Work underload

4.11 Work underload implies that a job is not sufficiently challenging and therefore fails to maintain the ongoing attention or interest of the employee. This includes targets that are set too low, together with repetitive work routines that can result in boredom and under-stimulus. Certain workers, eg emergency service personnel can face particular aspects of work underload in that they may often have to deal with long periods of relative inactivity, interspersed with periods of high demand and, on occasion, crisis action.

Case studies

4.12

Emergency services

Frank and Dave were firemen based in Nottingham. They had both been in the service for some years but were each affected, in their own way, by the unpredictable nature of the job. Although there were frequent call-outs, less than 40% of these turned out to be genuine fires in which their professional skills were needed. Much of their time was spent attending to routine maintenance work (that was nevertheless still vital to the efficiency and safe running of their vehicles, machinery and equipment). Consequently, their jobs imposed on them an intermittent start-stop work demand in which their physical and mental resources were called upon to give peak output during emergencies and then to return to 'waiting mode' for long periods.

However, Frank was a keen gardener and had turned his house and garden into a showpiece of flowers and shrubs. Any spare time he had at the fire station was spent assiduously learning the Latin names of all his plants. He was a sought after speaker at his local horticultural club but more importantly, he was always in control at work.

Dave, on the other hand, was entirely different. Although he had been in the job for the same number of years as Frank, he had never managed to adjust to the stop-start rhythm of the work and was only really happy when speeding along the road with the siren blaring and blue lights flashing. Afterwards, having returned to the station (and particularly the following day), he would invariably become taciturn and withdrawn. Deprived of his essential stimulus, he would answer in monosyllables until the bell next rang for active duty, which could be hours, or even days, later.

Monotonous tasks may involve repetitive movement and the lack of stimulus can slow concentration and reactions. This can lead to a declining output, poor judgement and an increased incidence of error or accident. Where it is possible to incorporate variety into the work routine, this can often lead to increased performance levels and improved productivity.

4.13

Manufacturing

Mary works on an assembly line at a TV monitor manufacturing facility and is paid a piecework rate. Her job is to press a switch to

(cont'd)

check that the picture tube has power. She has no ability to influence either the speed of the line (and hence the pace of her work) or the stultifying monotony of the repetitive job that she is paid to perform. She is not given any opportunity to change stations with another operator, there is no stimulation of any sort and by mid-morning each day she suffers from a headache that tends to last for the rest of the day.

Jill, however, works in another company, with similar hours but a different routine. She works two days in one department, then two days in another. By doing so, she meets new people and has the benefit of variety in her work that helps to dispel any monotony. Jill enjoys her work and is usually bright and happy as a consequence of her job satisfaction. Unlike Mary, she rarely has a headache and invariably looks forward to her working day.

Potentially stress-inducing work patterns

Shift working

4.14 Shift working has been around for a long time, although it was originally limited to those in the military or emergency services. Then, in the late 1880s, industrialists (such as Henry Ford in America), began to realise the benefit of continuous working schedules. If they ran their operations around the clock, equipment would be fully utilised and both production and profits maximised with little proportional increase in overheads. This became particularly evident in certain manufacturing operations such as steel production and iron foundries, where shutting operations down at night and re-starting them in the morning, would have been not only expensive and inefficient but in many cases, impossible.

Today, many different manufacturing facilities work on a 24-hour basis, as do some in the food and service industries. It is now commonplace to find retail supermarkets, service stations and grocery stores that are open 24 hours, seven days a week. The latest growth area for shift working can be seen in the call centre industry, where some information and products can be obtained or ordered around the clock. Then, of course, there are the various types of employment sectors that by their very nature have always employed shift working, such as the police, catering and transport sectors.

Although shift work may be good for business, it can take its toll on employees and their families. For parents working atypical hours, shift work may have both advantages and disadvantages. On the positive side, parents may be able to work opposite, or complementary, shifts to one another, so that there is always someone at home and they are therefore spared the expense of outside childcare. However, shift working usually limits family time together and can easily put relationships under strain.

The scheduling of shift work can also be difficult for both the company and the worker. The least troublesome schedule is one that has been prepared by management, after consultation with the workforce. This leads to greater co-operation, less cause for complaint and, consequently less stress.

The stress of night work

4.15 Due to the disruption of the circadian rhythm and the impairment of normal sleep patterns, night work has been shown to have the following effects:

- Fatigue is a common problem and can lead to greater susceptibility to illness, poor performance and low motivation. Chronic tiredness often leads to an excess of caffeine consumption and its ensuing potential health hazards (see CHAPTER 9).
- Being overly tired can also induce a short attention span, making it difficult to concentrate and increasing the possibility of errors or job-related injury.
- Some night workers complain of loss of appetite, resulting from the non-availability of suitable food and the requirement to eat at unsociable hours. This can very often lead to indigestion and other stomach problems.

Research by the University of Surrey in 2001 suggested that working irregular hours depresses the immune system, thereby reducing the body's natural resistance and making the body more vulnerable to disease. Workers on such shifts are also more likely to develop heart disease, type 2 diabetes and sleep disorders. Despite all the well-researched, negative health effects, a fifth of employees in industrialised countries now work irregular hours, often involving shift and overnight working.

Night work can have detrimental effects both on the worker and on other family members – disrupting social life and weakening family cohesion.

Social isolation can therefore be a real problem for night-workers and can lead to loneliness, introspection, and in some cases, psychosomatic illness and, increased stress levels.

Accident rates, both at work and when commuting, can also be affected by tiredness and fatigue. The particular danger period is the one that occurs towards the final third of the shift schedule, shortly after the transition from night to day work. This is when our biological systems of resistance are at their lowest ebb. It is of little surprise, therefore, that many disasters of recent years have occurred during night shifts, eg Chernobyl, Bhopal and Exxon Valdez, being notable examples.

Case study

4.16

Night work

John works nights for a manufacturing company. His wife also works (but during the day) and they have two children. As a result of working nights for the past eight months, John often takes sleeping pills to help him sleep after his shift. Unfortunately, by the time his wife returns home, it is almost time for him to leave again for his night shift, so they rarely have time to sit and talk. Their social life has suffered, and John feels very isolated, particularly as there is no one really to talk to at work. Added to this, the work is monotonous and because the factory canteen is closed at night, John tends to eat fast food that he buys on his way into work, where he then drinks countless cups of coffee to keep him awake.

Obviously, John's lifestyle is not ideal and his high intake of junk food has led to him being considerably overweight, which in turn leaves him breathless. He recently visited his GP who confirmed that his blood pressure was too high and advised him to exercise more, see a dietician and lose weight.

On his day off, John saw David Evans (the company's Occupational Health Adviser) to complain of chest pains. On examination, it was confirmed that John needed to re-assess his lifestyle. They went through his daily schedule and structured a lifestyle programme to suit the demands of the job, including a daily, 20-minute exercise programme to help John's overall fitness and stress levels together with a reduced fat and low salt diet. They also looked at ways to effect changes in John's shift work pattern that might serve to minimise the root causes of his stress.

To help John feel less isolated, David recommended that he make an effort to see more of his friends who also work night shifts. John left, feeling more in control and confident enough to discuss these issues with his manager.

Working at home

4.17 Many people enjoy working from home but for others, it can pose problems. It is suggested that the ideal balance for home-based workers is to spend at least a half or one day per week in the office, but if this is not possible, then it is essential to keep regular hours and to meet people outside of the 'home-office' – hence the development of new tele-working 'communities' and breakfast/ lunch clubs.

Those working without daily, peer group support, can generally be divided into four distinct groups, which, although sharing certain characteristics, have different advantages and disadvantages. These groups are:

- Home workers on fixed contracts, eg self-employed sales representatives, etc.
- Employees who are employed by the company but work from home as opposed to from the firm's office premises.
- Self-employed professionals who see clients, or offer consultations from home, such as private doctors, counsellors and health therapists.
- Freelancers who bring in their own work, such as journalists, graphic designers, authors and researchers etc.

Common environmental/ personal characteristics appertaining to these four groups include:

- Isolation and lack of daily peer group interaction such as meeting around the photocopier or coffee machine for a chat.
- A tendency to become demotivated leading to a lack of concentration, low productivity and missed deadlines.

It has been recognised that only individuals possessing a particular temperament may be best suited to home-working. Particular strengths include flexibility, self-motivation, good time management as well as being a self-starter and able to work without supervision.

A strong network of supportive friends is considered beneficial, if not essential, to retaining good mental health. Where people miss the stimulus of inter-personal relationships, they may, over time, become depressed and withdrawn. In this context, it should be remembered that colleagues at work not only spend a substantial amount of time talking to each other about their work but also discussing such topics as their social lives, families and even subjects as television 'soaps'. This frequent interpersonal contact, sense of belonging and mutual concern is fundamental to human needs but can very often be missing when an individual regularly works from home.

Company employees using their home as an office, should therefore be aware that this method of working requires a change in outlook, including:

- A clear understanding of the overall objectives of the (now virtual) team of which they are a member.
- A definition of their responsibilities, role and place in the hierarchical order of the departmental or company.

- Periodic face-to-face personal contact in order to minimise social isolation.

Freelancers and the self-employed, on the other hand, need to:

- Join professional organisations, eg a local Chamber of Commerce, professional societies or business groups
- Be proactive in initiating co-operative alliances and mutual ventures.
- Arrange meetings to maintain contact with colleagues and clients, eg business luncheon clubs.

Overall, the type of temperament and skills required to be a good home worker should include the following characteristics:

- Commitment to the job.
- Not easily distracted.
- Good time management.
- Self-discipline.
- Able to communicate well, by telephone, email, fax and letter.
- Ability to network.
- Flexible attitude to working hours.
- Willingness to be adaptable.

Case study

4.18

Home working

Annie returned to work after her divorce, when her children were at full-time school. She re-trained for a year, studied for a degree, and then worked from home as a self-employed home tutor. Initially she used the family sitting room but soon found the intrusion into living space to be difficult. Eventually she moved, and one of her requirements for the new home was a self-contained space situated near the front door that could be utilised as an office, in order that clients did not have to walk through the living area of the house.

However, after the stimulus of university, Annie found it difficult to accept that on many days, the only people she communicated with were her clients. She missed the interaction inherent in university life. She also found it difficult to motivate herself sufficiently to keep records up-to-date, block out lessons, read new material and market herself by phoning potential clients. Another irritant was that friends and neighbours tended to take advantage of her and appeared not to take her work seriously, eg by asking her to pick their children up from school if arrangements fell through, or to let in building contractors, etc.

On days when there was no work for her to do, it could be difficult to get out of bed, whereas at other times of the year, especially close to exams, the workload could be extreme. As a self-employed worker, it was difficult to turn down work and therefore very easy for Annie to over-extend herself. By contrast, other periods were unbearably quiet and it could be difficult to cope with the lack of control over her workload.

Annie eventually joined a gym to ensure that she went out regularly and because she enjoyed the sociability of the class. She could go during the day, rather than being restricted to weekends when the gym was busier and so she benefited from a cheaper off-peak rate. Annie found it important to keep in contact with friends regularly, although this resulted in higher expenditure on meals, cinema, etc.

On the whole Annie enjoyed her new lifestyle and the freedom to plan each day. Initially she found it difficult to cope with the highs and lows of work but this became easier as a pattern began to emerge.

As a home worker, it was important for her not to become isolated and to accept that she had to make an effort to go out and meet people. She joined a professional association and often travelled to London to attend lectures. This helped her to network on both a professional and private basis, as well as keeping her up-to-date with the new information that was essential to her work and professional discipline.

Hot desking

4.19 Various organisations follow a practice whereby desks (and work stations) are provided on a 'first come, first served' basis. Consequently, staff may have no permanently allocated space of their own. Desks need to be cleared at the end of each working session in order to be available for use by the next person. However, employees frequently find this arrangement to be both impersonal and detrimental to their overall productivity levels as they tend to feel a loss of identity and self-worth.

Another example of the impact of 'e-working' is 'hotelling' ie working for periods of days, or weeks, out of a hotel bedroom using a portable computer. The disadvantages of this type of working include social isolation from both family and colleagues and poor working conditions that are not conducive to optimum performance.

Case study

Gina had a promising career in the sales department of a major telecommunications company. Even though she spent much of her time 'on the road', she enjoyed her regular visits to head office and the opportunity to discuss her sales with other members of her team. One day, and without any consultation, her company introduced a 'hot desking' policy as a means of saving space and reducing overhead costs. Consequently, her usually friendly office environment was eventually replaced by a large, open-plan, open-access concourse; and rather than friends, her colleagues quickly became competitors for available desk space. Few colleagues had the time or inclination to speak to Gina during her office visits. The number of her former friends at work reduced as a result and so too did her commitment to the company.

Short-term contracts

4.20 It is commonplace in today's business world for employees to accept that, upon completion of their own short-term contracts, they may well have to re-apply for their own jobs, in competition with external candidates. For many, this represents a new uncertainty, and one that can cause high levels of financial worry and emotional strain.

No longer is there a realistic expectation of a 'job for life'.

New graduates coming into the work arena have to accept that their career path will have many changes of direction during the course of their working life. It is those individuals who have never known anything other than one particular employer, one location and one boss, who may find these changes harder to accept.

Short-term contracts are particularly prevalent in specific industry sectors such as catering, tourism, car sales, retail sales and certain sections of the IT sector.

Other organisational stressors

4.21 In addition to demand and control, work overload and underload and the various potentially stress-inducing work patterns described above, there are also many other organisational stressors, such as:

- staff assessment procedures that are either too frequent or over-critical;
- inadequate staffing, relative to workload;

- too many permanently unfilled posts;
- poor inter-departmental co-ordination;
- insufficient job training;
- inflexible working patterns;
- inconsistency in style and approach;
- consistent rumours of company takeover or downsizing;
- too frequent policy and/ or procedural changes;
- over-emphasis on competitiveness.

A number of these are dealt with in the following sections.

The role of the individual in the organisation

4.22 An employee's role in the organisation is one of the seven primary risk factors that have been identified by the HSE as requiring assessment. Role conflict, ambiguity and changing roles often contribute greatly to stress, indicating that clear, role definition is a priority.

An individual's role in the organisation should be clearly defined and understood by both employer and employee, and the expectations placed upon them should be reasonable and non-conflicting.

Nevertheless, individuals can and do experience 'role conflict' or 'role ambiguity'. There can be some confusion about these terms, which on occasion seem to be used interchangeably. An explanation of the HSE definition is outlined below.

Role conflict and role ambiguity

4.23 Role conflict occurs when an individual has to face conflicting job demands. These are commonly caused by:

- receiving work and/ or instructions that are not identical, and from more than one source;
- being asked to perform tasks that are perceived as being outside of normal duties;
- having to work in a way that demands two different behavioural patterns, such as working to tight deadlines whilst contemporaneously having to perform long-term strategic tasks.

Role ambiguity

4.24 Role ambiguity occurs when an individual is not in possession of a clear picture regarding the scope or aim of their work. Typically, they might not:

- have a clear understanding of the objectives;
- fully understand what their co-workers and managers expect from them;
- understand the scope and responsibilities of the job.

Common causes of role ambiguity that are experienced, include:

- moving job, or position, horizontally within the organisation;
- moving into the organisation from a similar position outside;
- being promoted to a newly created post;
- being given responsibility for additional staff;
- an individual being uncertain as to where they fit into a new organisational structure with no clear lines of communication.

Relevant questions to be asked in relation to the role of employees should include the following:

- Has the role been clarified and defined either verbally or in writing to both the individual and his/ her colleagues?
- If there has been a job change, has this been fully discussed with the employee in advance of the new post being accepted?
- Has the person's role been explained to (and discussed with) their new team regarding responsibility level and authority structure, prior to the taking up of the position?

Any concerns need to be voiced at an early stage so that expectations, on all sides, are clarified. Open lines of communication need to be maintained so that potential problems can be addressed and any necessary interventions introduced as early as possible.

Case study

4.25

Role conflict and role ambiguity

Susan works in the marketing department of a large department store and one of her responsibilities is the production of the company's quarterly magazine for account customers. The HR department recently decided that the company had now grown to a size where it would benefit from an internal staff newsletter and so, because of her expertise, Susan is drafted onto the team to get the project off of the ground.

At the initial meeting to discuss the newsletter, it emerges that no budget has yet been allocated for its production but Susan is asked whether she would be happy to look after the writing, design, etc, as this is 'only really an extension of her existing role'. Feeling unable to say 'no', Susan accepts the new task. In doing so, she has not only added to her workload; she is now effectively working for two departments rather than one, and on a project for which no metrics or evaluation criteria have yet been established.

Susan finds that the production of the staff newsletter takes up far more of her time than she had anticipated. In view of the fact that both it and the customer magazine are quarterly publications, she finds that she has less and less time to attend to her personal life and has periodic bouts of self-doubt that occasionally border on panic. She rapidly begins to see the newsletter as a 'millstone around her neck' and resents the fact that she agreed to take on the responsibility for it. Her boss (in marketing) has little interest in the newsletter project; while HR's major interest is simply that it is published on time. Having previously enjoyed her job in marketing, Susan now feels overworked, somewhat exploited and undervalued. This is compounded by the fact that she feels unable to speak to anyone or complain about the situation, since she does not want to be seen as being unable to adequately cope with her workload.

Changing roles

4.26 In the quest for ever-increasing productivity (and lower overhead costs), many firms and organisations expect employees to do 'more for less' in terms of assistance and in the setting of unrealistic targets. When given additional responsibilities, staff will usually require appropriate support and training. When these are lacking, then employees can feel they are no longer being treated as people, but merely as human resource items in the competitive race for market share. This is particularly apposite when they lack either any degree of autonomy or clear directions from management.

Responsibility without authority

4.27 There are also instances when individuals are given greater responsibility but without the commensurate authority with which to implement their given goals and/ or targets.

Case study

Leslie was a bank manager for one of the large clearing banks. He started 25 years ago as a clerk and having studied hard for his AIB certificate was eventually rewarded with his own branch. He had discretion to grant loans up to £25,000 and would 'hire and fire' his own staff. Unfortunately, with the advent of corporate mergers in banking and insurance, control became more centralised and progressively taken away from local branches.

As a consequence, Leslie was obliged to seek authorisation before granting any loans over £15,000, (or for more than a 12-month term). His authority and role were clearly diminished and he felt he had once again become 'just another clerk' – now with an important title but little or no authority. With the loss of job satisfaction he gradually became more frustrated and unfulfilled and these emotions eventually resulted in stress. In the event, he turned to alcohol to ease his frustration. After some months, he became a serious drinker who needed 'a little something' about midday to see him through the afternoon and then 'something' again on the way home.

The ultimate consequence to Leslie was the loss of a worthwhile career with an early retirement on a reduced pension. The bank lost a valuable employee whom they had trained over many years, at substantial cost, and who could conceivably have remained a senior employee with at least 15 more years of service to give.

Responsibility for others

4.28 When managers are required to make decisions, they will (hopefully!) find themselves comfortably in control of the situation. According to the demand and control model described earlier (see 4.8), they should not therefore find the decision-making process stressful. However, when taking decisions (the consequences of which will impact, directly or indirectly, upon individuals for whom they are responsible), they may experience stress-related symptoms related to the potential outcomes of their decisions, notwithstanding that they are ostensibly 'in control'.

Such potential for stress can be minimised when the decision-making process is shared among colleagues such as in committees or teams. However, persons who work in high risk professions (such as the emergency services) are often required to make spontaneous decisions, in the light of immediate circumstances, and will therefore not have the opportunity for prior consultation with others.

Other role-based stressors

4.29 Other organisational stressors that impact on the individual include:

- too little, or too much, responsibility;
- insufficient, or no, participation in decision making;
- a lack of support from management;
- onerous time pressures;
- having an uncertain or ambiguous role;
- the compulsion to conform to the culture of the organisation;
- an under-utilisation of personally acquired or inherent skills.

Relations within the organisation

4.30 Among the other issues that may contribute to workplace stress are those that concern relationships with management, colleagues and subordinates. As discussed in CHAPTER 1, every individual has their own perception of the actual cause(s) of their stress, but however subjective that perception, each person's feeling of stress must be respected, if it is to be managed satisfactorily.

Where an employee has poor relationships, either with his/ her boss, peer group or those under their supervision, this may lead to poor communication possibly resulting in anxiety, resentment or even anger (see CHAPTER 5). These reactions will invariably be reflected in levels of work performance. Issues concerning personality clashes can be better addressed if there are specific communication systems in place that facilitate feedback (see 4.33). It is desirable, if not essential, that concerns be raised as they arise, within a confidential (safe) framework and this can be assisted by a trained HR facilitator – enabling problems to be aired openly and in a constructive way.

Other stressors affecting relationships within the organisation include:

- unsympathetic (or frequently unavailable) management;
- social or cultural differences;
- difficulties in delegating responsibility;
- personality conflicts;
- bullying, racial or sexual harassment;
- unrealistic goal setting.

The way the organisation is managed

4.31 Given the frequency of mergers, acquisitions and buyouts, many employees are increasingly faced with organisational change, new methods of working and, all too often, redundancy – all of which are potential stressors.

Organisational stressors that are commonplace, include:

- insufficient staff to complete a workload to a specified schedule, with the result that employees feel overloaded;
- lack of communication or co-operation between departments, resulting in delays, frustration and poor work-flow;
- insufficient training, leading to poor performance, decreased motivation and low self-esteem;
- little control over workload, or rigid working procedures, giving rise to resentment that there is no facility to allow constructive input for possible improvements to existing practices;
- inadequate time being allowed for adjustment to organisational changes caused by restructuring.

All these stressors, (plus others as illustrated in Figure 6), may be further exacerbated by the way the organisation is managed. If, for example, there is inconsistency in management style and approach, or if procedures are frequently changed, workers will tend to feel insecure and/ or apprehensive. Similarly, where excessive competitiveness is part of the culture, tensions between colleagues very often escalate. This can cause anxiety and a sense of failure and consequent de-motivation – the very opposite of that which was intended. This will be particularly problematic if it results in the creation of a 'blame culture', an environment where employees are more interested in highlighting the failings of others, rather than in celebrating their mutual achievements.

'Crisis management' is another very frustrating and stressful way of working and it is often the case that, when looked at retrospectively, a crisis could have been averted had existing communications been improved. Stress is also often experienced by staff who are financially dependent on regular, but excessive, overtime and the ensuing problems to social and family life that this can bring. (There are, in this, similarities to the problems caused by shift work as discussed earlier in this chapter).

Poor management styles

4.32 The style and methodology, with which management tackles day-to-day issues, is clearly important in order to preclude them from becoming problems. At all levels, there will usually be a pressure to perform in

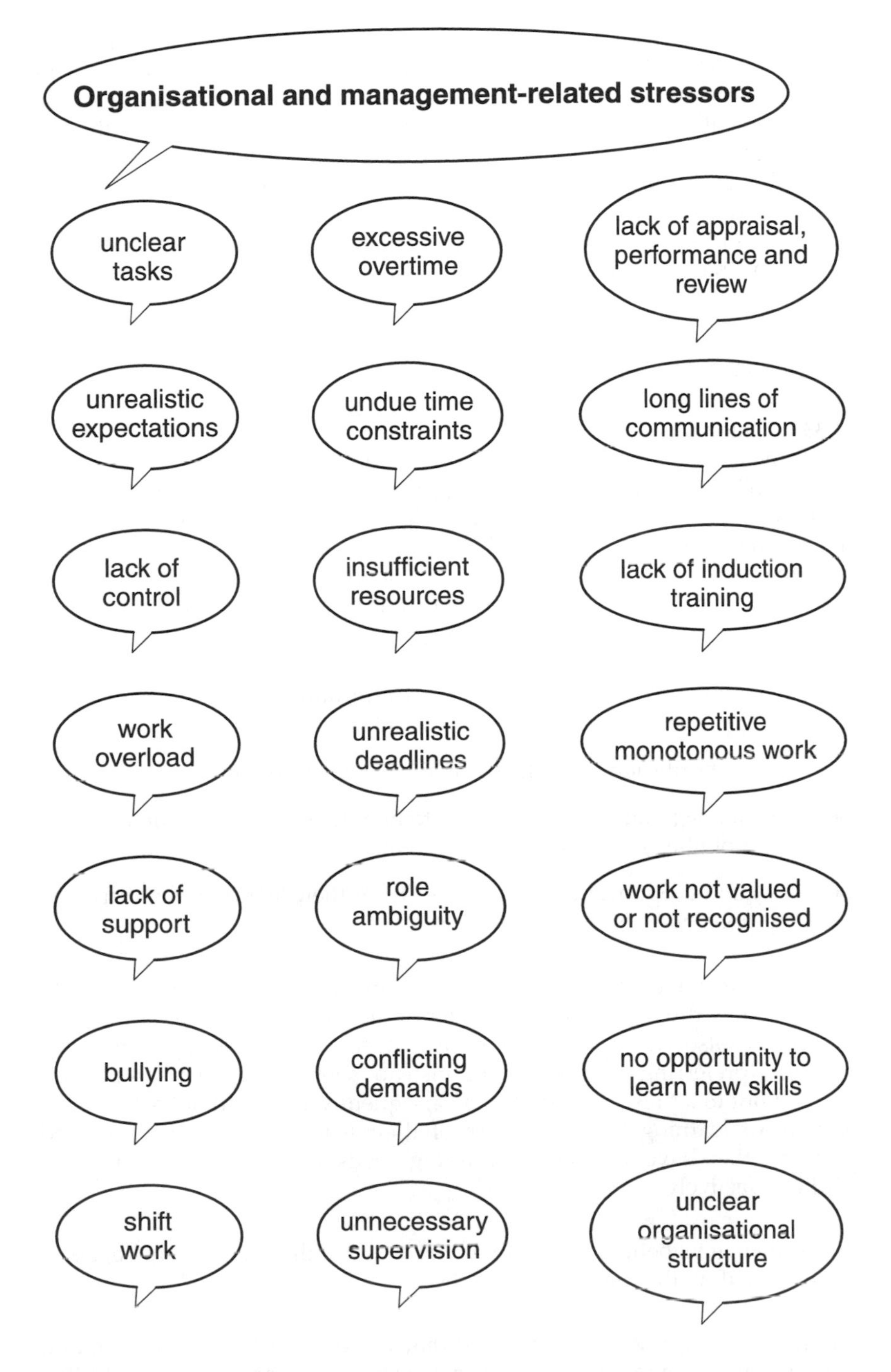

Figure 6: Examples of organisational and management-related stressors

respect of workloads and targets. However, this pressure, if prolonged, can be stressful. Of course, the reality is that there is always the opportunity for stress to occur, as a consequence of the power and authority of the management position, no matter what management style exists within an organisation. It is incumbent, therefore, that the particular management style that is employed, achieves the appropriate balance between consultation and control. In situations where work is delegated, for instance, training and support should be sufficient to pre-empt the emergence of problems that could well have been foreseen.

Poor communication skills

4.33 Unfortunately, there are managers who still believe that actively listening to their employees is a threat to their authority and weakens their ability to issue directives as by doing so they would establish (in their own mindset), too close a personal contact. They may also tend to believe that managing people is solely and exclusively about work-related issues, whereas, in fact, it is important for them to appreciate that individuals may have home-related problems and that these can conceivably exacerbate work-related issues or vice-versa.

Managers need to be proactive in communicating with their teams and need to appreciate that:

- an authoritarian attitude is not always appropriate;
- the managerial skill of 'active listening' to employees, is an essential part of effective people-management;
- obtaining optimum performance (from their team), is the duty and responsibility of every manager.

There are managers who use bi-annual appraisals as the only times with which to talk directly to every member of their staff, on a one-to-one basis. There are also managers who refuse to see appraisals as a two-way process, preferring to view them instead solely as a mandatory task carried out to a prescribed format. Consequently, they will not achieve the benefit of learning how members of their team are coping, or indeed whether they have any specific training needs in order to carry out their duties effectively.

Some managers believe that listening is a 'soft skill'. That it may be, but it is a soft skill with a hard edge attached.

Active listening involves indicating that the speaker is being heard, and subsequently being offered feedback. The manager who uses 'active listening' as a management tool will invariably generate, in his or her team, an ethos of responsiveness, loyalty and co-operation.

Some managers (or team leaders) are, unfortunately, 'stress carriers' and create, rather than reduce, the incidence of stress in their department. In such cases, it is not unknown for employees to look forward to those times when their managers are out of the office, in order that they can get on with the job in hand.

Lack of training

4.34 Adequate induction procedures and training are essential for all new employees and also in situations where new working practices, or system changes, are being implemented.

Case study

In July 1999, Beverley Lancaster, a housing worker employed by Birmingham City Council, was awarded £67,000 damages, after her employer admitted liability for personal injury caused by stress. Beverley had worked for the council for 26 years in a variety of posts, before being forced to retire on grounds of ill-health. For the first 21 years she had an exemplary work record but then, when her post was abolished, she was transferred to another job. Despite repeated requests, Beverley neither received any training for the job (that would have included how to deal with members of the public in potentially violent circumstances), nor did she receive the administrative support given to other staff working in the same area. Eventually Beverley found the levels of stress unbearable and had no option but to give up her post. (See also CHAPTER 2.)

Lack of support

4.35 In this, and foregoing chapters, we have made frequent references to 'support'. This term includes ongoing, day-to-day backing from supervisors and colleagues as well as constructive feedback from line managers, in addition to any required technical support. As we have seen (in the demand and control model at 4.8), without adequate support from colleagues and managers, employees are likely to lose motivation and commitment – with consequent adverse effect on performance.

Case study

In January 2000, Worcester County Council admitted liability and offered their employee, Randy Ingram, £203,000 for the prolonged stress which had forced him to take premature retirement at the age of 39. Ingram had successfully worked as a warden, managing travellers' sites over a two-year period. However, when the site management was transferred from the district council to the county

(cont'd)

council, there was a marked difference in the support he received. When the residents became aware that the warden no longer received the support of the district council, a minority took advantage of this and subjected the warden to violent and abusive behaviour. Ingram was the third warden on the site to suffer from stress and he increasingly felt powerless and isolated. The stress and subsequent depression affected his home life and necessitated hospital treatment as an in-patient, prior to his enforced retirement on grounds of ill-health.

Career development (or lack of it)

4.36 There is little doubt that any individual's self-perception is primarily influenced by the success (or lack of it) in the development of their career. This usually becomes central to how they judge themselves and the level of their self-esteem.

However, if their chosen career path is not progressing as they would wish, they will often blame their lack of progress on the way that the organisation is managed, the opportunities that are, or are not, available and the perceived fairness of promotion within the organisational structure. It may well be, of course, that the primary factor influencing their career progress lies more within themselves than elsewhere.

Issues related to career development can also be influenced by the trend to more flattened, organisational structures. Within a hierarchical arrangement, where individuals usually receive regular promotion, it is relatively easier to be satisfied with career development. However, in a flattened or 'banded' structure it is often more difficult to ensure personal advancement.

Various key stressors experienced by individuals in relation to their career development, include:

- *Under-promotion* – people being over-qualified for the position they hold. They may have originally accepted the job as a temporary measure in order to make ends meet but run the eventual risk of high stress levels due to boredom, frustration and lack of stimuli.
- *Thwarted ambition* – an individual's expectation may exceed the job they are doing. They may have been in the company for many years and then have to accept the fact that someone younger and possibly less experienced, is promoted over their head.
- *Job has insufficient status* – often, the importance of the work carried out by an individual is given insufficient recognition and the position becomes, therefore, undervalued and consequently under-rewarded.

- *Over-promotion* – employees having being promoted, realise they have insufficient training and may be under-qualified for the new position and their increased responsibilities.
- *Insufficient resources* – employees feel that they have been given inadequate tools (or time) to deal with the tasks that they are required to perform.
- *No parity with others who do similar work* – individuals may feel frustrated and angry at receiving that which they perceive as being insufficient reward for their endeavours.

Fear, uncertainty and change

4.37 Fear can be another important stressor in today's workplace, especially fear of unemployment, debt or failure. Individuals will rarely admit to experiencing stress-related problems as a result of these (sometimes irrational) fears, and will certainly not wish to be viewed by their employers as being weak or unable to cope. Nor do they wish to be considered potential candidates for the next round of redundancies.

Many people now live to work, rather than work to live.

Fear of the future

4.38 If people are not kept well informed about what is happening within their organisation, there will inevitably be a process of rumour, based either on misinformation or incomplete facts. However, they should they be made aware as to how a particular position or change may affect them personally, then any misinformation being circulated will lose its potential to cause anxiety.

Of course, fear of the anticipated effects of a change in company policy can rapidly circulate around the office especially when anyone becomes anxious regarding their future job security. Such a scenario is not uncommon and makes it the specific responsibility of organisations to be as clear as possible regarding imminent changes in company policy or methods.

Uncertainty and change

4.39 The anticipation of the consequences of organisational change can frequently be more negative in impact than the actual change itself. Rumour and conjecture, combined with inaccurate information of changes elsewhere, may all contribute to the generation of a disproportional fear of change and a consequent, heightened resistance to it.

This may apply particularly when individuals can personally see no logical reason for any modification to the status quo. A restructuring or reappraisal of our personal role can have a significant impact on our self-perception. Whether we perceive the change to contain either threats or opportunities can have a profound effect on us. It is recognised that many managers, and employees, find their organisation's handling of such matters, to be inadequate.

The effect of change on morale and productivity

4.40 Change is acknowledged to be an area that can also have a decided effect on both morale and on individual stress. However, it is less well appreciated that it also has an effect on productivity. Although the aim of most organisational change is to increase business efficiency, the process itself may actually lead to a drop in productivity, making the stated goal more difficult to attain within the timescale envisaged.

In a publication by L Clarke entitled *The Essence of Change* (1992), the relationship between productivity and change is illustrated. Typically during the change process, productivity will follow a similar downward trend and level out and start to rise as the employees go through the acceptance stage.

Job insecurity

4.41 The strains of job insecurity can also indirectly affect safety. The threat of job losses has been known to reduce motivation, and in turn, employee compliance with health and safety rules. These changes in attitude can often be linked to an increase in accidents, where employees, who have to juggle the conflicting demands of production, quality and safety, cut corners in order to hit their production targets. This is especially true if workers are worried that they may lose their jobs, and are therefore not actively motivated to work in strict accordance with health and safety regulations.

Redundancy

4.42 Redundancy is an emotive word. Fortunately, many companies are now moving away from large-scale redundancies and instead towards long-term planning of staffing levels through processes of natural wastage such as early retirement, etc. However, no one stratum within an organisation is ever completely immune from the implications of revised manning levels and the threat of redundancy, because, in the final analysis, employees are too often perceived as an expendable commodity that can (easily) be replaced.

The real cost of staff cuts can, however, be immense. When procedures are handled badly, (as is sometimes the case) the ensuing repercussions of redundancies can sometimes threaten the infrastructure of even the largest company.

A redundancy programme can leave in its wake a trail of despondency and insecurity, plus a weakened morale in those departments affected.

Of course, redundancy does not begin and end with the selection of which particular individuals are scheduled to be 'let go' in a company's downsizing exercise. The process is far more complex and its execution demands specific managerial skills if the detrimental, knock-on effects are to be avoided or minimised. The consequence of such negative effects can easily upset the positive atmosphere within the workplace that is a prerequisite in the maintenance of high productivity and good quality control.

The double-edged sword

4.43 A major error, often made by management, is to consider that only those staff who are actually being made redundant will be adversely affected. However, staff cutbacks can create unrest for all personnel, both those remaining in employment as well as those destined to leave. Remaining staff may be faced with altered job status or responsibilities, additional workloads, modified relationships with peers and supervisors, and possibly even changed working conditions, location or even remuneration. They may also be resentful of the attention and pay-offs sometimes given to those leaving.

There is little doubt that the insensitive and unsympathetic handling of redundancy can cause chaos. The absence of clear communication can initiate an atmosphere of fear in which conjecture and rumour run free, and the anticipation of the consequences of change can reverberate through all levels of the company, possibly generating more negativity than the direct impact of the change itself.

Even if a cutback or change in the company involves just one or two individuals out of a larger workgroup or unit, everyone will need to be informed of the proposed changes in order to foster a feeling of security, maintain group co-operation and establish an awareness of the restructured organisation.

Some of the insensitive methods that have been used when handling redundancy notices have included:

- Employees being handed a 'brown envelope' and given just 24 hours to clear their desks.

- A company running a video requesting employees to pack their belongings should their name not appear on the screen.
- Personnel being asked to be present at their desks while names are read out over a Tannoy system.
- 'Personalised messages' on in-house emails informing individual personnel whether or not they will still have a job tomorrow.

In essence, it is the long-term effects of how a redundancy programme is managed that will be longest remembered. Handled insensitively, it may take a considerable time for management to re-establish confidence within the company, whilst the knock-on effect can leave a legacy of a low-motivated, negative and possibly hostile workforce, which is obviously counter-productive.

The indirect effects of staff redundancies can be felt throughout an organisation and can seriously affect long-term morale and the individual perception of career advancement and opportunity.

Expectations of managers

4.44 Among the many responsibilities imposed on managers during a redundancy process, are to:

- make selections based on fair (and legal) criteria;
- ensure confidentiality until official announcements are made;
- make preparations for the news to be broken to employees;
- deliver the news to those staying, and leaving, with sensitivity;
- maintain and foster a team spirit during times of change or restructuring;
- be adequately prepared to deal with the differing reactions from the workforce;
- ensure that the process does not damage long-term business prospects with either customers or suppliers.

These responsibilities require mature, management skills. Redundancy is a time of personal crisis for those involved, and managers may need to receive additional training in order to manage the situation correctly. Without these skills, they may experience feelings of inadequacy and be unable to cope with the additional stress involved. This could give rise to a loss of confidence and a sense of failure, in which case it may be the manager who could be next in line for redundancy.

Sometimes there is a belief, (albeit usually unsubstantiated), that poor top-level management has led to the necessity for staff cuts. Managers may be placed in the invidious position of appearing to be straight and open with their staff whilst actually harbouring negative feelings relating to company policy, and it is these feelings that need to be addressed.

The act of selection for redundancy, itself, can cause conflict where managers may have to overcome feelings of guilt when deciding who stays and who goes. They may even disagree with the selection criteria, or feel frustrated over the timing of the redundancies, or indeed at having themselves been delegated to serve the notices.

Redundancy can be traumatic for the individual(s) concerned and can induce feelings similar to those of bereavement.

Bearers of bad news may experience measurable stress which can lead to sleep disturbance, and it is therefore considered important that they receive peer group support (and training) in managing their own stress levels.

The actual word 'redundancy' can also be termed downsizing, rightsizing or restructuring but, in the end, it still spells job loss. This loss can bring with it a plethora of emotions to those affected including shock, hurt, disappointment, rejection, confusion and despair, which in turn may lead to an identity crisis as a result of the loss of status.

The aftermath

4.45 For those employees who remain in the organisation, there can be a sense of 'survivor guilt' that manifests itself in feelings of awkwardness, embarrassment or an inability to communicate easily both with their colleagues who are staying as well as those who are leaving. The often sudden realisation that paternalistic organisations no longer exist, can be a shattering emotional blow for many.

When the initial shock has subsided, those employees under notice to leave, may feel anger and resentment, in addition to feeling a degree of isolation. They may be disappointed and bewildered by the embarrassment they appear to be causing to their colleagues who, owing to their own feelings of guilt at having been selected to stay, find communication difficult.

No matter how sound the reassurances given by a manager that the company's decision was objective and without personal bias, very few people will accept this. Being made redundant is personal and the effect of the decision will be taken as such.

Case study

4.46

> **Redundancy**
>
> After ten years continuous service, Paul, a Senior Sales Representative, was given notice of termination of his employment contract by an HR Manager in a cold, abrupt and insensitive manner. He was not offered any support to help him to cope with what for him was an unexpected, traumatic and life-changing event. He seemed, however, to accept the news in a resigned manner and was asked to report in the following morning to clear his desk. The next day he failed to go in to the office although his wife said, subsequently, that he had left home at the usual time. The following day, Paul's body was found by a neighbour, (when out walking his dog), hanging by a rope from a tree, in a nearby wood.
>
> The manager who had terminated Paul's employment was distraught. Work in his old office became extremely difficult, with people exhibiting shock and disbelief. At his funeral, a week later, many of his colleagues who wished to attend were refused permission by Paul's widow and family. They considered that the company was to blame and were angry in their grief. In the subsequent days that followed, Paul's manager and members of his team ultimately had to come to terms with the trauma of their colleague's death and the possibility that it might have been averted with better and more sensitive management procedures. In particular, the wholly inadequate support given to Paul upon his unexpected receipt of notice of termination.

Other stressors related to the way the organisation is managed

4.47 In addition to the many organisational stressors described in detail above, other stressors relating to the way the organisation is managed include:

- lack of consultation and/ or employee participation in decision making;
- budget constraints and inadequate staffing;
- a rigid, inflexible hierarchy;
- unnecessarily harsh disciplinary procedures.

Technology and equipment

4.48 In instances where organisations are investing in new technology, this can create its own particular stressors. The now universal use of email, mobile phones and other forms of 24-hour communication can put employees under increasing pressure to deliver immediate results, in addition to ensuring that they are accessible at all hours. The introduction of new technology and systems to the work environment requires employees to adapt quickly to new methods of working and to ensure efficient use of new equipment.

Unfortunately, some organisations make insufficient allowances for the impact of new technology and refuse to accept the changes resulting from them as a valid reason for poor performance.

There are certain forward-thinking companies who provide 'buddies' (a support team usually trained in counselling skills to help with the implementation of new technology and to pre-empt any potential feelings of inadequacy in operatives or staff (see CHAPTER 10). Research by ISMA[UK] (International Stress Management Association) in 2000, found that for many people, new technology can reduce stress and improve their quality of life, but for others, technology simply adds to their stress levels often for the reasons outlined below.

Information overload

4.49 The increasing volume of email that many have to deal with can include a mix of business and personal correspondence, newsletters, copies of other peoples' emails, etc (not to mention unsolicited messages or 'spam'). There is an expectation that all emails should be read, usually within 24 hours and this requirement can cause stress. To ignore the task today will doubtless entail double the work tomorrow!

Ordinary 'snail mail', by comparison, does not bring with it the same sense of urgency, nor does it interrupt our working lives, uninvited, throughout the day.

Certainly, both managers and staff say that having to cope with ever-increasing volumes of email affects their work performance and often their home lives as well. 'I have to get into work at least an hour earlier each day in order to read all my email', said one City executive. Copying emails to colleagues who really do not need to receive them is another common source of complaint.

These comments are reinforced by the results of a study from more than 800 managers published in February 2000 by the Institute of Management. The study found that for the first time, keeping up with emails had entered the top ten list of stressors affecting working life. Nearly a quarter of people interviewed said that having to deal with excessive emails had caused them stress in the previous 12 months. This was in stark comparison to research conducted by the Institute in 1993, in which 'new technology' was not a stressful subject for UK managers, who were more concerned about office politics and incompetent senior management.

Flame mail

4.50 'Flame mail' is jargon for an electronic version of a very old problem, ie harassing or bullying behaviour by a member of staff within an organisation, which is often done anonymously.

As internal email systems and company intranets proliferate, harassment and bullying behaviour are emerging as increasing problems Many incidents involve not only racist or sexist taunting but also straightforward aggression and are quite common experiences via the computer workstation. Bullies are attracted by their ability to upset, and exercise power over, colleagues by remote access. The ability to get a cheap laugh by emailing throughout the organisation is also too tempting for many who would otherwise be more circumspect (see CHAPTER 6).

Under existing legislation (the *Health and Safety at Work etc Act 1974* and the *Protection from Harassment Act 1997*) employers are required to take reasonable steps to prevent employee harassment. Where that harassment is not based on race, sex or disability, there is still a duty of care by an employer towards its employees. Employers must act firmly to prevent such misconduct.

However, it is not sufficient to be reactive after a problem has come to light. Companies need to issue clear codes of conduct about emails and the use of electronic networks, in order to pre-empt abuse of the system. This should clearly state company regulations concerning who is authorised to access the internet, the private use of the company's computers, and include instructions to comply with the provisions of the *Data Protection Act 1999*.

Staff should be reminded that once created, it is virtually impossible to completely delete an email, as it will still exist on a server, somewhere, even when erased on the originating workstation, and may be retrieved, months or even years afterwards. Such information has been known to be used as evidence in court or other proceedings.

Technology overload

4.51 Employees can often be faced with sudden technological change, without receiving the necessary support and/ or training from their organisation and are often expected to become their own administration centres – with the computer being the central core.

Gone are the days when employees always experienced daily face-to-face interaction with their colleagues – a form of communication that is fast becoming obsolete. Email may offer greater efficiency but is without any personal contact – often being sent from one corner of the office to another by people who prefer to send even simple messages electronically rather than going over to ask the question in person.

The personal approach is fast disappearing from modern office environments and with it the employee's sense of contributing towards cohesive team work – a factor that can have a profoundly negative effect on job performance.

Physical environment

4.52 A working environment needs to be designed with adequate consideration for the individual employee. An environment that is uncomfortable, unsafe or unhealthy, apart from the possibility of breaching the provisions of health and safety legislation, can cause stress, or add to any existing stress caused by any other psychosocial factors (see Figure 7). This is because our physical surroundings, in terms of noise, lighting, smells and all the other stimuli that bombard our senses, can affect our mood and overall mental state, whether or not we find them consciously objectionable.

Unfortunately, employees invariably have little influence over the design of their own workspace or equipment, despite the impact that this may have on their health and performance at work.

Poor design of working space can also have the effect of discouraging essential communication between colleagues.

In order to minimise stress to the individual, all elements within the workplace should ideally be compliant with recommendations regarding ergonomic design, ranging from the adjustability of a computer terminal chair, to the correct glare factor on the terminal screen.

Access to health centres and/ or rest rooms, whether in-house or external, are invariably of benefit to employees, and can provide physical relief from stress, provided that the organisational culture allows time for staff to make use of these facilities. Having access to fresh air always features high on any individual's wish list, but with climate-controlled buildings and permanently closed windows, is often impossible. Some problems may be alleviated by the prudent use of glass partitioning and the correct specification and location of light sources. In addition, the provision of available-to-all 'breakout' spaces, can prove to be beneficial.

The majority of workplaces are now designated as non-smoking areas. This puts specific pressure on staff who still wish continue to smoke and consequently have no option but to stand outside the front door (unless there are set-aside smoking areas). On the other hand, many organisations now see the value of providing more employee services and facilities that are directly accessible by their staff, eg back or shoulder massages at their desk (see CHAPTER 11).

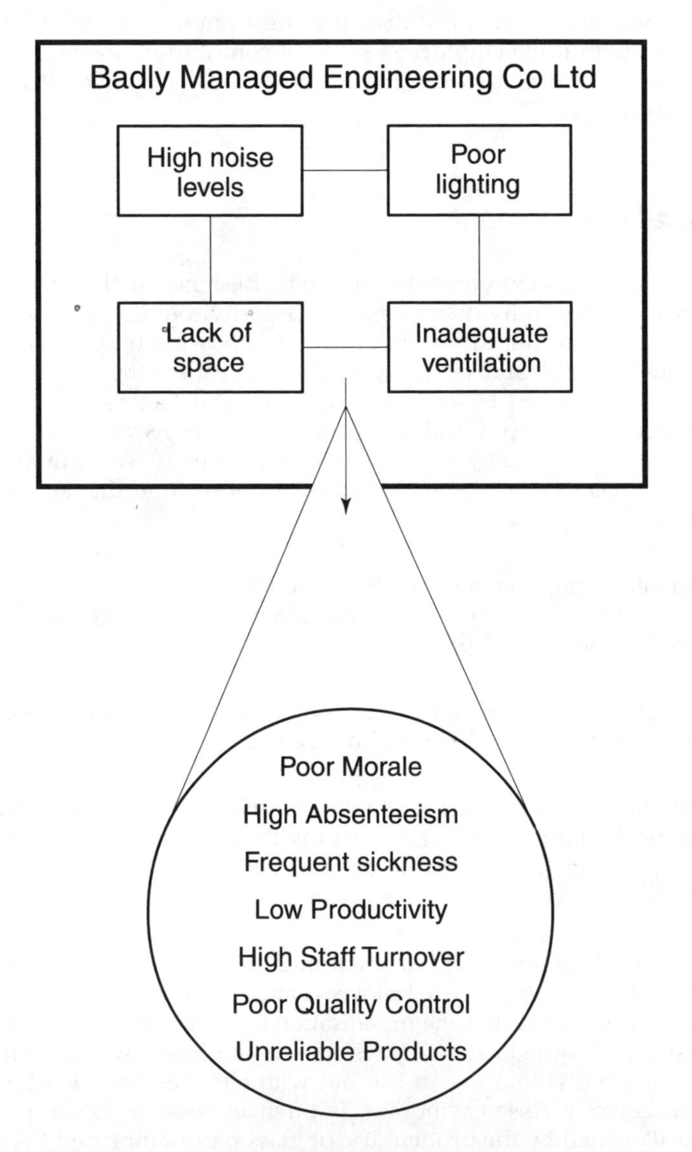

Figure 7: Environmentally engendered stressors

Noise

4.53 High levels of noise can cause impaired hearing. Even lower levels can interfere with communication and, particularly if prolonged, cause anxiety, irritability and tension, increased fatigue and reduced efficiency. As well as causing symptoms of mental ill-health, workers exposed for long periods of time to high noise levels appear to have a high incidence of allergies, respiratory and digestive disorders, in addition to musculoskeletal and heart problems (Cox (1993)).

Other stressors relating to the physical environment include:

- insufficient space;
- lack of privacy;
- badly designed open-plan office space;
- extremes of temperature;
- toxic fumes and chemicals;
- poor lighting;
- poor ventilation;
- insufficient fresh air;
- glare;
- dirty work areas;
- lack of personally identified space, eg storage;
- lack of recreational facilities;
- overcrowding;
- ergonomically unsuitable furniture;
- inadequate toilet and washing facilities.

Travel

Stress and driving

4.54 We have already discussed in CHAPTER 1 the differences between type 'A' and type 'B' personalities, and it is interesting to note that some type B individuals can also demonstrate signs of type A behaviour when they are behind the wheel of a vehicle! Pressure is perceived differently by different people and, from a driver's point of view, this may be determined by several factors that, taken together, will result in a particular attitude or mode of driving.

Road rage

4.55 'Road rage' is the term used to describe the feelings experienced (by a motorist towards another road user) of extreme anger and hostility coupled with a desire to retaliate against the perceived guilty party.

When driving, we may often experience anger as an automatic reaction to a stressful situation. Such situations can be caused by frustration at driving conditions encountered during the journey where increasing congestion is often to blame. However, when the cause of that frustration is the action of another driver, eg dangerously changing lanes and forcing another driver to brake, the reaction will be one of aggression. That reaction can lead to an out-of-control feeling of hostility that may well motivate us to cause injury (or insult) to the (apparently) offending driver of the other vehicle. However, as we are well aware that violence is a criminal offence, we will usually be content with the employment of verbal abuse (possibly combined with a belligerent gesture, made, of course, from within the safety of our own vehicle!).

On occasion, a driver, believing that he or she has been insulted or demeaned, will actually resort to physical violence or 'road rage'. Such action has already resulted in a number of documented fatal incidents where aggrieved drivers have, in fact, employed violence by means of shooting or knifing, resulting in the other driver being killed.

There is no reason, of course, to suppose that such rage is markedly different to any other rage, other than that driving is a frequent occupational activity with most people, particularly in urban areas. The incidence of road rage being a manifestation that reflects that situation.

Many cases of road rage, appear to be the consequence of a simple, inadvertent driver error or momentary loss of concentration, that is, unfortunately perceived by another driver to be an example of aggressive driving. This might be the apparent 'refusal' to allow another vehicle into 'our' lane of traffic on the premise that 'we were here first and I'm damned if he's pushing in front of us'.

Of course, any particular driving style on a given day may be influenced by additional factors such as a difficult day at the office or a nagging spouse. These pent-up feelings of frustration may only surface when the individual concerned is subjected to additional stress that may serve as a trigger to release his/ her frustrations that are then directed at another driver.

Other contentions include the hypothesis of instinctively aggressive behaviour as an automatic response related to defensible space or a perceived overcrowding or personal safety issue. In the former, the car is viewed as an extension of personal space and any attempt to intrude upon that space is an act of invasion or territorial aggression. It should also be noted that such a notional, defensive space will usually be automatically extended to an area close to the vehicle and on all sides and any attempt at intrusion, into these areas, can initiate tension.

In many instances, where a driver threatens another's 'territory' by the act of 'cutting in' or even overtaking, then the other driver may feel it incumbent upon him or her to carry out a manoeuvre to re-establish either the status-quo or superiority in terms of speed or size. Such action may merely be a blast on the horn or a flashing of headlights to show disapproval or to vent feelings of annoyance. In other cases, the aggrieved driver may try to recover 'lost territory' and deliberately try to cut-back in or re-overtake. Normally, that may the end of the incident and both drivers will continue on their respective ways, either satisfied in the knowledge of reasserting perceived dominance or fuming at his frustration. However, it is by no means unknown for one driver to follow another, in an attempt to stop the other vehicle and remonstrate with or even to physically assault the driver. The common thread in all of these examples, is stress.

Points to minimise stress on the road

4.56

- Ensure that you know where you are going and how to get there.
- Avoid congestion, where possible, and traffic blackspots.
- Remember to check your vehicle before a long journey, including oil, fuel, water, lights, indicators and tyres.
- Leave adequate time for the journey with allowances for delays.
- Anticipate actions, and reactions, of other drivers and road-users.
- Adhere to legal road restrictions including vehicle speed.
- Do not allow yourself to lose concentration by being distracted.
- Do not use a mobile telephone when driving.
- Be awake and aware at all times. If you are tired – don't drive.
- Be tolerant towards other road users.
- Adopt a mindset that allows you not to become frustrated and stressed.
- Accept that there are situations, on the road, over which you have no control.
- Ensure you take adequate rest breaks.
- Do not take medicines containing sedatives before driving.

Commuting

4.57 Not surprisingly given the above, commuting by road can be a serious stressor for many employees, leading to distracted drivers, low concentration and resultant poor driving performance. When drivers are on familiar commuter routes, this rises as concentration falls.

Ease of access and adequate parking facilities are also important considerations. Disputes over parking often contribute to stress in (already) busy people. Those who travel as part of their job, or who live in places that lack adequate public transport, can be helped by good parking facilities and/ or transport laid on by the organisation. In instances where little can be done to actually improve journeys, then facilities should be made available, if required, for people to relax before they settle down to work.

Of course, commuting can have positive effects for families who want a better quality of life away from city centres but it can also put pressure on relationships. Long distance commuters may have to leave early in the morning and return home late at night, with the consequence of less quality time spent with families.

Case studies

4.58

Rail travel

Jane, an Advertising Executive, lives in Bournemouth and travels to London by InterCity train every morning from Monday to Friday. She is usually able to get a seat and uses her journey to teach herself French, from a tape in readiness for her holiday in Paris. She is in full control of how she manages her nearly two hour's journey time and puts this to good use.

In the alternative, she could have resented the fact that she has to make such a long, daily journey into work and allowed herself to become stressed by counting down the minutes to her arrival time at Waterloo and constantly worrying about timetables and underground connections.

4.59

Air travel

Although air-travel opportunities may seem very appealing and glamorous to many people, all modes of travel can be stressful, and travelling by air is no exception. Traffic snarl-ups on the roads or at airports, delayed or cancelled flights or trains, can all increase stress levels when not anticipated and well managed.

According to Dr Thomas Stuffaford, the medical correspondent for The Times, 'too much travel is associated with an increased rate of coronary heart disease and high blood pressure. The upset to the body's natural clock causes by crossing time zones and the tiredness that comes with having sleep patterns altered, plays havoc with digestion and metabolism. Blood sugar is raised and, in the long term, so is the risk of developing type 2 diabetes and high cholesterol levels. The sleeplessness, dehydration, rich food and cramped conditions on board, contribute to jet lag. Constant flying affects the levels of hormones, in particular cortisol, which helps to determine the body's response to stress and indirectly controls carbohydrate metabolism. Too much air travel and the busy executive become a near cortisol bankrupt – quite unfit to negotiate and, in time, prematurely aged. There are increased cancer rates among regular flyers; whether this is a pathological response to flying or a reaction to lifestyle is unknown' (The Times, 20 January 2002).

4.60

Flying

Adrian, a City broker who regularly has to travel to New York on business, initially found the idea of 'club class travel'to be exciting, but soon found it to be tiring and monotonous. He says, 'It's just another plane, another taxi and another hotel in another city. When I am there, I could be in New York, Sydney or back in London, it makes no difference. In addition, since the terrorist attack in New York on 11 September 2001, there is now much increased security attached to air travel, that it adds additional stress to the journey. What used to be a pleasurable trip is now a much more serious affair. Plus the fact, that due to company cutbacks, I'm no longer able to use club class. Consequently, much of the previous enjoyment of travelling has, for me, simply disappeared'.

Individual concerns

Bullying, harassment, discrimination and victimisation

4.61 One of the most important and serious stressors in the workplace is exposure to violence, prejudice, discrimination, harassment and bullying, which is why this particular subject is covered in depth in CHAPTER 6.

Case study

Marion was a senior secretary at a large firm of London solicitors. Four months ago, her department head was replaced with a woman who, for some unknown reason, took an instant dislike to her. As a result, Marion tried to keep out of her way but this was easier said than done. After some weeks and many sleepless nights, Marion

(cont'd)

plucked up the courage to complain to the new head of department regarding her workload, which she regarded as excessive. She was met with an uncooperative and ill-tempered response. She was told, 'Either you carry out the work I give you or I will have you transferred out of this department!'

Marion started to suffer from stomach pains, that her doctor said were symptoms of irritable bowel syndrome and suggested she was under stress. Over the next few weeks, the pain gradually became more severe and she was eventually diagnosed with ulcerative colitis. She received treatment that helped, but in the end Marion asked to be transferred to another department and in the process lost a degree of seniority for promotion. Unfortunately, she had little choice.

As a postscript to this episode, the departmental head is still in her post, her behaviour remains unchanged and, in the absence of Marion, she now bullies another member of her staff.

Home/ work interface

'Presenteeism'

4.62 Maintaining a proper balance between work and home life is an increasing challenge for many employees. How do we maintain a proper balance and what does it look like? Is it a 40-hour working week? Is it the amount of time an individual spends thinking about work – whether or not they are there? What about employees who take their work home with them, or who make sure that they are contactable even whilst on holiday?

Where there are a number of different stresses emanating either from home or from work, each may serve to exacerbate the other. In contrast, effective stress management, be it in the workplace or at home, will serve to lessen the adverse impact and improve our resilience, and, as a result, our mental and physical health. This is why so much emphasis is now being placed on the importance of a correct work-life balance.

Traditionally, society has rewarded those of us who show up for work every day regardless of circumstance. For years, perfect attendance has stood as a criterion of loyalty and commitment. As a result of this, and as mentioned earlier in the chapter, more and more of us are living to work instead of working to live. We often work at weekends and whilst on holiday, with many people finding it hard to take time off without feeling guilty.

All too often, organisations encourage these feelings by creating a norm of long working hours. Employees are made to feel guilty if they leave the office or workplace on time. Some organisations inculcate this mindset in staff as the cultural norm. Clearly by so doing they impact on the external activities that employees would otherwise be able to undertake such as reducing the 'quality time' they could be spending with their families and friends.

Most 'long hours' workers feel that they have an unsatisfactory work-life balance, with 56% saying that they dedicate too much of their life to work. Two-fifths of those working more than 48 hours per week report that working long hours has resulted in arguments with their spouse or partner in the last year, and the same proportion feel guilty that they are failing to pull their weight on the domestic front (The Chartered Institute of Personnel and Development (CIPD) (2001).

Presenteeism is an unhealthy attitude adopted by employees, to remain at work when most everyone else has gone home or to feel guilty about taking their annual holiday entitlement. For some employees, it is a case of 'don't forget your mobile phone and your laptop', rather than 'don't forget your passport and your sun screen'!

In certain instances, presenteeism can be a symptom of home-related stress, when the employee prefers to be at work, rather than at home.

Some people choose to work long hours to avoid their partners or families, whilst others may work longer hours to make up for other inadequacies in their private lives. Whatever its cause, presenteeism is effectively the opposite of absenteeism and is now believed to be just as detrimental to business. In many cases the erroneous belief is that such visibility improves the chances of keeping one's job and of future promotion. However, the results of this behaviour are often high levels of stress, psychosomatic illness, depression and poor performance – adding to the very risk (ironically) of job loss that the employee is so anxious to avoid.

Case study

4.63

Presenteeism

John, an accountant, wears a suit to work but feels the need to keep an extra jacket on his chair at all times, so that his Finance Director cannot easily check on him to monitor when he is in or out of the office. It is not that he is not pulling his weight, but every day the Finance Director works a twelve-hour day, from 7 am to 7 pm – and John cannot keep up.

It may be the culture of the organisation to work these long hours, but the organisation must appreciate that just because an individual is at work and ostensibly behind his or her desk, this does not mean that they are being productive or working at optimum performance – often the contrary is true. Senior management should examine the

(cont'd)

existing culture and identify whether or not it actually works to the benefit or to the detriment of their employees and the organisation.

The long-hours culture

What constitutes 'long hours'?

4.64 The EU Working Time Directive became law in the UK on 1 October 1998. Its main requirements include a maximum working week of 48 hours averaged over a rolling 17-week period; weekly and daily rest periods of 24 hours and 11 hours respectively; a guaranteed four-weeks paid leave per year; and protection for night workers.

To put the implications of this into perspective, the International Social Survey Program in 15 OECD countries (March 2002) found that 35% of British employees work more than 40 hours a week; eight out of 10 reported being stressed at work; one in four stressed employees planned to leave their work within the next 12 months; and a major cause of work stress was long working hours. Just five months later (29 August 2002), the Department of Trade and Industry's (DTI) Work-life Balance campaign and 'Management Today' revealed that one in five workers want a better work-life balance, and that there had been a steep rise in the number of people work excessive hours over the last two years.

Despite the legal requirement, perceptions about the number of hours that constitute 'long hours' can vary according to the type of work and what is considered to be the norm in a particular organisation. The extent to which long hours cause a problem for employees can also be influenced by the amount of control they have over their working time, with individuals who work long hours on a voluntary basis, being less likely to be affected.

Most individuals can work at an excessive pace for a short period of time, with out adverse effect when this is continuous, it can lead to ill-health and 'burnout'.

Case study

4.65

Long hours

Giles, an experienced Marketing Manager, joined a financial company in the City of London. His contract of employment and conditions of service clearly stated that his working hours were '9 am to 5 pm, Monday to Friday, with some overtime if required.'

> A year later he was working most evenings until 7 or 8 pm and taking work home every weekend. One Monday morning, his wife Lisa telephoned the Marketing Director to inform him that her husband had suffered a heart attack on Saturday evening and was now in the Cardiac Care Unit of the local hospital. She had been told that Giles was lucky to have survived and that although he should make a good recovery, he was likely to be off work for the next three months.
>
> Lisa wanted to know why Giles had been working such long hours. She had found his contract of employment, and questioned why he was told his hours would be 9 am to 5 pm. She also said that she was taking legal advice.
>
> Prior to this incident occurring, Lisa was incensed that on two occasions in the previous year Giles had cancelled their holiday arrangements. On the first occasion he had too much work to do and he told Lisa that he had to be available in the office to complete a vital deal. On the second occasion, his holiday was cancelled at short notice when his boss told him that he had only just remembered he had booked three weeks' holiday himself and they both could not be out of the office at the same time.

Excessive working hours

4.66 As the above case study suggests, the most extreme consequence of working excessive hours continuously over a prolonged period, can be sudden death. However, physical and psychological fatigue is much more commonplace. It is the physical fatigue that weakens the immune system and leads to an increased risk of heart disease, sleeping difficulties, sexual disorders, gastric disturbances, headaches, backaches, dizziness and weight problems.

There is also a range of psychological and behavioural problems associated with excessive working hours including apathy, depression, disorganisation, feelings of inadequacy and constant irritability.

In Japan, sudden death from a heart attack or stroke brought on by overwork is known as 'karoshi'. Karoshi first emerged as a problem in the 1970s, coinciding with widespread job cuts and a resultant increase in individual workloads.

Case study

4.67

> **Karoshi**
>
> Nobuo Miuro was simply getting on with his job when he suddenly keeled over and died. It had been an extremely busy few weeks for the interior fitter from Tokyo; he had been struggling to get a new restaurant ready for its launch and had been putting in excessive overtime. The day before he collapsed he had worked from 11 am until 4.30 am the next morning, but had managed to snatch a few hours' sleep before starting again. However, when Miuro, 47, tried to pick up his hammer and nails again, he suddenly fell ill. He died a week later. The coroner returned a verdict of 'karoshi' – death by overwork. However, there are signs that some Japanese companies are introducing 'no overtime' days, meaning that for one day a week, workers get to actually go home when they have completed their contracted hours (The Guardian, March 2001).
>
> Karoshi takes place in Britain as well.
>
> In January 1994, a junior doctor, Alan Massie collapsed and died in Warrington District Hospital at the end of an 86-hour week. He had worked seven of the previous eight days including two unbroken spells of 27 hours and one of 24 hours. The British Medical Council estimates that junior doctors frequently work more than 100 hours a week – equivalent to 16 hours 40 minutes a day for six days out of seven! (Observer, 10 April 1994).

The hidden costs of long hours working

4.68 The EU Working Time Directive ('WTD') focuses the attention of employers on the issue of long-hours culture. Some employers welcome the impetus this is providing to address problems about which they were already concerned. Others are worried about the cost of complying with the regulations. However, the WTD is not the only reason for employers to implement initiatives to tackle long working hours. Many are also recognising that although it may be necessary to work long hours in the short term, this is most inadvisable in the long term.

With technological advances, new ways of working, and against a backdrop of job insecurity, employees feel the need to justify their existence in terms of the hours they are seen to be working – and hence their indispensability. As we have seen, long hours do not necessarily equate to productivity or high levels of performance. Similarly, while working long hours may be viewed by employers as a way of obtaining the maximum benefit out of limited resources, in the longer term these effects are not sustainable.

Organisations must take firm action, senior staff need to set an example and insist employees go home on time and actually take all holiday entitlement. Where long hours are interpreted as demonstrating commitment, the example set by managers, combined with peer pressure, can and does generate a presenteeism culture. Compulsive 'presenteeists' should be shown how to plan their workload more effectively, and so avoid the issues described above.

Achieving a proper balance between work and home is increasingly being accepted as a critical factor in influencing the well-being and motivation of employees. Long hours have a negative impact on individuals and the organisations for which they work and ultimately on the national economy in general.

Key learning points

4.69 In this chapter, we have tried to demonstrate the many and varied causes of stress in the workplace. This account is not intended, however, to be a comprehensive catalogue of disasters waiting to happen, but rather to indicate where potential sources of trouble may lie if left unrecognised and unattended.

Many of these issues are either rectifiable or surmountable, and one of the principal keys to effective stress management is to maintain an awareness of where stressors may occur (or are occurring) and so be ready to address them prior to adverse effects manifesting themselves.

- The initial step in beginning to combat stress is recognising that there is a problem in the first place.
- Stress awareness and the related training are of benefit to *all* employees – enabling them to cope with stress at work and in their personal lives.
- Home and work-based stresses can both feed off and reinforce each other.
- Stress mapping can be a useful tool for exploring and resolving issues that cause stress.
- There needs to be a correct correlation between the work demands made on an individual, his or her ability and the amount of control over working practices that are available to them.

(cont'd)

- Both work overload and work underload can lead to stress.
- Shift work and night work can be inherently stressful and may lead to an increased risk of accidents.
- Home workers may feel isolated and require structured support.
- 'Hot desking' and short-term contracts bring their own particular pressures.
- Role conflict, ambiguity and changing roles all contribute greatly to stress.
- Inadequate communication is one of the most common mentioned organisational stressors.
- Management style needs to achieve a balance between consultation, support and control.
- Managers often need more training, particularly in communication and people skills.
- Dealing with redundancy brings its own particular specialist training requirements.
- Careful attention needs to be paid to the planning of physical workspace in order to ensure that staff are comfortable and motivated, and thus more likely to perform to their maximum potential.
- The introduction of new technology, if not approached in a planned and gradual manner, can add to stress levels.
- Organisations can mistakenly encourage a culture of 'presenteeism', in which employees feel the need to be seen to be working at all times.
- A correct work-life balance is essential to good health and efficient performance
- All travel, whether by road, rail or air, can be inherently stressful and it is recommended to take simple measures to mitigate or avoid it, altogether.
- Flying, by commercial or private airline, affects hormone levels that impact on metabolism. As above, as with other modes of travel, common sense methods can limit its effects.
- Finally, workplace stress is not something that if left alone will go away of its own accord. It can only be tackled through a process of identification, intervention and management and not through short-term initiatives or one-off 'quick fixes'.

5 The Effects of Stress on an Organisation

Introduction

5.1

> 'Intolerence of groups is often, strangely enough, exhibited more strongly against small differences than against fundamental ones'.
> *Sigmund Freud*

With the advent of new working practices, the causes of stress in today's workplace can be many and varied. In this chapter we will look in detail at the effects of stress on the organisation as a whole, and how the impact of stress on organisational performance can be used to highlight adverse conditions in the workplace.

When analysing stress, it is necessary to appreciate the highly individual nature of the response (a subject we will examine in depth, in CHAPTER 7).

Merely because one person might experience stress in response to a specific form of pressure does not indicate that everyone, or indeed anyone else, will necessarily be affected.

Similarly, where two people are subject to the same stressor and both experience a stress reaction, their individual responses will vary. One, for example, might experience mood swings and irritability, while the other might suffer from headaches, anxiety, panic attacks or a change in sleep pattern. However, regardless of the variation in response, the stress resulting from it can have adverse effects on well-being and overall work performance.

The most obvious and easily quantifiable effect of stress is its impact on sickness absence. However, there are also cumulative and often insidious effects on the organisation that will inevitably result from prolonged and excessive pressure on the workforce (see Figure 1 below). These include one or more of the following factors, all of which are detrimental to organisational performance and profitability:

- High levels of absenteeism.
- Increased staff turnover.
- Low productivity.

- Poor morale.
- Decreased worker commitment.
- Increased frequency of accidents.
- Poor industrial relations.
- Imperfect customer relationships.
- Difficulty in retaining or recruiting good managers.

We will now look in detail at these and other stress-related issues, starting with the costs to the organisation of work-related illness.

The costs of stress

5.2 An organisational culture that imposes excessive pressure on employees is not conducive to efficient and cost-effective working.

The Health & Safety Executive ('HSE') is one of many organisations that have published research into the costs of stress in the workplace and of sickness absence in general. Examples of their, and other recent published research, includes the following statements:

- A total of 33 million working days were lost to illness in 2002, with a further seven million lost because of injuries (HSE (2002)).
- The number of people taking time off work because of ill-health has doubled over the last decade (HSE (2002)).
- Absence from work due to sickness, costs UK business nearly £11 billion a year, but the estimated cost to society is nearer £23 billion. Employees take on average 7.8 days sickness absence per year, with the highest absence rate in the public sector at 10.2 days. The majority of sickness absence is short-term (5 days or under), although 40% of days lost are due to long-term absence of more than 20 days (CBI (2002)).
- When asked about the aspects of their work that cause them stress, the stressors cited by workers in the UK included too much work (62%), deadline pressures (58%), aggressive management and/ or poor communications (49%), an unsupportive work environment (43%), and problems with maintaining an acceptable work/ life balance (42%) (International Stress Management Association (2002)).
- 'Common causes of stress include lack of job security and control and work overload' (European Agency for Safety and Health at Work (2002)).
- 'The problem [stress] is thought to cost the EU at least 20 billion Euros a year in lost time and health costs' (European Agency for Safety and Health at Work (2001)).

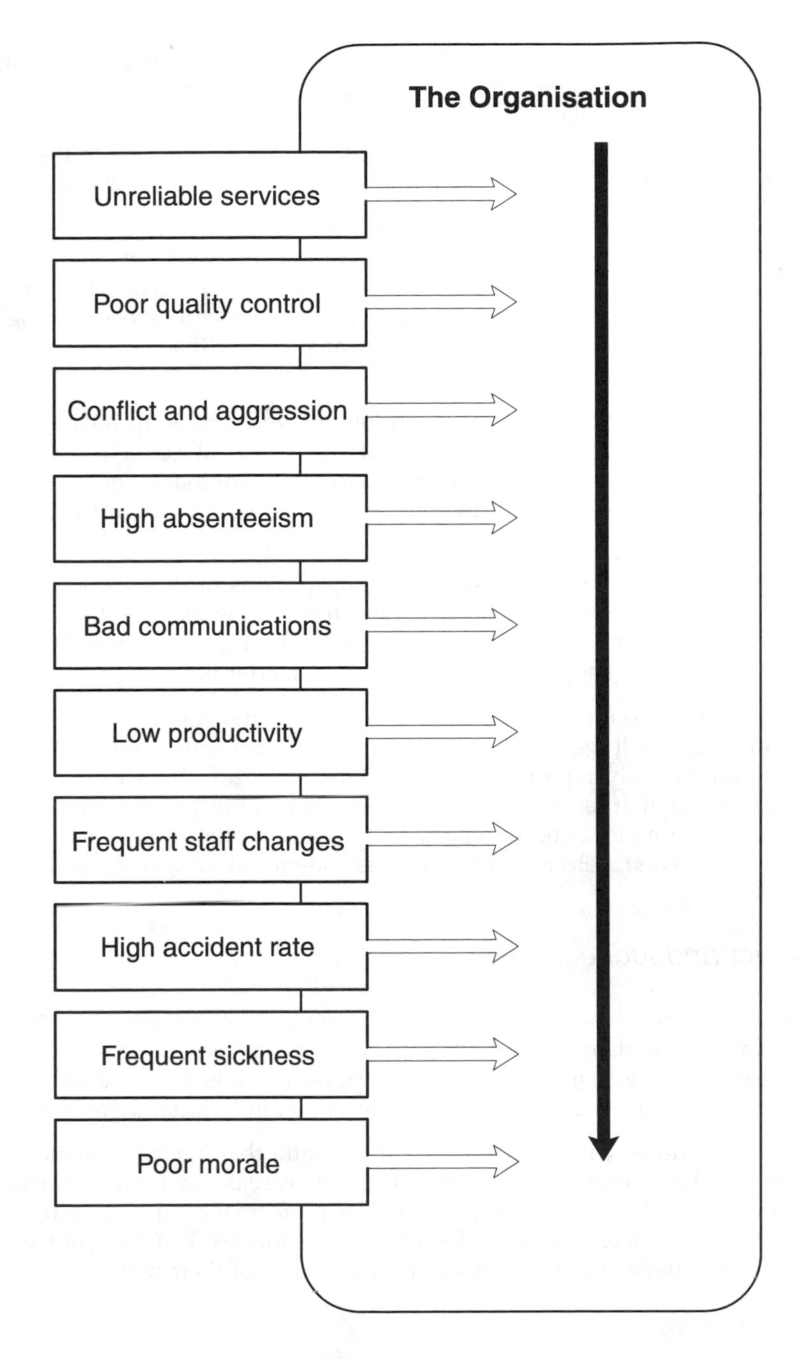

Figure 1: Effects of employee stress on the organisation

- Stress results in 6.5 million lost working days each year through sickness absence – costing UK employers £370m and society as a whole £3.75 billion (HSE (2001)).
- 'Stress may become the most dangerous risk to business in the 21st century' (Association of Insurance and Risk Managers (2000)).

The impact of sickness absence is felt not only in lost time and lost production. The morale of those staff required to cover for absent colleagues (see 'short-term sickness absence' at 5.9 below) will also tend to be low, especially where it is perceived that the absence has been brought about by organisational pressure, unrealistic deadlines or an overly aggressive management style.

A frequent characteristic effect of prolonged pressure is an individual's inability to maintain normal patterns of sleep. The subsequent tiredness may eventually deplete coping resources to the extent that judgement may be impaired, and both the individual and those working with them may be put at risk.

Working under excessive pressure for long periods of time increases the risk of error and the incidence of accidents at work and these may ultimately impact on the ability of a department to meet its production or service deadlines and/ or its quality control standards.

Taking into account these issues, the consequential cost of stress to the organisation will very often have commenced prior to an affected employee actually reporting sick or taking leave of absence. In fact, it could well be that the costs of their stress-related poor performance might have gone unnoticed and unreported for a period of months, (or conceivably even years), before an employee is diagnosed with a stress-related illness.

Direct and indirect costs

5.3 The previously quoted figures by the CBI and others of a cost of hundreds of millions of pounds attributed to sickness absence for UK industry as a whole, may be difficult to appreciate. It is therefore meaningful to evaluate these figures in the context of an individual organisation.

In order to do this, it is necessary to understand that the consequences of stress in the workplace are reflected in an increase in both direct and indirect costs. These result from the under-performance of those employees who are unable to satisfactorily manage the level of pressure with which they have to contend in the normal course of their work.

Direct costs

5.4 Direct costs are primarily those concerned with human resources and levels of productivity but may also include contingent claims for compensation, for stress-related illness, employer negligence or wrongful dismissal.

- Human resources costs (ie those that are employee/ staff-related) include absenteeism, lateness, as well as staff turnover and the consequent costs of recruitment and training.

- Performance-related costs result from the influence of stress on employee morale/ behaviour at work, including the quality and quantity of production, increased incidence of accidents, repairs due to negligence and/ or bad maintenance, raw material wastage and inefficient use of other supplies and/ or services.

The potential legal consequences of stress to the organisation, may include compensation awarded by the courts as a result of litigation, including associated legal fees, both of which can be substantial. For example, a financial adviser, employed by the Pearl Assurance Company, obtained a six-figure compensation award for work-related stress based on his second nervous breakdown (AMICUS (2002)).

Indirect costs

5.5 Indirect costs occur as a result of the loss of motivation, low morale, communication breakdown, weak decision-making, poor working relationships and lost business opportunities.

Low morale and motivation can result in variable performance that occurs when employees are constantly faced with demands that exceed their resources. High staff turnover, in particular, is related to low morale and a consequent reduced quality of working life.

Communication breakdown results from a reduction in the frequency and/ or availability of contact between employees and managers, as well as that between employees themselves. The incidence of 'role conflict' or 'role ambiguity' can also affect inter-communication and the successful completion of tasks and activities (as discussed in CHAPTER 4).

Excessive pressure on employees can cause impaired decision-making that may damage team-working and lead to animosity amongst members. This can contribute to potential conflict, whilst any deterioration in employee relationships is likely to be an unwelcome distraction from the work in hand.

Finally, while good management has the ability to counter and minimise the impact of environmental and other hazards; an organisation with a stressed workforce will be at a distinct disadvantage because, over time, it will have depleted the energy needed to maintain a competitive advantage.

Table 5.1: Direct and indirect effects on the organisation of employee stress-related ill-health

Direct effects of stress-related ill-health	*Indirect effects of stress-related ill-health*
Absenteeism	Retraining costs
Occupational sick pay	Increased recruitment costs
High staff turnover	Low morale
Use of Agency staff	Poor quality control
Increased overtime	Poor communications
Early retirement pension payments	Increased cost of production
Compensation awards	Loss to company image
Increased administration	Lost production
Insurance premium loading	Late deliveries
Legal and professional fees	Indifferent customer care

Employee performance

5.6 It may be considered to be a good sign to see an employee sitting behind their desk, but merely being present, affords no indication that an individual is achieving his or her optimum performance level.

Managing employee performance by regular appraisal and monthly target-setting, is one way of identifying the quality and quantity of work being carried out. Where an individual appears to be spending unduly long hours in the office or regularly taking work home, then it is appropriate to enquire as to the reason why the workload cannot be completed during normal office hours. It should be established whether this situation is due to work overload or because there appears to be a capability issue regarding time management, job knowledge or work ability. Either way, a management intervention is required in order to remedy the position.

Another damaging effect of prolonged stress, on work performance, is the extent to which individuals may eventually find that their sustainable powers of creativity and rational thinking have been compromised or weakened.

Early indications of stress in others, include a perceptible change in individual behaviour, or an exacerbation of an existing personality trait. In circumstances where individuals are under excessive pressure and consequently have difficulty in coping, they are likely to exhibit behavioural changes and attitudes that are generally out of character.

Initially, such changes may be minor and possibly inconsequential, such as occasional disagreements with colleagues. However, where more extreme symptoms are observed, such as social withdrawal or a pronounced over-reaction to ordinary events, then the individual should be approached by their manager in order to endeavour to ascertain the reason(s).

The following are just some of the warning signs that give an indication that stress may be adversely effecting the organisation.

- Increased absenteeism and/ or frequency of minor sickness.
- Reduction in work performance, without any apparent reason.
- Tendency to lose concentration on the job in hand.
- Increase in workplace errors, accidents or 'near misses'.
- Change in timekeeping habits, for example persistent lateness.
- Evidence of increased drinking or alcohol/ drug abuse.
- Working long hours without any apparent need (although this may be indicative of problems outside of work, ie that being at work is preferable to being at home).
- Seemingly withdrawn before and during meetings (possibly with frequent trips to the washroom) or failing to make any contribution to the subject under discussion.
- Taking an irritable, or aggressive stance, with business associates, customers and/ or suppliers.
- Receiving increasing complaints from clients or customers.
- Avoidance of face-to-face contact (eg emailing someone who is sitting in the next office).
- Loss of ambition, for example failing to apply for promotion.

Individual, personal responses to stress can also include:

- Lack of attention to personal appearance.
- Emotional instability, eg closeness to tears, exhibiting anger or hostility.
- Irrational behaviour.
- Frequent confrontation with others.
- Absenteeism (as discussed at 5.8 below).
- Tendency to suffer from frequent colds, flu and minor illnesses.
- Negative thinking.

- Mood swings
- Poor decision-making.

Managers need to be aware of the strengths and weaknesses of their teams and be adept at recognising when staff, under their supervision, are under-performing or behaving unusually. Early recognition of frequent changes in mood and/ or behaviour will enable appropriate support to be offered and performance to be reviewed before any serious health or safety issue arises.

Case study

5.7

Employee performance

Julie worked for a printing company and she was a bright and conscientious member of the team. She was well suited to her role, which was to ensure that print jobs were scheduled and despatched efficiently and on time.

Unexpectedly, the Production Manager for whom Julie had worked for over the last two years, announced that he had been head-hunted by one of the company's competitors. Rather than replacing him, and because Julie was doing so well, the company decided instead to give overall responsibility for production management to the Finance Director, whilst leaving day-to-day scheduling up to Julie.

Problems soon arose within the department, as Julie felt that the Finance Director was constantly over-critical, and was only interested in the occasional job that went out late, rather than the hundreds that were dispatched on time.

As the weeks went by, the situation deteriorated. Julie felt unable to make even the slightest mistake. The consequences were that, at night, she would often lie awake for hours worrying. Eventually, she started to phone in sick on Mondays, because she simply could not face the stress of the week ahead.

Julie's performance began to suffer. She found it more and more difficult to concentrate on the task of scheduling incoming print work, and all her previous enthusiasm for her work evaporated.

Events finally came to a head when one of the firm's more 'difficult' clients phoned to complain about a typographical error in the brochures they had received. The brochures had also been damaged in transit, and Julie was held responsible. Being unable to cope with the pressure any more, Julie burst into tears. Feeling unsupported

> and angry she told her manager exactly what he could do with his job, before clearing her desk and leaving the building.
>
> There was a happy ending, however. Julie was offered a job at the printing company that her previous manager had left to join, which she accepted and soon found herself, once again, to be happy and contented in her work.

Sickness absence

5.8 The CIPD 'Employee Absence Survey (2002)' cited stress as being the most common cause of long-term absenteeism. With sickness absence costing employers an average of £522 per employee per year (an average of 10 lost working days), there are good reasons to look closely at the root causes of absenteeism and, where possible, provide early intervention to support employees in regaining their health.

Ongoing short-term absence is often found to be the result of factors such as ergonomically unsuitable working conditions, 'sick building syndrome', or other outstanding stress-related issues which, if properly identified, can be dealt with efficiently.

Short-term absence

5.9 Short-term absence is usually defined as a period of absence of less than 10 consecutive working days, and will usually be as a result of the employee suffering from a minor medical condition.

Persistent short-term sickness, however, is one of the most common problems that employers have to face, and is an unrecoverable overhead cost. Arranging temporary cover when an employee is off sick may not always be viable, and in many instances will be disruptive and costly. Many employers therefore adopt the approach of persuading existing employees to cover for absentees on an *ad hoc* basis.

Whilst this policy may work in the short-term, when applied over longer periods of time, it puts pressure on existing staff, as they struggle to do their own work in addition to that of an absent colleague or workmate.

The effect of this on staff morale can be damaging and counter-productive. Staff frequently feel resentful, if required to do two jobs, often within the same time-scale and for no extra remuneration. The situation may be further compounded when the absentee employee returns to work and is met with resentment from those who have had to cover for them during their absence.

Long-term absence

5.10 Long-term absence is defined as any period of absence in excess of 10 consecutive working days. Such absence, particularly where it is stress-related presents a different problem for employers. In the short-term, they may feel able to cover an absence internally, as discussed above. In the longer term, however, it may be necessary to recruit temporary staff who will normally require induction training and who may not necessarily fit in well with existing teams. Temporary staff will obviously increase the salaries and wages bill, in addition to the payment of costly agency fees.

After a long-term absence, a phased return to work will most certainly be recommended, with possible training needed to support the employee 'back into work'. Where rehabilitation is not an option, the costs of premature retirement due to ill-health will need to be taken into account. Stress, as can be seen, has a quantifiable impact on health, safety and individual well-being, as well as on the operational and financial performance of the organisation as a whole.

Attendance patterns

5.11 The link between stress and absence is so well proven that statistics on non-attendance are often used as an indicator of stress 'hot spots' within an organisation. These figures may also be used as a control in order to measure the effectiveness of any stress management in interventions.

In the analysis of attendance patterns, any extended periods of sick leave will immediately be apparent. Obviously, a stress-related illness or injury already sustained by an employee cannot itself be undone. However, positive steps can still be taken by actively managing the return to work of the employee, and to minimise the risk of any identified stress recurring.

Of even more importance is the monitoring of short-term absences that may well be the first sign of excessive pressure. Typically, absences that tend to fall into a pattern, eg if an employee is off sick every Monday, or absences that are linked to particular operational requirements such as reporting periods, are the ones that should be identified as likely to be stress-related.

It is important to initially look at the pattern of absence, rather than the reasons that may be given for it.

Stress is typically under-reported as a reason for absence, especially in the early stages, with various alternative reasons being given by those concerned, such as colds, back pain, migraine or general fatigue.

This under-reporting can occur for a number of different reasons. For example, it may be that the individual has not recognised that they might actually be suffering from stress, or possibly that they are reluctant to admit, either to others or to themselves, that this is the real problem. There is often a stigma attached to stress that reflects a perceived inadequacy or inability to cope. This exacerbates the problem by creating an artificial barrier to its identification and management.

Absence management

5.12 A successful absence management policy will ideally create a culture that will enable any individual to admit more readily to stress-related ill-health, without feeling that their future employment or career prospects may be damaged. Clearly, the earlier that specific sources of stress are identified, the sooner appropriate action can be taken to reduce the poor attendance that often ensues.

In order to establish a level of control over sickness absence, and to implement an effective policy, it is necessary to analyse the data on employee absence over the previous 12 months.

This will involve the following types of records:

- The number of days lost in any one year.
- The number of employees taking leave of absence.
- The average length of absence per employee.
- The names of the employees and the department(s) with the worst, and best, record of absence.
- Are there any identifiable patterns of absence?
- Does the data show that absence is influenced by age, gender, the number of years in the job or seasonal variations?
- How many employees take their maximum paid sickness entitlement in a year?
- Which is the greater proportion of those who take sick leave or other absence during the year ie workers, staff or management?

It will then be necessary to assess the reasons for the various types and frequency of absence, including the following:

- Is a particular job too stressful or too boring?
- Is the work dangerous or does it require too much physical effort?
- Is the working environment unsuitable?
- Is management weak or over-aggressive?

- Is morale poor?
- Is there a culture of taking days off at particular times?
- Do working practices lack organisational support?
- Is the work location difficult to travel to by car or public transport?
- Is there a general lack of incentive and motivation?

When all this information has been collated and analysed, it can then be used to devise policies and procedures in consultation with staff representatives that should, when properly implemented, substantially reduce the incidence of absence. Of course, it should be borne in mind that any policy or procedure designed to reduce the causes of absence will take time to implement and results are unlikely to be achieved overnight.

Capability issues

5.13 The issue of capability is referred to in the provisions of the *Employment Rights Act 1996*. In instances where an employer has sufficient reason to dismiss an employee who is under-performing as a result of capability, it is necessary to follow proper procedures including an investigation of all the circumstances and the effects of the under-performance on the organisation.

These investigations should properly elicit any factor regarding the health of the employee. For example, it is necessary for the employer to know whether there are any underlying reasons for ill-health that may have influenced the under-performance. Where work-related stress is given as a reason for poor performance (and/ or a possible compensation claim), it will be necessary for the employee to show that the employer had no or insufficient procedures in place to monitor workplace stress and/ or failed to investigate the complaints from the employee.

It is possible that an employee's inability to carry out their required role may be caused by their own professional limitations eg that they are insufficiently qualified or trained, or that they are employed in a job to which they are not well suited.

Whatever the reason, an inability to perform their tasks to the required standard will eventually produce its own stress response. It is important to therefore ascertain, therefore, whether this response emanates from the individual's personal inability to perform their tasks, or from the unreasonable expectations of the organisation, eg in terms of the skill levels necessary, or the time constraints allocated, to required tasks.

Many organisations have procedures for addressing instances of under-performance and such issues can often be dealt with within these. However, it should be a line manager's responsibility, through continuous appraisal, to identify any capability problem and, where appropriate, arrange training and set clear performance targets.

Person-job fit

5.14 When evaluating capability issues that may be contributing to work-related stress, it will be helpful for managers to consider the following points in respect of individual members of their team:

- Are they in the right job? It should be confirmed that everyone has the appropriate training, experience and ability.
- Are they working with the right people? Do their duties and responsibilities properly interconnect with those of the rest of the team or should they be in another team or department? It is worthy of note that some workgroups become so well established that they develop into cliques that tend to make newcomers unwelcome. It is essential, therefore, that the role and duties of any newcomer be communicated prior to joining the team of which they will be a member.
- Do they have the right working conditions and access to resources in terms of tools, equipment, environment and time?
- Are they rewarded appropriately? Do they get paid the industry rate for the job, ie parity with other workers doing the same work?
- Do they understand their role and responsibilities? It is essential that these are communicated effectively and feedback obtained.
- Are they sufficiently in control of their own work? Does the organisation encourage employee input regarding work scheduling and methodologies?

Poor communications

5.15 As we saw in 5.2 above on the costs of stress, research commissioned by the ISMA (International Stress Management Association) in 2002, found that 49% of people who reported experiencing stress at work said that this was the result of aggressive management and/ or poor communications; and 43% said that it was caused by an unsupportive work environment. This is a reflection of the fact that some managers are unfortunately promoted into managerial roles with insufficient experience and training, in handling and understanding others.

Many stress-related problems occur as a consequence of the inability of management to recognise the signs and symptoms of stress and to deal with its cause(s) effectively.

Sharing information, ensuring that employees know what is happening and what is expected of them and listening and responding to what they are saying, are all important skills in the management of stress. Information should be disseminated appropriately to the correct people and at the optimum time, in order not to contribute to issues of uncertainty and misinformation.

Low morale and productivity

5.16 Employee morale is essential to the success of any organisation. In relative terms, a workforce with high morale will usually be well motivated. Staff and employees will usually look forward to the challenges that each day brings, endeavour to give their best at all times for the good of the organisation and for themselves, and will tend to be conscientious and loyal.

By comparison, employees who have to work continuously under excessive pressure will feel devalued or perceive themselves as receiving inappropriate recognition whether in terms of praise or money – attitudes that can quickly lead to demotivation. This can have a negative effect on colleagues and contribute to valuable energy being directed into seeking employment elsewhere.

The effects of unremitting pressure and low morale will usually translate into decreased productivity and consequent cost implications for the organisation, although these problems might initially not be perceived as being stress-related.

Staff who are under continuous pressure will be difficult to manage and are likely to see themselves as victims of the system. There is an established correlation, in organisations, between low morale and the incidence of stress.

Negative thinking also influences low morale and vice-versa, reinforcing a pessimistic attitude within the group.

Such demoralising attitudes have been known to spread rapidly throughout departments and, in some cases, even influencing the outlook of entire organisations.

Examples of how poor morale can affect an organisation can be seen within the service sector, where it can quickly bring about declining sales – that, in turn, will further influence motivation. At this point, employees lose interest and commitment to their work and customers who, if they have a choice, will often take their business elsewhere.

The effect of change on employee morale

5.17 It is a significant fact that organisational change, whether it be negative or positive, when managed badly or ineffectively, can cause stress by requiring adjustments in established working practices. The introduction of change can seriously affect the short-term morale of those affected, and care is therefore needed with the management of this process.

Lack of recognition

5.18 Unfortunately, many employees feel under-valued and feel their work input is unrecognised, for example: 'I feel like I'm just a number in the organisation. My manager never acknowledges me, doesn't smile or say good morning. I came into work at the weekend and okay I was paid overtime, but what I didn't get was even a "thank you" for all the extra effort I'd made. Let someone else do the hard graft in future. I'll just work my hours and go home on time'.

These are the words of an employee who is not being managed effectively. If managers fail to heed these comments, then dissatisfaction can escalate to anger and end in demoralisation, anxiety and resignation. Managers need to be proactive and give support and acknowledgement for a job well done, which in return, brings about satisfaction, enthusiasm, raised self-esteem and a sense of pride and commitment.

A sense of fairness, purpose and trust are basic processes that drive the mechanisms of demand/ control and effort/ reward – the foundations upon which every organisation stands. However, if perceptions of unfairness and the resentment that ensues, are allowed to translate into stress, then this can result in adverse outcomes to health, with the accompanying negative effects on performance described earlier.

Loss of trained personnel

5.19 High rates of staff turnover can prove to be expensive for a company – increasing recruitment and training costs, reducing efficiency and disrupting other workers. In particular, in instances where a high level of technical or professional expertise is required, the loss of key personnel within a department can make the difference between profit and loss.

Employees who find themselves working continuously under excess pressure, are likely to seek jobs elsewhere. This means that both the knowledge and the skills they have acquired during the time with their company will be lost – adding to unrecoverable, overhead costs. Recruitment costs for key and senior personnel, in particular, are notoriously high – with fees for a single employee typically costing the equivalent of one month's salary, (as at 2002) ie £3,333 in the case of a £40,000 per year, middle manager in the South-East of England.

Identifying and recognising responses to pressure within teams, is extremely important if valuable resources are to be retained. A management culture that has in place procedures that enable signs of stress to be detected and then managed efficiently is, therefore, essential within all sizes, and types, of organisation. Labour turnover is particularly high in fields such as hotel and catering, nursing, call centres and tourism, where too little acknowledgement of service given.

Case study

5.20

Customer service

John had worked for a van hire company for just over ten years. He had established an excellent rapport with the firm's clients and was customer-orientated. Repeat business was good and the organisation had built its reputation on personal service. The company, however, was taken over and John was then given responsibility for the car fleet as well as the van fleet, effectively nearly doubling his workload. John became frustrated because he was unable to give the level of service that he had in the past and he started to become brusque with customers. Due to increasing pressure, he failed to give the friendly, courteous service for which the company had become known and, eventually, customers began to take their business elsewhere.

Accidents and workplace safety

5.21 Another indication of stress in the workplace is a lack of concentration that frequently results in lowered productivity (eg as a result of reduced dexterity in manual operations) and a possible increase in workplace accidents ensuing from a decreased vigilance in safety procedures.

While non-fatal accidents at work are generally on the decrease, (HSE 'Safety Statistics Bulletin 2000/01'), each week, in the UK, an average of five people are killed at work and nearly 3,000 are also injured (according to the Health & Safety Commission in 2002).

It is believed that many work-related accidents occur as a result of employees working under excessive pressure. As we shall see in CHAPTER 7, individuals under continuous pressure may turn to drugs or alcohol, ostensibly for relief. (The former includes recreational or prescription drugs, or over-the-counter medications – some people may even become addicted to painkillers). All of these substances can reduce reaction times, weaken functionality and contribute to workplace error.

In the case of employees who are required to drive vehicles or handle machinery as part of their job, the effect of any impaired ability, both upon themselves and others, cannot be overstated. This is the rationale for organisations in high-risk industries including power, chemicals, transportation and haulage already undertaking programmes of random testing for drugs and alcohol.

Clearly, it is important for managers to strictly observe, and implement, the policy of their organisation regarding the misuse of alcohol and drugs, and to take appropriate action should they become aware of an adverse change in behaviour of any person, within the workplace.

Professional boundaries

5.22 In instances where relationship boundaries, within the workplace, are not clearly defined, the position can very often result in misunderstandings that can cause undue pressure to a person, for example, a female secretary who receives unwanted attention from her male manager. This may not necessarily be a case of sexual harassment, more a breaking of professional boundaries through over-familiarity. Similarly, a young male manager, newly promoted to lead a predominantly female team, may not have either the experience or the communication skills to know within which boundaries he should work. Using the wrong language to a team member could, for example, leave him open to criticism of favouritism or harassment, putting stress on both him and the employee – an issue that can be just as difficult to deal with for female managers leading predominantly male teams.

In our multi-cultural society, it is also necessary to recognise, respect and, as far as is practicable, accommodate, cultural differences and customs. Individuals can easily (and quite possibly unwittingly) cause stress and offence by words or actions that may breach cultural boundaries. For example, attitudes to personal space, formality of language and attitudes to dress, are all areas that may need to be addressed, while respect must also be paid to non-Christian holy days and differing dietary requirements.

Conflict and aggression

5.23 Conflict between colleagues, and between managers and subordinates, is a well-documented cause of stress. Managers, under pressure to meet their targets or budgets, will not always think before they act or speak. Their attitude to those around them may well appear to be confrontational and may possibly be perceived to be one of bullying.

Where team communication breaks down due to conflict between members, the resultant effect on both productivity and morale, can be pronounced.

The manner in which individuals communicate with each other, is often a good indicator of their levels of stress. Short, abrupt sentences, accompanied by aggressive body language, will often indicate conflict within teams, whilst stressed individuals will also appear to have lost their former sense of humour. Either one, or both, of these behavioural characteristics can easily distort and exaggerate an otherwise minor incident and turn it into one of conflict.

Conflict can also result as a consequence of the differing perceptions of a situation by the parties involved, leading either one or the other, to the conclusion that the choices available will inevitably involve an incompatible outcome. A simpler definition is that conflict can result from a clash of opposing ideas or forces.

When conflict is dealt with constructively, people can be stimulated to greater creativity that can lead to a wider choice of actions and more beneficial outcomes.

Invariably, it is not the conflict that is the problem, but the correct management of it.

Conflict becomes a problem when it is avoided, or is viewed as a win/ lose situation. When these latter responses are invoked, conflict can become a source of stress and distress.

Examples of conflict in the workplace include:

- Personality clashes.
- Unhealthy competition.
- Rivalry between employees/ departments.
- Perceived injustices.
- Industrial disputes.

If conflict is not adequately addressed, there will be a negative effect on both the organisation and the individuals involved. Symptoms produced by unresolved conflict include:

- Anger and aggression.
- Anxiety.
- Retaliation.

- Depression.
- Insomnia.
- Psychosomatic disorders.
- Low morale.
- Arguments/ resentment.
- Proliferation of grievances.
- Increased staff turnover.
- Litigation.

This list is far from exhaustive. However, the fact remains that to ignore conflict is an abdication of duty and responsibility and can have legal repercussions.

Managing conflict

5.24 In order to manage conflict effectively, all parties need to understand the nature of the different types of conflict that can occur in the workplace.

The first is conflict in regard to methods of working or conditions of work. These disagreements regarding issues within the organisation can usually be resolved through discussion and compromise.

The second type of conflict is one based on personalities, in which there develops a personal animosity between one or more members of a group. This dislike may be for a specific reason, but more often is based on an emotional reaction that may be caused by frustration, anger or even some unspecified reason. However, because there may be no issues of substance involved, problem solving is far more difficult, particularly as, in many instances, neither party has an interest in finding a solution. Indeed, they may possibly have a vested interest in exacerbating the cause of conflict.

As Charles Handy states in his book *Understanding Organisations* (1999), 'Neglected, conflict is like weeds, it can stifle productive work.'

Case study

5.25

Conflict and aggression

When Anne started her new job at her local factory packing vegetables, she was a little apprehensive but anxious to please and to work hard. The work involved taking washed vegetables from a conveyor belt, weighing them and packing them into polythene bags, which were then labelled and heat-sealed by an automatic machine. On her second day, Anne became involved in an argument with another girl on the packaging line, who appeared to be jealous of Anne's bubbly personality. From then on, things deteriorated with 'cat calls' and shouting between the two girls, until eventually the packing department supervisor became aware of the situation and transferred Anne to a different shift. That simple intervention was sufficient to remedy a situation that might otherwise have escalated out of control and seriously disrupted the work of the department.

In instances where conflict cannot be resolved within the team then it may well be necessary to use the services of a mediator, or other third party, to avoid the situation escalating and to find a solution, where possible, that is acceptable to all parties.

Violence

5.26 Violence within the workplace continues to be a major problem. Unless addressed immediately, it can cause damage to the victim(s), and to morale, increase absence, tarnish the organisation's image and may incur substantial expense to the organisation, in terms of legal and professional fees in any possible litigation and/ or prosecution.

A TUC survey in 2002 of 5,000 workplace, safety representatives, found violence ranked within the top five workplace complaints and was reported by nearly one third of all workers.

The *Management of Health and Safety at Work Regulations 1992 (SI 1992/2051)* (now revoked and replaced by *SI 1999/3242*) imposed the need for employers to identify any hazards within the workplace. Under the *Reporting of Injuries, Diseases and Dangerous Occurrences Regulations 1995 (SI 1995/3163) (RIDDOR)*, which came into effect on 1 April 1996, violence was finally, and significantly, classified as a notifiable hazard.

Despite this legislation, however, when the TUC published its report 'Violent Times' in 2002, it revealed an 'epidemic of workplace violence', with one in five workers likely to be attacked, or verbally abused, in the workplace. Those most at risk were workers who dealt with the public, particularly nurses, security staff and care workers. One in three nurses, for example, had been violently attacked or abused, during the year; and 70% of teachers believed that violence in schools was increasing. The report also showed that the most likely victims of workplace violence were young women (11% of women aged 25-34 had suffered a physical attack, compared with 6% of men in the same age group), and that the younger and less experienced an individual was, the more likely they were to be attacked.

The TUC's research reinforces the findings of the British Crime Survey 1997 (published by the Home Office), which identified the eleven highest-risk occupations as being:

- Security and protection workers.
- Nurses.
- Care workers.
- Public transport workers.
- Catering and hotel workers.
- Education and welfare staff.
- Teachers.
- Shop workers.
- Managers and personnel staff.
- Leisure service workers.
- Other health professionals.

However, it is important to recognise that an occupation considered to be low risk can change significantly dependent upon specific factors and circumstances, for example:

- Working alone and/ or working with money, is often a major risk factor.
- Working outside normal hours or in semi-dark or isolated areas, can increase the incidence of risk of violence.

Working alone does not necessarily mean that an employee is in danger, but other risk factors may contribute to vulnerability.

How violence is expressed

5.27 Clearly, violence is a major contributor to stress and injury. Violence can mean irreparable damage being sustained by the victim, both emotionally as well as physically.

To label particular behaviour as being violent, is a subjective concept, as one person's gregariousness can be another's aggression. However, if we feel unsafe or fearful of a person because of their behaviour, be it verbal or non-verbal, then we will react accordingly. Similarly, if we are afraid of a person because we perceive them to be violent, then we are unlikely to stop being afraid merely because we may be told that the person means no actual harm.

It is also important to realise that violence is not just about punching or kicking another person. Non-physical forms of violence include:

- Hate mail.
- Flame mail.
- Threats of violence.
- Verbal abuse.
- Bullying.
- Intimidation.
- Sexual innuendo.
- Stalking.
- Blackmail.
- Inappropriate language.

Verbal abuse and threats of violence, can be just as stressful, and as damaging, as the consequences of physical violence.

A number of different factors can trigger both physical and non-physical violence. These include:

- Fear.
- Perceived injustice.
- Frustration or impatience.
- Desire to dominate.
- Alcohol and drugs.
- Confusion.
- Self preservation.
- Boredom.
- Blackmail.
- Mental health problems.
- Extremes of temperature or other environmental factors.

Ultimately, people may exhibit such behaviours for reasons that we do not always recognise or appreciate.

Violence between workers is a growing phenomenon. A study in 1997 by the European Foundation for the Improvement of Living and Working Conditions found that 35% of European Union workers exposed to physical violence, were absent from work over the previous 12 months, compared to an average of 23% amongst workers in general. Incidents (including stalking, verbal abuse, sexual and racial harassment, physical violence and even murder) do happen, and these problems are far more prevalent than many managers realise.

Case study

5.28

Violent behaviour

Graham was suspended indefinitely from work, and subsequently dismissed by the school's head, Rosemary, for alleged fraud whilst working as a bursar at his daughter's school. Such was the embarrassment and shame he felt for his wife and daughter, that he went home that evening and took an overdose of paracetamol. He survived but, unfortunately, the major focus of his attention, during his rehabilitation, was Rosemary, his former boss. He became obsessively concerned with some form of revenge and became consumed with making plans to cause her harm.

After he recovered, Graham systematically stalked Rosemary by following her home every night for two years. He sent her offensive letters and telephoned her during the middle of the night. The authorities had difficulty in obtaining sufficient evidence to prosecute although they did issue him with a warning.

As a result of this threatening behaviour, Rosemary suffered severe anxiety and depression eventually leading her to resign from her position and move away. Her illness and anger were fuelled by the knowledge that the failure of the authorities to stop the harassment had meant that her tormentor had succeeded in forcing her to move away from the district where she had lived, and worked, for so many years.

(cont'd)

It subsequently transpired that Graham had apparently been moved from department to department by his previous employers because of his 'difficult' behaviour and had threatened various of his colleagues, including female employees, despite management warnings. Eventually he was dismissed. Records eventually became available that showed him to have been convicted of assault some years previously and to have served a term of imprisonment. None of this had been declared to the school authorities, who had been unaware of his unsavoury character and had, unfortunately, failed to insist on references before offering employment.

How workplaces can trigger violence

5.29 The American expert on conflict, Gavin De Becker, highlights the issue of inter-employee violence in his book *The Gift of Fear*. He says, 'The loss of a job can be as traumatic as the loss of a loved one but few fired employees receive a lot of condolence or support'.

De Becker revisits and retraces the work histories of violent people, and (similar to Graham, in the previous case study in 5.28 above), there are invariably, warning signs that employers deliberately or perhaps inadvertently ignore because they wish to avoid industrial unrest or litigation, or are just inefficient. Some employers may even provide incompetent or violent employees with excellent references, just to get rid of them. Yet the consequences of this can be very serious, and illegal.

Pre-employment screening evaluation

5.30 Employers are expected, by custom if not by law, to perform adequate screening of potential employees via a background investigation with previous employers and by taking up personal references. Typically, however, these background checks are often limited in scope by the provisions of the *Data Protection Act 1998*, anti-discrimination or other legislation.

During pre-employment interviews, previous relationships with employers, colleagues and family should be explored, as well as any criminal history. Simply asking the prospective employee how he or she feels about certain past employers may yield a wealth of information. The interviewer should also take into consideration any changes in the applicant's demeanour to certain questioning and, before reaching a decision, to possibly employ psychological testing to reveal any personality difficulties, interpersonal functioning or drug/ alcohol misuse.

While measures such as these may appear excessive, as the quoted TUC survey suggests, there are an increasing number of 'Grahams', and so it is important that managers are aware of the possibility of offering employment to someone who may be dysfunctional, violent or anti-social.

How to recognise a potentially violent individual

5.31 The box below summarises a simple two-minute exercise which will help to assess the potential risk of violence in the workplace.

Two minute risk assessment for violence

1. Do you know the person?

   Yes = 5

   No = 10

   The best predictor of future behaviour is past behaviour.

2. If yes, does the person have a history of violence?

   Yes = 10

   No= 0

   Don't know = 10

   If a person has a known history of violence then this must be acknowledged. Of course a person may have been violent as a result of a set of unique or exceptional circumstances which may never reoccur. Nevertheless, if known, a history of violence needs to be acknowledged. What is of more significance is the score attached to the 'don't know' response. If we are not aware of any history of violence we may assume the person not to be violent. This is not good practice. Until there is evidence that the person is not violent then the default assumption should always be 'If you are not sure, assume the worst'.

3. Has the person become verbally abusive or suddenly become quiet?

   Yes = 10

   No = 0

   If there is a noticeable change in someone's behaviour, this should cause alarm bells to ring, since such changes are indicative of unpredictable behaviour. Unpredictability should prompt caution. It is necessary to appreciate that verbal abuse is often a precursor to physical violence.

4. Has the person said that they intend to become violent towards you or a colleague?

(cont'd)

Yes = 10

No = 0

As obvious as this seems, many employees who deal with threats of violence on a daily basis can become 'immune' to the incident. Often an individual after threatening violence, then executes it, prompting 'surprise' from the victims, or statements such as 'I didn't think he/ she would attack me'.

5. Does the person have or appear to have a mental health problem?

   Yes = 10

   No = 0

   Mental health problems are often erroneously associated with violence. However, research suggests that a person who is expressing bizarre behaviour may prove to be unpredictable and a source of anxiety.

6. Is the person under or appear to be under the influence of drugs or alcohol?

   Yes = 10

   No = 0

   There have been numerous pieces of research that have confirmed the links between alcohol, drugs and violence. In this context, one of the most common reactions to alcohol and drugs is the loss of inhibition.

7. Is the person's body language hostile or aggressive?

   Yes = 10

   No = 0

8. Is the task being undertaken likely to cause the person to become angry? For example, are you giving bad news, or doing something that the person may find painful or distressing?

   Yes = 10

   No = 0

   One of the most common triggers of violence in the workplace is the extended use of negatives such as 'no' and 'never', whereby a request or demand is being refused. This is a situational factor that needs to be taken into consideration.

9. Are there the correct number of staff on duty and available to manage a violent situation should it arise?

Yes = 5

No = 10

The aim of this risk assessment is to balance the risk of violence against the safe systems of work that are in place, in order to minimise or even eliminate the risk of violence. Note that the correct number of staff might be two or three. This number may be more effective in dealing with a situation than, say, five or six staff.

10. Does your unit provide safety first for staff?

    CCTV = 9

    Personal alarm systems = 9

    Training in Aggression Management = 5

    No = 10

    It is important to understand that having no safety systems in place attracts a score of 10. Personal alarm system or CCTV are only one point less each. Training in aggression management scores 5.

11. Do staff feel comfortable about the situation?

    Yes = 5

    No = 10

    This is about intuition or 'gut feelings'. Although somewhat unscientific, many victims of violence have in retrospect, articulated views that something was not right. Hence, intuition does have a role and should not be dismissed as crude or primitive.

Scores

Between 13 and 34 = low risk of violence

Between 35–84 = medium to high risk of violence

Between 85–110 = high risk of violence

[W Brennan 'Zero tolerance to violence' (2000) *Health and Safety at Work* Vol 22 No 12]

Disciplinary issues

5.32 In extreme cases, an employee's response to a stressful situation may be to become angry, aggressive and to exhibit a loss of control. Such behaviour is unlikely to be and should not be tolerated in the workplace, and (where an organisation has 20 or more employees) this would normally be dealt with under the organisation's disciplinary procedure or in accordance with the guidance given by ACAS.

If disciplinary cases are mishandled or employee grievances ignored, then the performance of teams and individuals may suffer. Some employees may even feel they have sufficient cause to complain to an industrial tribunal.

Industrial relations

5.33 Where an organisation is perceived to be exposing its employees to undue or unnecessary pressure, it is likely to lead to poor industrial relations, and any subsequent negotiation to settle such a dispute may be acrimonious.

Where excessive pressure is an ongoing issue within an organisation, it is often the case that staff consultative representatives and/ or trade unions will become involved. In many cases of worker/ employee litigation against employers, this involvement has been an important factor in a successful outcome for the plaintiff.

However, on other occasions, trade unions have been successful in brokering an agreement between the organisation and the employee in order to achieve a negotiated settlement and obviating the need for any litigation.

Crisis and disaster management

5.34 While post traumatic stress is a subject that stands alone, it is equally important to understand the impact of crisis and disaster on both employers and their employees. Natural disasters such as flooding and fire, in addition to workplace abuse or violence, or the damaging effects of an accident, can leave employees traumatised and possibly in need of professional support.

In CHAPTER 8, we will look at how organisations can best prepare themselves for the effect of both natural and man-made disasters, and how to support those employees that have been directly and/ or indirectly involved in such an incident.

Suicide in the workplace

5.35 Although suicide is rare in most organisations, it is likely that some employees will have had a direct or indirect experience of the effects of a suicide of a colleague or friend. The suicide may not have taken place on-site, but nevertheless, the impact upon the workforce can be substantial, and it is important that employers are aware of what interventions to make, should such a situation occur.

The effect of a suicide on former team members can include guilt at possibly being part of the problem, anger, shock and a lack of comprehension because they were unable to anticipate the event or to prevent it from happening.

It is very difficult for an employee to return to the workplace after they have made an (unsuccessful) attempt to take their own life. When it is known within the organisation that a fellow employee has attempted suicide, this can create a general feeling of awkwardness and embarrassment. This situation can in turn result in the person who attempted the suicide, being isolated, at a time when they most need support, simply because people are hesitant about broaching the subject. This situation therefore needs to be carefully managed, both for the person returning to work, and for the rest of the workforce.

Although we will look in more detail at the issue of suicide in CHAPTER 7, it is appropriate here to look at some of the warning signs that can indicate that a person may be suicidal.

There are no definitive rules on how to recognise when someone is feeling suicidal. Given the opportunity, one person may want to talk through their feelings, whereas another may not. Individuals feeling suicidal do, however, often exhibit signs of anxiety and depression, particularly when their life seems, to them, to have lost all meaning. A secure, trusting environment is therefore needed, so that the person feels free to share their emotions and fears.

According to the Samaritans, individuals often show their suicidal feelings by:

- Being withdrawn and unable to relate to those around them.
- Thinking about methods of committing suicide, and possibly speaking of tidying up affairs, or giving other indications of ending their life.
- Indicating that they are feeling isolated and lonely.
- Expressing feelings of failure, uselessness, lack of hope, or loss of self-esteem; constantly dwelling on problems for which there seem to be no solutions; or expressing the lack of a supporting philosophy of life, such as a religious belief.

The risk of suicide will also be greater where there is:

- Recent bereavement or the break-up of a close relationship.
- A major change in health or in personal circumstances, such as redundancy, early retirement or financial problems.
- Painful incidents or problems leading to a dependency on alcohol or other drugs.
- A history of earlier suicidal behaviour, or of suicide in the family.
- Chronic or acute depression.

Fortunately, the incidence of suicide in the workplace is not a commonplace occurrence but where it does occur it can have devastating effects on those closely associated with the victim and the organisation as a whole and needs to be managed in a caring and open manner.

Key learning points

5.36 A company's success invariably depends upon the individuals who work for it. Stress is an extremely complex phenomenon that can affect individuals in many different ways and to differing degrees. The effects of sustained pressure, on employees, can severely affect the performance of an organisation to its eventual detriment and that of its end product or service.

- Stress costs UK businesses hundreds of millions of pounds each year.
- The most detrimental effects of stress include high levels of absenteeism, poor job performance, low morale, low commitment, increased incidence of accidents, difficult industrial relations, poor relationships with customers and possible litigation.
- The link between absence and stress is so well proven that non-attendance statistics are often used as an indicator of stress 'hot spots' within the organisation.
- The effect of stress on work performance is damaging to the extent that individuals suffering from high levels of it may eventually find that their powers of creativity and rational thought have been weakened.
- Where an individual is unable to perform their job to the required standard, this will eventually produce its own stress response.
- Many stress-related problems can be exacerbated as a direct consequence of management not having the required expertise to deal with them.

- Employee morale is vitally important to the success of any organisation. Low morale and lack of recognition by the employer will often lead to the loss of valuable trained personnel.
- Where relationship boundaries within the workplace are not clearly defined, this can lead to misunderstandings that can cause undue pressure.
- Conflict can be endemic within the workplace, and if not addressed will damage both the organisation and the individuals involved.
- In its most extreme form, workplace stress can result in violence, or even suicide, either within or outside the workplace.
- In the circumstance of endeavouring to rehabilitate an employee who has recovered from an attempted suicide and has returned to work, it is essential that the rehabilitation is handled carefully in what is bound to be a difficult and sometimes embarrassing situation. Colleagues and managers need to be aware of the necessity for a sensitive approach particularly in the early days.

6 Bullying at Work

Introduction

6.1

> 'You can kill a person only once, but when you humiliate him, you kill him many times over.' *The Talmud*

Bullying at work is undeniably a serious 'stressor'. Its emergence as a workplace issue is relatively recent, compared with our knowledge of other identified stressors such as excessive noise or shift work, having been brought to prominence in the UK in 1992 by Andrea Adams's book entitled, *Bullying at work: How to confront and overcome it.*

The systematic study of bullying in the UK began in the mid-1990s, and builds on work undertaken in Sweden, Norway and Finland in the preceding five years. In this chapter we look at the types of behaviour that can be characterised as 'bullying'; the impact that these can have on the individuals that are exposed to them; who is typically responsible; the measures that individuals and organisations can take in order to eradicate bullying behaviour and current (and anticipated) legislation related to bullying at work.

What is bullying?

6.2 Andrea Adams's definition of bullying is that it is 'the misuse of power or position to persistently criticise and condemn; to openly humiliate and undermine an individual's professional ability until this person becomes so fearful that their confidence crumbles and they lose belief in themselves'.

Such attacks on the individual may be irrational, unpredictable and demonstrably unfair. Despite this, bullying behaviour at work is still perceived in many organisations as part of an 'effective management' system, that delivers the required results.

A person in an ultimate position of power is unlikely to be successfully challenged. Furthermore, in many cases, it has been found that such a person ie a bullying manager does not possess the skills necessary to negotiate and compromise successfully. Notwithstanding this, there is also an assumption in some organisations that the more senior the position held, the less the requirement is for management training and professional development.

Most bullies are extremely self-orientated. The way in which they see themselves will rarely coincide with the view of others who are, or have been victimised by them. 'Just do as I say' will be seen, by some, as a justified command when given in a management role. However, while the assertiveness implied by this is used to urge employees into action and could be conceivably considered to have a constructive effect, the damning, condemnatory and often covert behaviour of the adult bully, invariably results in a destructive outcome.

Being bullied is an isolating experience. It tends not to be openly discussed, in case this increases the risk of further ill treatment. Those who are the prime targets, often feel ashamed to discuss it with colleagues in case their professional credibility is called into question. Even the mildest form of intimidation may be very disturbing for the target and as bullying or intimidation intensifies over a period of time, the proportionate effect on the victim can become severe.

Bullying behaviour may be confused with and viewed as:

- Personality clash.
- Attitude problem.
- Autocratic management.
- Poor management style.
- Harassment.
- Abrasiveness.
- Intimidation.
- Unreasonable behaviour.
- Victimisation.

The wider effects of bullying can extend well beyond the people who are directly involved. Listed below are some of the issues that may result from this type of behaviour in the workplace.

The consequences of bullying at work include:

- low morale and motivation;
- a tense and apprehensive workforce;
- staff feeling devalued and demoralised;
- reduced creativity;
- ineffective management;

- increased costs;
- declining productivity and profit;
- absenteeism and high staff turnover;
- loss of investment in training.

The problem of bullying in UK workplaces

6.3 Surveys relating to bullying at work show that around 15% of the UK's working population claim to be being bullied at any one time (UNISON (1997)), with around half the population claiming to have been bullied at some stage (Rayner (1997)). These estimates vary between studies, but rarely fall below 10% (Hoel and Cooper (2000)) and can be over 20% (Lewis, 1999). There are few sector differences, and little of significance can be found regarding age or gender in the UK. Most bullies are reported to be managers, and more men than women are reported as bullies – but as more men are in management positions, the gender issue disappears, leaving status as the single most important factor for 80% of reports of bullies.

What is curious in this 'equal opportunities' phenomenon is that bullying also seems to happen to the same extent regardless of where one sits in the organisational hierarchy. A study by Hoel and Cooper, for example, based on a consortium of employers, found rates of bullying to be around 10% at all levels throughout the organisations surveyed (Hoel and Cooper (2000)).

Some of the reasons why individuals submit to bullying are because they:

- are intimidated;
- do not know what remedial action is available to them;
- have no confidence in any such action being taken;
- possess low self-esteem;
- are vulnerable;
- may be apprehensive regarding the possibility of losing their job;
- are worried about the repercussions should they complain.

As Table 6.1 shows, bullying is a widespread and pervasive issue that tops the list of employee complaints.

Table 6.1: Top 10 complaints raised by employees (2002)

Harassment/ bullying	45%
Discipline	27%
New working practices	23%
Grading	22%
Discrimination	18%
Staffing levels	17%
Non-pay terms and conditions	17%
Pay	15%
Health and Safety	2%

[Source: 'Don't nurse a grievance: Resolving disputes at work' (2002) *IRS Employment Review* No 759]

Organisational factors

6.4 According to the trade union AMICUS in their publication *Bullying at work: How to tackle it*, the type of workplaces where bullying is more prevalent are those where one or more of the following factors exist:

- An extremely competitive environment.
- Fear of redundancy or downsizing.
- Fear for one's position of employment.
- A culture of promoting oneself by putting colleagues down.
- Envy among colleagues.
- An authoritarian style of management and supervision.
- Frequent organisational change and uncertainty.
- Little or no participation in issues affecting the workplace.
- Lack of training.
- De-skilling.
- No respect for others and/ or appreciation of their views.
- Poor working relationships in general.
- No clear published and accepted codes of conduct.
- Excessive workloads and demands on people.
- Impossible targets or deadlines.
- No procedures or policies for resolving problems.

Another area for delineation is the difference between bullying and that which is merely considered to be 'tough management'. Without doubt, some industries and professions, and certain organisations, encourage (or condone) 'tough' ways of working, and there can be a very narrow dividing line between these and that which is generally perceived to be 'bullying' behaviour.

Bullying may, unfortunately, be part of the culture of an organisation and seen as integral to a strong management strategy, ie an effective way of getting things done quickly and efficiently.

What are bullying behaviours?

6.5 As already stated, different people react differently to the behaviour they experience at work in so far as each individual has their own perception of events and will cope with them (or not) in different ways. This creates a complex problem for the study of bullying since it means that there is no definitive list of bullying behaviours.

The original work in the field (conducted by Heinz Leymann in 1996) used interviews that identified 'critical incidents' that targets felt had epitomised bullying behaviours for them. After several hundred interviews, Leymann generated a list of common behaviours, and more recent work has built on these methodologically sound beginnings. In general, the following signs demonstrate the range of behaviour that will typically be reported as bullying.

Early warning indications of being bullied include:

- 'This relationship is different to anything I've experienced before.'
- 'I am persistently got at for no good reason.'
- 'My work is forever being criticised even though I know that my standards haven't slipped.'
- 'I'm beginning to question my own ability.'
- 'I wonder if all these mistakes are really my own fault.'
- 'I don't want to go into work anymore. It's making me ill.'
- 'My supervisor is overbearing and constantly rude!'
- 'My boss is constantly ridiculing me in front of my team.'
- 'My manager is always putting me down at meetings.'

Attacks on professionalism

6.6 It is not unusual to hear complaints from individuals that their professional competence has been called into question as a result of disparaging remarks or criticism either from colleagues or managers, who they find to be undermining their efforts.

These attacks might be overt actions such as a public 'dressing down' for work errors, or might also include covert behaviour such as the circulation of rumours or gossip that appear to question an individual's ability. One problematic area is where this includes 'non-action', for example, not giving acknowledgement and/ or approval for a good piece of work, or not asking for an opinion from the person who is clearly the best qualified to provide that input.

These areas are very difficult for the targets of bullying to raise (as they question their own validity), and anyone receiving such accounts should refrain from discounting these often very subtle reports until it is possible to clarify the position. Small events themselves will often form a picture that is quantifiably different from the sum of the parts. The focus of attention should therefore be on the entire picture.

Examples of bullying behaviour

6.7 Bullies will typically:

- Make life at work constantly difficult for their targets.
- Make unreasonable demands: constantly criticising.
- Insist that their way of carrying out tasks is the only way.
- Shout at victims, publicly, in order to get things done.
- Give instructions and then subsequently change them, for no apparent reason.
- Allocate tasks which they know that the person is incapable of achieving.
- Refuse to delegate when appropriate.
- Humiliate their targets in front of others.
- Block promotion, refuse to give fair appraisals or refuse to endorse pay increases or bonus awards.
- Exclude the victim from meetings or other legitimate business activities.
- Constantly make attacks on the professionalism or personal qualities of their targets.

Personal attacks

6.8 In addition to attacking a person's work role, 'bullying' behaviour may also include actions and statements that are intended to undermine people personally, for example where someone has an interest that is easy to ridicule (such as amateur dramatics or train spotting), or supports a different football team to the majority of the other people in the department.

Personal comments may also extend to physical characteristics, such as making fun of an individual's height, weight, clothes or hairstyle. It is clearly inappropriate for these comments to be made within a work context, particularly as in most areas, very few individuals are employed on the basis of their appearance. Such personal attacks which may also be rooted in a lack of respect for someone else's culture, can undermine a person's standing at work and become the focus of attention, rather than the quality of their work.

Isolation

6.9 Being 'sent to Coventry' is a well-known English phrase describing the enforced social isolation of an individual within a group. It is more often applied in a workplace setting (although not exclusively so) and requires enormous courage for any one group member to break ranks with their 'bullying' colleagues and risk the consequent ridicule and rejection. Once these situations happen, for whatever reason, they are very hard to stop.

Social isolation and its effects should not be underestimated. It is reminiscent of the playground and the lonely child who is not part of a playgroup, and can be just as miserable and humiliating an experience for adults as it is for children, or more so, for it can jeopardise their livelihood.

Overwork

6.10 Overwork, as a category, can sometimes raise eyebrows, in an environment where long hours are often the 'norm' (see CHAPTER 4). However, studies by UNISON have found that people are able to distinguish between overtime, caused for example by a temporary lack of staff, unusual pressures of productivity or simply busy periods; and overwork in a bullying sense (UNISON (1997)). In the latter context, the overwork is seen as being the imposition of highly unrealistic deadlines where people are effectively deliberately 'set up' to fail. This may also appear as 'micro management', where every dot and comma, bolt, nut and screw is checked so often that incompetence or inability is deliberately implied.

Destabilisation

6.11 'Destabilisation' is a catch-all category which focuses more on the processes and outcomes than the actual targets of bullying, themselves.

People who are deliberately destabilised feel that they have lost control over their work environment and, as a result, have ceased to be able to carry out their duties in a relaxed manner without being threatened. Instead they live from day-to-day as they fight to regain a position of normality, often unsuccessfully.

Workplace behaviour, such as obvious inconsistencies in the allocation of rewards, unequal enforcement of working standards, withholding of privileges, changing objectives without warning or breaking of agreements invariably lead to extreme discomfort for the individual(s) concerned.

To summarise, behaviour that can be identified as bullying includes overt action such as yelling and shouting; covert action such as rumour and gossip; and non-action such as deliberately failing to include individuals in discussions, thereby isolating them, or a more general inaction that adversely impacts on their situation, or security, at work.

A 'bullied' individual can be vulnerable to targeting in various ways, including harm to their personal or professional position, through overwork, social isolation or the deliberate reduction of the level of control over their working environment.

Case study

6.12

Bullying at work

Jim, a Sales Executive, had worked harmoniously with his team for ten years. Staff turnover was kept to a minimum, and sickness levels were usually well below average.

However, when Jim's Line Manager retired, the position was taken by Mary, a former colleague. What then ensued was six months of extreme discomfort and harassment as Mary systematically stripped Jim of his authority, ridiculed him in front of staff and unreasonably increased his workload. When Jim sought help, Mary questioned his professional ability.

Jim became nervous, over anxious, confused, and his health began to suffer. Mary's style of leadership was not person-centred, and her constant references to the need for staff changes in the department resulted, in Jim's anxiety turning to panic.

Constant paging by Mary, being set unrealistic tasks while being offered no support, plus a wall of silence, eventually resulted in Jim experiencing disturbed sleep patterns, chest pains and violent headaches. His deteriorating health, however, brought no sympathy from Mary. When stress was eventually diagnosed, Mary nevertheless continued to undermine his authority. Then, after one particularly disturbing incident, Jim was forced to conclude that he was the victim of psychological bullying over which he had no control. Unable to retain his composure, he sought further medical help but, finally, broke down completely.

Off sick for six months, Jim was eventually forced to take early retirement. At his exit interview he unequivocally expressed his concerns about Mary. He believed that by highlighting the problem, action would be taken to ensure that Mary's unacceptable behaviour be curbed. His comments were, regrettably, not taken seriously by his employer, and no action was, in fact, taken. This allowed Mary to continue in her uncommunicative and aggressive manner, which resulted in a further staff member leaving and another suffering a nervous breakdown.

Despite this catalogue of damning evidence that highlighted Mary's shortcomings as a manager, she still remains in post with the same employer.

The effects of negative actions at work

6.13 During Professor Leymann's studies of severely affected targets of bullying, he established that they could be shown to exhibit symptoms that would normally be associated with Post Traumatic Stress Disorder (PTSD) or Prolonged Stress Duress Disorder (PSDD). He also postulated that bullying at work contributed to adult suicide (Leymann (1996)). Fortunately, many people do not experience such a traumatic effect, but we must be clear that such extremes do exist (Tehrani (2001)).

Effects of bullying behaviour on the individual

6.14 A full discussion of the different types of harm resulting from bullying behaviours can be found in Rayner, Hoel and Cooper *Workplace Bullying* (2002). In general, bullying can be seen as a 'psychosocial stressor', and the effects are those which would normally be associated with psychological stress. It is also important to remember that whether or not people actually label themselves as being bullied, they are nevertheless likely to suffer the effects shown in Table 6.2.

Table 6.2: Physical, psychological and emotional effects of bullying behaviour on the individual

Insomnia	Anxiety
Headaches	Irritability
Increased blood pressure	Panic attacks
Skin rashes	Depression
Nausea	Lack of motivation
Sweating, shaking	Lack of confidence
Stomach/ bowel problems	Feelings of isolation
Back pain	Hopelessness
Change in eating habits	Suicidal feelings

Often the victims of bullying will choose to leave their jobs rather than to stay and allow themselves to continue to be bullied fearful that complaining will merely exacerbate the negative behaviour towards them – an issue that we will look at later in the chapter (see 6.22 below).

Identifiable patterns of response in the individual

6.15 One area that remains to be investigated is differences in reaction and harm, and how these relate to the patterns of bullying. For example, a study in 1997 of a large sample of UNISON members found that one third of 'bullied' people claimed that the whole of their workgroup was subjected to the same behaviour, whereas at the other end of the spectrum, around 10–15% of people reported being singled out and victimised. These two experiences obviously differ. In the former, the bullying behaviour is relatively unlikely to be 'personal', whereas when an individual is singled out, then it would be logical for the target to regard this behaviour as 'personal'. As yet, there is little or no systematic data on how these processes affect people, either in terms of the nature of their reaction, or their ability and choice of strategy to deal with the situation.

Who does the bullying?

6.16 In the UK, bullies are reported to be in managerial positions in around 80% of cases (UNISON (1997); Rayner (1997)). However, some environments such as hospitals have quite complex management hierarchies, and therefore in certain organisations, reports that identify the bully, show lower rates of bullying by direct line managers.

It should also be noted that around a third of respondents to surveys (who label themselves as being bullied) claim that there is more than one bully. Such a scenario provides added complexity for investigators, since exploring such incidents is commensurately more difficult. Table 6.3 shows replies to a questionnaire from respondents who work in a complex management setting, ie civilian staff in the police service.

Table 6.3: Responses to UNISON police section questionnaire

People who had labelled themselves as bullied were asked to 'Please identify who has been bullying you'.

	Type of setting respondents worked in:			
Person(s) instigating bullying:	On their own	With other civilians	With other police officers	*Total*
Line manager	13	9	4	26
'Manager' (x1)	10	5	8	23
'Managers'	12	5	4	21
Police office manager (x1)	9	0	4	13
Police officer managers	0	0	2	2
Police officer	18	6	4	28
Co-worker (civilian)	8	4	0	12
Co-worker (police officer)	2	0	2	4
Total	72	29	28	129

[Source: 'UNISON police section members experience of bullying at work' (UNISON (2000))]

Note: The table shows numbers of people reporting. Managers are specified in 85 of the 129 cases (66%) which is relatively low – most other studies provide a figure of around 80%. Police officers are identified in 55 of the 129 cases (43%). There is ambiguity in that many managers are not identified specifically as police officers or civilians.

This picture of the association between bullying (in the UK) and hierarchical power is repeated in Australia (McCarthy, Sheehan and Kearns (1996)), but not in the Scandinavian countries where only 50% of 'bullies' are cited as managers, the rest being co-workers (Einarsen and Skogstad (1996)).

Can the organisation itself be 'the bully'?

6.17 A not insignificant point is that in some circumstances the organisation itself can be perceived as being 'the bully'. This contention has had very little exploration as, in general, it has not yet been the subject of staff questionnaires. However, an important piece of work by Liefooghe and MacKenzie Davey (2001), found that people in call centres described feeling undermined, humiliated, isolated and over controlled.

These particular workers sit at computer terminals in organisations which provide (remote) information, help or advice to telephone callers, and can be situated, possibly hundreds of miles away. Staff blamed the way they felt on the design of the software and performance management systems that are implemented to control their work and output, rather than their actual supervisors or managers. In one organisation, for example, staff were allotted a specified number of minutes that they were allowed to be away from their terminal in order to visit the toilet, and their bonus could be affected if the time taken away from the terminal was in excess of the time allotted.

Other sections of this book have examined how organisations can be seen to employ methods of working which (albeit inadvertently) can be causes of stress. Is it therefore possible that the organisation can be identified as a 'bully'? Such a notion has typically been resisted on the grounds that no employer would knowingly condone obvious bullying behaviour. However, the role of the organisation in a bullying environment, is gaining validity, for example, as a result of organisations choosing to ignore a bullying 'culture' of which they have been made aware, or in systemic situations such as those exemplified by some specific call centre environments. (It should be noted however, that there are some excellent employers in this field).

The role of culture at work

6.18 There is little doubt that the organisation can play a part (in terms of the messages that it does or does not send out) in relation to its potential complicity in the incidence of bullying at work. Unless 'bullies' are made aware that their actions are deemed to be unacceptable and that they will be brought to task if they continue to engage in such behaviour, then the organisation will be seen (by its employees) as specifically condoning bullying in the workplace.

Early work in the UK identified the fact that where people held the opinion that 'bullies could get away with it', then this was one of the key reasons for bullying to be present in their organisation (Rayner (1998)). Of course, it is quite possible that bullies within a particular organisation cannot, in fact, 'get away with it', but if staff think they can, then a belief system will be set up that underpins and reinforces the incidence of bullying. This attitude will also contribute to staff being unwilling to report the situation because they do not believe that any remedial action will be taken, even if they did make the effort to report the situation.

If there is one single message for organisations to understand regarding bullying at work, it is that 'doing nothing' is not seen by staff as a neutral act – it is perceived to be the organisation positively condoning bullying at work. In order to avoid such a judgement, tangible action needs to be seen to be taken, meaning that even if the organisation is not perceived as 'the bully', the level of action taken (or not taken) against bullying can influence the impression that staff have of its collusion in, and therefore acceptance of, bullying behaviour.

To summarise, (within the UK), we have a reasonably clear picture that a majority of 'bullies' are reported to be managers or supervisors who exhibit unacceptable behaviour towards subordinates; with the minority being co-workers or colleagues (it is very rare to come across cases of 'upwards' bullying, with a typical incidence of only 1%).

There are certainly instances where there is a reluctance by organisations to take anti-bullying measures, and where this is the case, they are seen as being complicit in condoning the behaviour. In this latter situation, management therefore has a crucial role to play in containing and eliminating the incidence of bullying at work.

Bullying cultures can also exist that rely on group pressure for people to conform. Little systematic evidence has been collected on this dynamic, but anecdotal evidence (Rayner et al (2002)) points to strong work cultures as contributing to the continued existence of bullying at work.

What can individuals do about bullying at work?

6.19 Possible actions available to those being bullied:

- confront the bully;
- approach the bully's immediate superior;
- contact the human resources department and/ or occupational health;
- Involve a trade union;
- speak to a colleague;
- stay and do nothing;
- leave the job.

The UNISON survey (2000) of police section members (see 6.16 above) asked those who had successfully combated bullying what actions they had taken.

A wide range of actions was reported, ranging from reacting with aggressive hostility, to taking the 'bully' aside and gently explaining the effect of their actions. What was common to these 'success stories' was that all actions were taken quickly, ie on the first or second occasion. It would seem that once bullying has become established and entrenched as an 'acceptable' pattern of behaviour, then it is extremely difficult to stop or modify.

Combating bullying when it starts

6.20 If individuals were to follow the findings from the UNISON survey, the advice would be to confront the bully *immediately*, and in a direct but low key way that does not escalate the situation, ie so that the bully is not insulted and does not wish to exact revenge. This seems to be quite sensible advice. De-escalation of emotion is a principal aim, and consciously treating others with dignity and respect at all times is fundamentally beneficial to good relations (both in and out of the workplace). In this way, it may be possible to resolve the situation quickly, without it escalating into a position that can be damaging not only for those directly involved, but also the section and/ or department in which they work. All our knowledge of this topic points to the nature of conflict escalation, whereby parties often end up in formal proceedings and then questioning how the situation could have deteriorated so far in such a short time.

When bullying is established

6.21 If an individual is in a situation where they consider that they are subject to an 'established' pattern of bullying, the very real difficulty of bringing about substantial change should not be underestimated. Staff often report that informal complaints are met with 'nothing' as a response (Rayner (1998)). While confidentiality practices may preclude clear and immediate responses being given to complainants, communication to indicate that something is being done would clearly be welcomed by staff who have expressed concern over a bullying situation. This view applies not only to personnel departments, but also to trade union representatives as well.

Leaving the job

6.22 It is estimated that 25% of people who have been bullied, eventually leave their jobs and seek work elsewhere. This represents an enormous drain on any organisation, in terms of disruption and replacement costs. Whilst there is little hard evidence, it is suspected that many of these people leave quietly, without ever making a formal complaint (Rayner (1997)). It should also be noted that over 20% of people who directly witness persistent bullying to others change their jobs as well, possibly because they want to avoid similar treatment, and so take a proactive approach to ensure their own future work security.

Possible results of management interventions:

- Nothing (this is often the most likely).
- Complainant labelled as a trouble maker.
- Allegations overruled.
- Dismissal threat to complainant.
- Bullying worsens.
- Complainant offered a sideways job move.
- Bully disciplined.
- Bullying stopped.

Case study

6.23

Confronting bullying at work

Christine worked as a commercial IT trainer. Having been a contractor for a telecommunications company, she was offered a permanent position – a role she had happily settled into. However, two weeks after accepting her revised job status, Bob, a new manager arrived. His style was not so much a 'new broom' as a policy of someone committed to the strategy of 'divide and conquer'.

Christine's confident manner, together with her loyalty to her team, unfortunately made her an immediate target. The new manager did not so much 'move the goal posts' as keep them 'constantly floating'. Whenever Christine got close to completing a project or assignment, he revised the objectives and added to the task.

Christine went from being a happy, content and conscientious person to dreading the time that she had to spend in the office. Worse still were the twice-weekly face-to-face meetings, always in private, with no witnesses, that resulted in a gradual whittling away of her confidence and self-esteem. Christine talked to her colleagues with whom she had an excellent working relationship, but gained the impression that most of them thought she was bordering on paranoia – even though some entertained vague doubts about Bob.

Eventually, Christine approached the human resources ('HR') department, who had noticed the change in her demeanour. A meeting was subsequently arranged to discuss the worsening

(cont'd)

relationship, at which Bob attempted to change the theme of the meeting and focus instead on Christine's work performance. Christine was ready for this, and after a short time, the rest of the meeting was spent trying to set objectives for their future working relationship.

Christine found the meeting so stressful that four hours later she was admitted to hospital for emergency treatment for severe chest pains. Her condition was of a physical nature, but how far the stress of the situation at work had contributed to it, was difficult to ascertain. However, Bob's bullying subsided for a couple of weeks only to restart again with added vigour.

Christine decided that enough was enough. She could no longer withstand being bullied at work, a situation that was making her ill, and made a personal decision to leave the company at the first available opportunity. The end of her probationary period was near, and so when the meeting between herself and Bob to discuss the end of her probation was arranged, it was her turn to have her say.

On the Friday when she was summoned to the meeting. Christine was ready – her profile of a serial bully in hand. She waited for the moment when her future employment with the company was to be discussed. It was then that she announced that she had no intention of staying with the company, and with both the HR manager and Bob looking shocked by her response, she was given the opportunity to explain her reasons.

Christine was calm and professional. She talked through all the points she had prepared, and ended by saying, 'Bob, you are a bully and I will no longer submit to your behaviour'. She left with her head held high, her dignity intact, and a tremendous feeling of relief.

From then on, Christine decided to return to being self-employed – she would no longer allow herself to be the target of a bullying manager. Despite the professionalism with which she had handled the situation, Christine still had times when she would entertain doubts about herself, but she knew in her heart that what had happened had not been her fault.

Six months later Christine received a phone call from another member of her former team, who had been made redundant with a proviso that they would receive an enhanced payoff if they discontinued with the grievance procedure they had instituted against Bob. For Christine, this was an added justification that it was Bob's unreasonable behaviour that had caused her to leave her job.

What can managers do?

6.24 It is important to be aware that most targets of bullying have two main aims – to keep their jobs and for the situation to return back to 'normal'. These apparently simple goals can get lost in the distractions of defensive positioning regarding possible legal claims and worries about future action to remedy the situation.

If the perpetrators of bullying behaviour can be talked to quietly and succinctly regarding their actions and the harm that they are causing, then this could be the ideal solution. However, such interactions are difficult to handle, and it is essential for those in authority (who have to deal with these complaints) to make sure that they feel comfortable with managing the situation effectively. Quite often, managers do nothing simply because they do not know what to do.

It is important, that this type of informal dispute handling becomes part of management training, so that managers can perform this facet of their job effectively. Talking to staff without escalating the situation, gently pointing out that intent is not an issue, and ensuring that all staff feel in a safe environment at work, is a major aim and, as such, an essential management skill.

With a potentially high staff turnover from the targets and witnesses of bullying leaving their jobs, it is in everyone's interest to expose and deal with these situations. For managers, this might mean the need to keep a close eye on aspects of employee behaviour in addition to the usual performance appraisal. Being sensitive to negative changes in the workplace and keeping a lookout for staff who appear to be nervous or frightened, can clearly help in the detection and elimination of workplace bullying.

It should be borne in mind that managers need to create a mechanism to enable staff to 'whistle-blow' if they feel the need to do so. The role of witnesses speaking out (or not) is suspected to be another factor contributing to bullying at work being seen as the 'norm' (Tehrani (2001)). If an individual is a witness, it is important for them to have a safe place to express their concerns – possibly through existing management reporting procedures; central services such as human resources; or through an external party such as a trade union or staff representative.

Managers may often hear reports from colleagues or other professional staff about individual members of their teams, possibly regarding bullying further down the hierarchy. While it may be tempting to always defend their own staff, it may be very unwise to do so in the case of bullying at work.

Many people present different 'faces' to different people, and managers are typically likely to be presented with the 'best' aspects of staff who may not be so rational, reasonable or tolerant towards others in the organisation. This is why it is so important to treat accusations of bullying carefully, to suspend judgement until after inquiry and not to pre-judge the situation. Calmness and level-headedness, as in other situations, will save considerable time and effort in the long-run.

What can organisations do about bullying at work?

6.25 It is clearly important that employers recognise the impact that bullying can have on the organisation as a whole, as well as on individual employees. The implementation of a bullying policy indicates that the organisation takes the issue seriously, and provides a mechanism for dealing with complaints, both informally and formally.

Developing a policy on workplace bullying

6.26 It is important to have a formal policy and procedures in place to deal with the issue of workplace bullying and/ or harassment. The policy should include a clearly written, and user-friendly statement on the organisations view of the types of behaviour that may be interpreted as bullying. Statements contained in the policy need to assure staff that the organisation takes these types of adverse behaviour seriously.

The trade union, AMICUS, recommends that a policy on bullying should:

- have commitment from the top, and be jointly drawn up and agreed by management and trade unions;
- recognise that bullying is a serious offence;
- recognise that bullying is an organisational issue;
- apply to everyone throughout the organisation and at all levels;
- guarantee confidentiality;
- undertake that anyone complaining of bullying will not be victimised;
- be implemented as early as possible.

The complaints process should be appropriate to bullying at work, and employers must properly investigate individual cases and the aftermath with sensitivity for all parties concerned.

There is frequently less clear evidence related to bullying than might be found in other cases of discrimination such as sexual or racial harassment. Investigators need to look at situations where a large number of small and, taken on their own, insignificant actions are the basis of a reported incident – but which when taken together, may reveal a discernible picture of consistent bullying behaviour. The system therefore needs to be able to cope with such patterns of evidence, and allow for proper investigations whereby people connected to the parties involved can also be safely interviewed. One third of 'bullied' UNISON members reported that the whole of their group was treated in the same manner (UNISON (1997) and (2000)), so it is often possible to find corroborating evidence.

Informal procedures for dealing with bullying and harassment allegations

6.27 An informal procedure can be a useful way of resolving complaints quickly with confidentiality maintained. AMICUS suggest that the problem can often be resolved at this stage, if the following facts are made clear to the alleged bully:

- that their behaviour is contrary to the organisation's policy and procedures related to bullying;
- that their behaviour must comply with the organisation's required standards of policy, procedure and codes of conduct;
- that they must realise the impact their behaviour has on others;
- that they must modify their behaviour and that the situation will be monitored accordingly;
- what will be the consequences should their behaviour not be satisfactorily modified;
- that the discussion is informal and confidential, at this stage.

However, in instances where no action is taken until formal complaints have been registered, then the staff involved are often in an irretrievable situation of escalated conflict. Many people are unwilling to take formal action, and this might be a contributory factor in the high exit rates that are associated with instances of bullying at work. Undoubtedly the most effective intervention is the training of managers to help them ensure the fast and effective resolution of disputes between their staff.

Making a formal complaint

6.28 A formal complaint should be brought in accordance with the organisation's policy on bullying and harassment, or where such a policy is not in place, under its existing grievance procedure.

The complaint should be made in writing and should give as much information as possible regarding alleged incidents including date, time and place, along with the names of any witnesses. The document should be signed and dated and handed to the designated person within the organisation responsible for dealing with such issues.

It is likely that a bullying and harassment policy will be invoked alongside an organisation's grievance procedure, and organisations should take great care to protect both the alleged perpetrator and the complainant, whilst a full investigation is carried out.

Investigations

6.29 Most policies allow for the informal resolution of a complaint, and this route should be fully explored before a formal investigation is undertaken. If no informal resolution can be reached, then it will be necessary for a formal investigation to be undertaken.

This process, depending on the nature of the complaint, may be carried out by individuals from within the organisation (but not in direct contact with the parties mentioned) or by an external investigating body.

Once an investigation is complete, the findings (in a written report) will be put before the commissioning manager, and it is at this point that a decision will be made as to what further action should be taken. Where allegations are proven, it is most likely that action will be taken using the organisation's standard disciplinary procedures. Where allegations are unfounded, then the organisation will need to consider what mediation may be required to ensure that the alleged perpetrator and the complainant are able to return to work and interact effectively together.

The role of the investigator

6.30 Investigators need to be trained and their actions seen as fair. It is imperative that they are totally impartial. Their role will be to interview witnesses, the complainant and the alleged perpetrator, collate evidence from all those involved, and evaluate whether or not there is a case to answer It may also be appropriate for investigators to make a recommendation for future actions relating to training and changes to organisational policy and procedures.

During the investigation, both parties need to be treated with sensitivity, and managers would do well to remember that being accused of bullying can be a devastating experience. Staff members who are alleged to have been involved in bullying may need as much support as those who bring the complaint.

Investigators can encounter complex situations, it is recommended that training in the gathering of evidence is undertaken by all investigators. Complexity will increase dramatically if more than one 'bully' is implicated, and investigations can also uncover long-term patterns of negative treatment, which in turn can be distressing for the investigators themselves. It is therefore good practice to have two people working together on any alleged case, plus a nominated person who they can call upon for help, if necessary. Smaller organisations may wish to secure the services of external agencies to help with, or to conduct, the whole investigation.

Appeals

6.31 Within the policy and procedures, there needs to be some form of appeal process. This will accord with general 'good practice' in relation to other unacceptable behaviour, eg racism and sexism, that organisations have had to deal with for some time. As with incidents of racism and sexism, it is possible that some staff might need to be dismissed where bullying is proven. While such situations are always difficult to handle, so long as the process has been (and has been seen to be) fair, then the effect on remaining staff is likely to be positive.

Malicious accusations

6.32 Many organisations are worried about malicious claims. While bullying can be hard to establish, it can also be hard to defend. As stated above, it is crucial that policies that deal with bullying mirror good practice, as in instances of 'whistle-blowing', malicious accusations need to be treated seriously and may therefore be considered a dismissal offence.

Monitoring

6.33 It is considered advantageous for an organisation to formally monitor levels of feeling amongst staff regarding the topic of bullying, and for larger organisations this can probably best be done as part of annual or bi-annual staff survey. Smaller organisations may wish to engage external agencies for the gathering of such information so that staff members are clear that they are giving information in confidence, and anonymity should always be an option. In this way, the organisation can monitor staff attitudes and feelings, and track the success (or failure) of their actions.

Exit interviews

6.34 These can be a remarkably effective way of ensuring that problems to do with bullying are brought to the surface. Essentially, exit interviews ask people who are leaving (or have left) the organisation to appraise their experience of their period of employment, in a candid manner – with an assurance that any comment will in no way affect any reference given. As the employee is leaving or has already left, they should have nothing to lose by commenting on issues they might not have revealed at an earlier time. Clearly, confidentiality is still of importance, but the value of evidence from exit interviews can be very good and serve to identify specific problems that might otherwise remain unknown.

Listed below are examples of the types of questions related to bullying that might typically be asked during an exit interview.

Exit questionnaires

6.35 Some questions to be included:

- Have you been bullied in the last six months?
- Have you ever been bullied?
- Have you ever witnessed bullying behaviour?
- Is there anything you think we should know about how staff members treat each other in your former department/ unit?

Back to basics

6.36 The real opportunity for organisations in dealing with bullying at work is to identify it informally and at an early stage. This means that 'good' standards of behaviour need to be clearly established and that managers are trained to deal with early concerns expressed by staff.

Key learning points

6.37 Bullying is a pervasive problem in UK workplaces, with around 15% of people at any one time labelling themselves as being bullied. This figure can be doubled if the label of 'bullying' is removed, and individuals are asked instead about persistent unacceptable behaviours that they experience at work. As described earlier, such behaviours range from overt actions such as shouting, through to covert actions such as undermining, and non-actions such as failing to assign credit for positive work.

- Bullying at work can happen anywhere and to anyone regardless of gender, age or hierarchical position.
- The effects of bullying can be severe, in extreme cases resulting in suicide or significant mental breakdown.
- Around 25% of people who are bullied leave their job because of it, and many do so quietly and without fuss; and in addition, around 20% of witnesses leave their jobs. Together, such exit rates place a very high cost on any organisation.

- In the UK, a majority of bullies are reported to be managers (80%).
- The organisation's action (or inaction) taken in relation to bullying will often be judged by staff, with inaction by organisations being seen as condoning bullying at work.
- Individuals should attempt to stop bullying behaviour becoming a pattern by bringing the negative effect of the bully's actions very quickly to their attention, in a non-confrontational way.
- Once bullying behaviour has become established, it is much more difficult to remedy.
- Organisations need to have formal policies and procedures in place to give employees a structured route to make legitimate complaints about bullying and harassment at work. However, the most effective way to challenge bullying is to train managers and supervisors not to engage in any behaviour that can be construed as bullying themselves, and to be able to deal with complaints regarding others early, informally and without escalating the situation.

The way forward

6.38 Following on from the above, summarised below are some of the most useful steps that managers and organisations can take to help combat bullying behaviour.

- Watch the workplace for a change in atmosphere among staff – particularly when cheerfulness turns to virtual silence.
- Adopt an holistic approach to bullying that is problem solving and not punitive.
- Focus on 'soft' skill development.
- Use stress/ culture audits to identify problem areas.
- Define and identify what is considered to be unacceptable behaviour.
- Introduce agenda-free meetings to provide a platform for troubled staff.
- During exit interviews, include the specific question, 'Have you experienced bullying within this organisation?'
- Ensure that all levels know that bullying behaviour is unacceptable and could possibly lead to dismissal.

APPENDIX
An Overview of Employment Law and the Human Rights Act 1998

6.39 The following is a guide to current and forthcoming legislation relating to harassment. The information was compiled by Jason Galbraith-Marten for the Andrea Adams Trust.

Legislation

I. The *Race Relations (Amendment) Act 2000* came into force on 3 December 2001. The Commission for Racial Equality has produced a Code of Practice (published in May 2002) dealing with the duty to promote race equality imposed by *section 2* of the Act (substituting *section 71* of the *Race Relations Act 1976*) and the setting up of race equality schemes. Note that the general duty encompasses three elements:

(i) eliminating unlawful racial discrimination;

(ii) promoting equality of opportunity; and

(iii) promoting good relations between people of different racial groups.

II. The Code suggests that in order to meet this duty, public authorities should:

(i) identify which of their functions and policies are relevant to the duty or in other words affect most people;

(ii) put the functions and policies in order of priority, based on how relevant they are to race equality;

(iii) assess whether the way these 'relevant' functions and policies are being carried out meets the three parts of the duty;

(iv) consider whether any changes need to be made to meet the duty, and make the changes.

III. The Department for Work and Pensions and the Department of Trade and Industry have both recently published new race equality schemes.

IV. The *Fixed-term Employees (Prevention of Less Favourable Treatment) Regulations 2002 (SI 2002/2034)* came into force on 1st October 2002.

The Regulations provide that 'A fixed-term employee has the right not to be treated by his employer less favourably than the employer treats a comparable permanent employee:

(*a*) as regards the terms of his contract; or

(*b*) by being subjected to any other detriment by any act, or deliberate failure to act, of his employer'.

V. The recently amended version of the Equal Treatment Directive (adopted by the European Council and Parliament on 18 April 2002 and coming into effect in 2005) contains a new definition of sexual harassment. The definition encompasses any situation where 'any form of unwanted verbal, non-verbal or physical conduct of a sexual nature occurs with the purpose or effect of violating the dignity of a person, in particular when creating an intimidating, hostile, degrading, humiliating, or offensive environment'. The definition is not dissimilar to that of the notion of unwanted harassment as it has developed in UK law (see below).

VI. With similar effect as the directive mentioned above, the Race Directive (adopted on 29 June 2000 and coming into effect on 19 July 2003) defines racial harassment as occurring whenever 'unwanted conduct related to racial or ethnic origin takes place with the purpose or effect of violating the dignity of a person and of creating an intimidating, hostile, degrading, humiliating or offensive environment'.

VII. Extending the concept of harassment beyond merely sex, race and disability, the *Dignity at Work Bill* (a private members bill), provides a right to dignity at work by prohibiting harassment, bullying or conduct which causes alarm or distress. The Bill received its third reading in the House of Lords on 29 May 2002 and now has a reasonable chance of becoming law.

VIII. In the circumstances, it is unlikely that the amended Equal Treatment Directive will require much amendment to UK law.

Unlawful discrimination

IX. It is now clear that harassment/ bullying on a prohibited ground amounts to unlawful discrimination. Importantly, what amounts to unlawful harassment is to be viewed from the perspective of the victim, and not merely objectively.

X. In the case of *Reed v Stedman [1999] IRLR 299*, Ms Stedman resigned because she found working for her male manager intolerable because of a series of sexual remarks, actions and innuendoes. Her claim of sex discrimination was upheld even though no single incident was in itself enough to constitute harassment. What was important was that the series of events could be looked at as a whole in upholding the Employment Tribunal's decision. Morison P gave some useful general advice on the meaning of harassment, and in particular what is 'unwanted conduct':

(*a*) A characteristic of harassment is that it undermines the victim's dignity at work and constitutes a detriment on the grounds of sex; lack of intent is not a defence.

(*b*) The words or conduct must be unwelcome to the victim and it is for her to decide what is acceptable or offensive. The question is not what (objectively) the tribunal would or would not find offensive.

(*c*) The tribunal should not divide up a course of conduct into individual incidents and measure the detriment from each; once unwelcome sexual interest has been displayed the victim may be bothered by further incidents which, in a different context, would appear unobjectionable.

(*d*) In deciding whether something is unwelcome, there can be difficult factual questions for a tribunal; some conduct (eg sexual touching) may be so clearly unwanted that the woman does not have to object to it expressly in advance. Alternatively, there are 'grey' areas where the conduct concerned would normally not be objected to but because it is for the individual to set the parameters, the question becomes whether that individual has made it clear that she finds that conduct unacceptable. Provided that that objection would be clear to a reasonable person, any repetition will generally constitute harassment.

XI. Although a complaint of 'harassment' carries the implication of conduct persisting over a period of time, there is no requirement that this be so. A single act of sufficient seriousness will be enough.

In the *case of Bracebridge Engineering Ltd v Darby [1990] IRLR 3*, Mrs Darby was assaulted by supervisory staff in the office of the factory where she worked. Her legs were placed around the body of one of the two men who were assaulting her, and one of the men placed his hand up her skirt and made an obscene remark. After a failure on the part of the management to investigate the incident properly she resigned and claimed in respect both of constructive dismissal and unlawful sex discrimination. She won on both complaints, receiving £3,900 compensation for unfair dismissal and £150 for sex discrimination. In the words of Wood P, giving the judgment of the court: 'We would deplore any argument that could be raised that merely because it was a single incident – provided it is sufficiently serious, I think one must say that it could fall within the wording of the subsection'.

The case of *Insitu Cleaning Co Ltd v Heads [1995] IRLR 4* concerned Mrs Heads who was subjected to a grossly offensive sexual remark ('Hiya big tits') by a manager, half her age. She established that this was unlawful sex discrimination. The

Employment Appeal Tribunal ('EAT') rejected the argument that an act could not be said to be 'unwanted' until it had first been offered and rejected, and also held that Mrs Heads had shown she had suffered the necessary detriment. The EAT took note of the difference in the ages between the perpetrator of the act and Mrs Heads, and also the fact that he was in a position of managerial responsibility over her.

In *Hereford and Worcester County Council v Clayton, The Times, 8 October 1996*, a fire brigade divisional officer had said to his watch, 'The good news is that you are getting someone else for the watch, the bad news is that it is a woman'. The finding by the Employment Tribunal that the making of this remark, even although not made to Ms Clayton, amounted to a detriment, was upheld by the EAT.

XII. Usually sexual harassment takes place between members of the opposite sex but this need not be so if a male employee treats another male in a sexually abusive fashion. This can be unlawful discrimination in circumstances where he would not have treated a female colleague in the same way.

XIII. In *Smith v Gardner Merchant Ltd [1998] IRLR 510, CA*, Ward LJ pointed out that not all forms of discrimination taking the form of sexual harassment will be gender-specific. The sexual harassment of a gay man may however be unlawful discrimination, if a gay woman would not have been similarly treated. In such a case, a comparison between how a hypothetical man or woman would have been treated is required. It is only when the sexual harassment happens to take a form which is sex-specific that the issue of how a comparison would have been dealt with becomes irrelevant. It is irrelevant, however, because it is obvious that a man would not have been treated in similar fashion, not because, as a matter of law the question of comparison does not arise. Not all cases of sexual harassment will fall into this category.

XIV. Leading on from *Smith*, the vexed question of whether discrimination on the ground of sexual orientation is prohibited by the *Sex Discrimination Act 1975* ('*SDA 1975*') is still working its way through the court system. The current position as set out in the decision of the Court of Session in *MacDonald v MoD [2001] IRLR 431* and of the Court of Appeal in *Pearce v Mayfield School [2001] IRLR 669*, would appear to be 'yes' and 'no'.

XV. The issue before the Court of Session was whether the MoD acted in contravention of *sections 1* and *6* of the *Sex Discrimination Act 1975*. The majority found that the term 'sex' as used in *section 1* was not ambiguous. It related to gender and *not* sexual orientation. In the circumstances the *Human Rights Act 1998* ('*HRA 1998*') was not required as a guide to interpretation. Furthermore whilst the *SDA 1975* might be deficient in not prohibiting discrimination on the ground of sexual orientation it could not be said that it was *incompatible* with the Convention. Lord Caplan could 'not see the 1998 Act as requiring the insertion of omitted provisions so as to enhance the legislation under consideration'.

XVI. In *Pearce* the Court of Appeal reached much the same conclusion but by a slightly different route. Contrary to the Court of Session, the Court of Appeal thought it was not permissible for the court to apply *section 3(l)* of the 1998 Act so as to give the *1975 Act* a different construction from that which bound the lower tribunals at the time of their decisions (applying *R v Lambert [2001] UKHL 37*). In respect of acts occurring before the coming into force of the *HRA 1998*, the Court of Appeal agreed that the term 'sex' within *section 6* of the 1975 Act related to gender and not sexual orientation.

XVII. In respect of acts occurring after the coming into force of the *HRA 1998* their lordships were equivocal. LJ Hale did observe that, had the acts complained of taken place on or after 2 October 2000, they may well have been capable of being a contravention of the right to respect for private life under Article 8 of the European Convention on Human Rights, particularly when read together with the prohibition of discrimination under Article 14. Note that in *MacDonald* it was conceded that the MoD, by requiring the applicant to resign after he had declared that he was homosexual, had acted in breach of Article 8 read together with Article 14, and that he was entitled to compensation in respect of that breach.

XVIII. Significantly, the House of Lords granted permission to appeal on 11 June 2002 in the case of *Pearce* on both the interpretation point and the retrospectivity point. Permission has also been granted in *MacDonald*, and the cases are likely to be heard together.

XIX. An action may well lie, therefore, against a public authority under *section 6* and *section 7* of the 1998 Act in respect of acts taking place on or after 2 October 2000.

XX. Note that the EAT held in the case of *Whittaker v Watson (7 February 2002) (Unreported)* that neither the Employment Tribunal nor the Employment Appeal Tribunal had jurisdiction to hear a submission as to the incompatibility of legislation with the European Convention on Human Rights.

XXI. Having taken regard of the above, it is worth noting the subtleties of this debate are soon to be rendered academic (in relation to the employment sphere at least). By 2 December 2003 the United Kingdom will be obliged to have in place legislation to prevent discrimination based on sexual orientation, following implementation by the European Commission of the Council

Directive 76/207/EC on the implementation of the principles of equal treatment for men and women as regards access to employment etc (the Employment Directive). Note that the Employment Directive also prohibits discrimination on the ground of religion and belief, disability and age.

XXII. A further attempt was made to overturn the decision in *Post Office v Adekeye [1997] ICR 110, CA* (no right to claim race discrimination in respect of acts occurring after the termination of employment) in the case *of D'Souza v Lambeth LBC (25 May 2001) (Unreported)*. In the EAT Morton I felt obliged to follow *Adekeye* but did so 'without relish' and specifically invited the Court of Appeal to reconsider it in the light of *Coote v Granada [1999] IRLR 452* (the right to claim sex discrimination in respect of acts of victimisation). The Court of Appeal had the chance to do so in the case *of Rhys-Harper v Relaxion Group [2000] IRLR 460* but upheld *Adekeye.*

XXIII. When the Court of Appeal came to consider the appeal in *D'Souza* it was (additionally) argued that the interpretation favoured *in Adekeye* and *Rhys-Harper* deprived the applicant of a fair hearing contrary to Article 6. The Court of Appeal disagreed pointing out that Article 6 does not come into play unless the applicant can first establish a substantive right. Article 6 extends only to disputes over civil rights and obligations which are recognised in domestic law; it does not itself guarantee any particular content for (civil) rights and obligations in the substantive law of the contracting states.

XXIV. The House of Lords has now granted permission to appeal in a clutch of cases dealing with the three forms of discrimination: *D' Souza* (race), *Relaxion* (sex) and the recent case of *Jones v 3M Healthcare Limited [2002] ICR 341* (disability).

Unfair dismissal

XXV. By virtue of *section 95(1)(c)* of the *Employment Rights Act 1996* an employee is taken to have been dismissed where he terminates his contract of employment, with or without notice, as a result of the employer's conduct. This is commonly known as constructive dismissal. The conduct relied upon must amount to a repudiatory, or serious, breach of contract.

XXVI. It is now well established that there is implied into every contract of employment a term of mutual trust and confidence. In the recent case of *Morrow v Safeway [2002] IRLR 9*, the EAT was called upon to consider the implications of a breach of the implied term. Mrs Morrow alleged that she had been subjected to harassment and unreasonable pressure by her manager in relation to the manner in which she worked. The tribunal held that the manner in which M's manager criticised her in public amounted to a breach of the implied term of trust and confidence, but that the breach was not sufficiently serious to amount to a repudiatory breach of M's contract of employment.

XXVII. Allowing the appeal the EAT held that a finding that there had been conduct amounting to a breach of the implied term would inevitably mean that there had been a fundamental or repudiatory breach of contract entitling the employee to resign. In other words any and every breach of the implied duty could lead to a claim for constructive dismissal.

Common law actions

XXVIII. In the case of *Waters v Commissioner of Police [2000] I WLR 1607* the House of Lords was of the view that an employer owes a duty of care to employees to prevent harassment. A cause of action arises if the employer knows that acts being done, or which might foreseeably be done to an employee by fellow employees, might cause him physical or mental harm and does nothing to protect him against such acts. Lord Hutton noted that a cause of action would lie in both contract and for the tort of negligence.

XXIX. Actions may also lie in trespass (including assault) and false imprisonment. In the case of *Khorasandjian v Bush [1993] QB 727* the Court of Appeal held, by a majority, that repeated unwanted phone calls amounted to a private actionable nuisance. The *Protection from Harassment Act 1997* creates a distinct tort of harassment, defined as a course of conduct causing alarm or distress to a person.

Human Rights Act

XXX. It is not yet clear whether the *HRA 1998* adds substantially to the protection already afforded to employees against bullying, as outlined above. Article 3 of the European Convention prohibits torture and 'inhuman or degrading treatment'. At first sight it might seem unlikely that Article 3 would apply in the workplace, but in the case of *Smith & Grady v MoD [1999] IRLR 734* the European Court of Human Rights accepted that unnecessarily intrusive investigations into the private lives of individuals suspected of homosexuality might contravene Article 3. The definition of sexual harassment contained in the amended Equal Treatment Directive (see above) refers specifically to 'degrading or humiliating treatment.'

XXXI. It is likely, therefore, that bullying treatment by one employee of another would amount to a breach of Article 3.

XXXII. However as such treatment would almost certainly amount to a breach of the implied term of trust and confidence (or to an act of discrimination if perpetrated on a prohibited ground) it is unlikely that a court would ever have to rely on the Article 3 right.

XXXIII. Article 8 guarantees the right to respect for private and family life, home and correspondence. This is not, therefore an absolute right. The notion of private life indicates an inner circle where the individual can live without state interference. The principle can extend, however, to business premises, including a person's office.

XXXIV. In the employment context, therefore, interception of correspondence, including email, telephone monitoring and searches (of lockers, common rooms etc) are potentially covered. Access to information about an employee, including medical information, is also potentially covered. The right to have and form social relationships and the protection of a person's reputation have also been held by the European Court of Human Rights to be covered by the right, so that, for example, a rule prohibiting office affairs might run counter to it. Covert surveillance, whether at work or away from it, is potentially covered.

XXXV. However, as the right is not absolute, if the employee has agreed in advance (usually through the contract of employment or workplace policies and procedures) to monitoring etc, such interference will be justified. So, for example, where an employer allows employees to use work email for private correspondence, the employer will not be able to monitor or intercept such correspondence merely because it emanates from the workplace. Where, however, the employer has issued an email and internet policy which prohibits or limits the use of private correspondence, such monitoring will be lawful (subject to the principle of proportionality). The case of *Halford v UK [1997] IRLR 471* is a good example of this in practice. In that case the European Court held that the interception of Ms Halford's phone calls, made from work, infringed her right to privacy under Article 8 as she had not been warned in advance that such calls might be intercepted.

XXXVI. Note that there is as yet no free standing anti-discrimination provision. Article 14 prohibits discrimination on any ground, but only in respect of the rights and freedoms set out in the Convention. A free standing anti-discrimination provision, the new protocol 12, is slowly being ratified by member states but is still some way off coming into force.

7 The Effects of Stress on the Individual

Introduction

7.1

'Business is a combination of war and sport'. *Andre Maurois*

In the previous chapters, we focused primarily on the causes and effects of stress within the organisation. In the case of individuals, the effects of stress can extend to cover virtually every aspect of physical and mental well-being. In this chapter, we will look in more detail at these effects, and what can happen when the pressure exerted on an individual, goes beyond their ability to deal with it.

Individual responses to stress

7.2 A useful way of looking at stress is to examine it in terms of the three key types of responses through which it is expressed:

- *Psychological* – how we think and feel.
- *Physical* – how our body reacts.
- *Behavioural* – how we perform and the signs we exhibit.

Psychological

7.3 The psychological symptoms associated with stress are relatively well known and most of us will have experienced them at one, or more, points in our lives. It is common, for example, to experience some anxiety before giving a major presentation or attending an important interview. This is normal and soon passes once the event is over. What is considered abnormal is the persistent feeling of anxiety experienced by many individuals who feel chronically stressed. This anxiety may manifest itself in irritation, impatience or anger and in many cases causes depression, a sense of hopelessness and other negative emotions.

Physical

7.4 In addition to its emotional impact, the adverse effects of stress on the physical health of an individual, should not be underestimated. There is considerable research to suggest that there are links between stress and serious illness, some of which may influence life expectancy.

Stress can weaken the immune system, leading to an increased susceptibility to frequent minor illnesses such as colds and flu.

When stress is prolonged, there is a tendency for minor ailments to develop into more debilitating conditions – with migraine, irritable bowel syndrome and chest pains, being amongst the most common.

Behavioural

7.5 However, it is not just the effects of physical illness that can impact on an employee's ability to perform their job. Stress invariably causes changes in individual behavioural patterns that may well prove to be disruptive, in the workplace. Anger and frustration can also be triggered when individuals are unable to cope with pressure and these feelings can, in turn, impact on productivity and morale.

A manager who is aggressive will fail to obtain the optimum output from his or her staff, who are likely to become resentful and possibly stressed themselves. In this respect, stress is infectious and may be passed on from one team member to another, in the event that appropriate action is not taken to identify and remedy the cause(s).

An intimidating management style or a culture of fear, are just two of the overriding factors in the incidence of workplace bullying. It is important to appreciate that stress can, in some instances, also turn victims into bullies themselves. Individuals can become aggressive when they become frustrated or 'out of their depth' and this can often result in bullying behaviour being exhibited, from one level down to the next, throughout a particular section of the organisation.

The normally composed manager, for example, may become frustrated and aggressive when put under pressure (supposedly to 'push' him or her to achieve the required results), and as a consequence, may find it necessary to apply similar pressure to others, at a subordinate level.

Stress 'carriers'

7.6 A stress 'carrier' is a person, very often in a managerial or supervisory position, who has the ability to cause stress in others by intentionally raising the anxiety level of all those around them, but without suffering any adverse effect on themselves.

Their behaviour is often domineering and others are reluctant to confront them and invariably feel a sense of relief when not in their company.

Case study

7.7

Stress 'carriers'

Bill, a Sales Support Executive, works in the rapidly expanding sales office of a large organisation. He manages his workload well but when Jennifer, his boss, comes into the office she deliberately imposes on him deadlines that are unrealistic, and seems to gain pleasure from intentionally increasing the working pressure throughout the department. However, Jennifer, herself, is confident of her method of management. 'People should be kept on their toes,' she says. 'If Bill feels stressed then he should get a quiet job picking up leaves. This is a workplace – not a park!'

Of course, anger and aggression are not the only behavioural symptoms of stress. Stress can also contribute towards an inability to concentrate, a weakening of essential focus and a difficulty in effectively managing time.

Whilst these effects may give some indication of the impact that stress can have on the individual, most people will only experience a limited number of these in response to any one particular stressor. The following, however, are some of the most commonly experienced warning signs and are listed under either physical, emotional or behavioural (although this table itself is by no means exhaustive).

Table 7.1: Some signs and symptoms of stress in the individual

Physical	*Emotional*	*Behavioural*
Palpitations, awareness of heart beating, chest pains	Mood swings	Susceptibility to accidents
Diarrhoea, constipation, flatulence	Panic attacks	Changes in eating habits
Indigestion	Morbid thoughts	Increased smoking
Loss of libido	Low self-esteem	Restlessness, hyperactivity, foot tapping
Muscle tension	Irritability	Over dependence on drugs and/ or alcohol

Physical	*Emotional*	*Behavioural*
Menstrual problems	Feeling of helplessness	Changes in sleep patterns
Tiredness	Impatience	Out of character behaviour
Breathlessness	Anxiety	Voluntary withdrawal from supportive relationships
Sweating	Crying	Disregard for personal appearance
Tightness in the chest	Cynicism	Loss of confidence
Skin and scalp irritation, eczema and psoriasis	Withdrawal into daydreams	Sullen attitude
Increased susceptibility to allergies	Intrusive thoughts or images	Clenched fists
Frequent colds, flu or other infections	Nightmares	Obsessive mannerisms
Rapid weight gain or loss	Suicidal feelings	Increased absence from work
Backache, neck pain	Paranoid thinking	Aggressiveness
Migraines and tension headaches	Guilt	Poor time management

Some individual coping mechanisms

7.8 In whichever way stress manifests itself in a particular individual, the most important consideration is that the person affected is able to recognise the warning signs in themselves and the fact that these may be stress-related. Only then can they take proactive steps to manage the problem, using the types of coping strategies which are examined in CHAPTER 11. Where stress is not managed, (see Table 7.1 above), it can result in adverse effects on both mind and body that, in extremes, can even lead to suicide.

Individual reactions and attempts to gain temporary relief from personal stress often include binge eating and alcohol, tobacco and drug abuse.

Smoking

7.9 The relationship between tobacco smoking and stress has for long been controversial. The basic paradox is that notwithstanding the fact that adult smokers maintain that cigarettes help them to feel relaxed, recent research shows that, overall, they actually feel more stressed than non-smokers.

According to Professor Andy Parrott of the Department of Psychology, University of East London (in a 1999 study), smokers often report that cigarettes help relieve feelings of stress. However, Parrot's research shows that the stress levels of adult smokers are slightly higher than those of non-smokers; adolescent smokers report increasing levels of stress as they develop regular patterns of smoking; and smoking cessation leads to reduced stress. Far from acting as an aid for mood control, nicotine dependency actually seems to exacerbate the problem.

This is confirmed by the daily mood patterns described by smokers, with normal moods during smoking and worsening moods between cigarettes. The apparent relaxant effect of smoking only reflects the reversal of the tension and irritability that develop during nicotine depletion, ie dependent smokers need nicotine to remain feeling 'normal'.

Many organisations have adopted a non-smoking culture. This can be a source of some resentment amongst co-workers, when smokers take a smoking break – even though this may only be for 10 minutes. Clocking out (where this is company procedure) can help maintain fairness but not all organisations have this system in place. In view of the recognition of the potential harm that smoking causes, an increasing number of companies have also implemented programmes to assist staff to either 'kick the habit' or to dissuade non-smokers from taking up smoking.

Alcohol and drug abuse

7.10 Alcohol misuse is a major issue affecting employers and employees in the UK, impacting on the health, safety and welfare of staff as well as on business productivity and profitability.

Factors that may contribute to stress (and subsequent alcohol or substance abuse) include:

- Constant workloads that are either too heavy or too light.
- Sustained long or irregular working hours.
- Under or over-promotion.

- Excessive pressure from colleagues or management.
- Intractable marital or domestic problems.
- Illness in self/ family or bereavement.
- Need to escape from feelings of inadequacy, fear, anxiety, guilt etc.

The stress of attempting to juggle a job/ career with family responsibilities has also been cited by Alcohol Concern as a major cause of 'binge drinking' in women. Their 2000 study found that females aged 16–24 were the heaviest binge drinkers, with almost half cramming their weekly alcohol consumption into just two or three days.

There may even be a genetic connection between stress and alcohol. According to recent research published, an abnormality in one of the genes associated with the body's stress response may predispose individuals to 'reach for the bottle' as their automatic response to stress (I Sillaber , G Rammes, S Zimmermann, B Mahal, W Zieglagaensberger, W Whrst, F Holsboer, and Spanagel Rin (May 2002) *Science*.

Alcohol and drug dependency

7.11 'Turning to drink' is a coping mechanism adopted by many people, apparently since time immemorial. According to stress researcher, Hans Selye, 'high levels of drinking are clearly more common in stressful occupations' (H Selye (1976)).

The world over, individuals turn to drink because they cannot deal with the particular stressors in their lives. Many (hoping that their problems will go away if they ignore them) try to lose themselves in drink, and are frequently in denial that they have a problem. When they need 'a little something' to help them through the day, they allow themselves 'permission' because they are going through a difficult time in their lives. Before long, they find that they also need 'something' to get them through the night and, then again, before going to work.

This, in turn, alters or modifies their behaviour. It is not difficult, therefore, to understand why individuals may be inclined to use alcohol, (or cannabis), in an attempt to mitigate the effects of stress. However, these can only ever be short-term fixes and not long-term solutions, because the problems the individual is experiencing are invariably still there the following morning. Indeed, it is often the case that someone who stops drinking may feel worse when the problem they tried to avoid has to be confronted again but with the added component of guilt. Contrary to fulfilling the hope of developing better coping resources, many become addicted to their 'false friends'. Unfortunately, long-term alcohol abuse also results in impaired liver function and consequent reduction in life expectancy.

However, not everyone, who uses drugs or alcohol, will have an identifiable dependency problem, nor will they necessarily be using such substances as a way of managing stress. In relation to this, the World Health Organisation has defined certain terms that help to clarify the severity of these issues and most professionals now work with these definitions.

- 'Dependence' is a compulsion to continue taking a drug (including alcohol) in order to feel good or to avoid feeling bad.
- 'Addict' is a drug user whose use of drugs causes them serious physical, social or psychological problems.
- 'Misuse' means using drugs in a socially unacceptable way.
- 'Abuse' means using drugs in a way that is harmful or quantifiably detrimental to the user.

Case studies

7.12 The following case studies illustrate how alcohol or drugs are often used in an attempt to combat the physical, emotional and behavioural effects of stress.

Case study 1

Ann is a 33 year old, departmental manager. She has two young children and when she returns home it is important for her to spend time with them, in addition to having to look after daily domestic chores. Her department has recently been under scrutiny and several identifiable problems have emerged. It is Ann's responsibility to resolve these difficulties, without (in her opinion) having the proper resources with which to do so.

Ann has always found that a glass of wine when she comes home helps her to relax and get through her domestic tasks. Recently she has found that she is drinking more and cannot wait to settle down in front of the television to finish off the bottle of wine, after putting the children to bed. Unfortunately, she now wakes often at 3 am, to find herself on the sofa with the blank screen of the television flickering at her in the darkness. Usually, she cannot be bothered to find her way to bed and finally wakes up in the morning feeling heavy-headed and unrefreshed.

When a friend suggested to Ann that she might be drinking too much, Ann disagreed as she was quite clear as to the benefits that she gained from it. By helping her to relax, she felt better able to cope with those areas of her life that she found increasingly difficult and stressful, when sober. She was also adamant that she could stop

(cont'd)

whenever she wished. Ann was not, (at that point), able or willing to see that her alcohol consumption was probably exacerbating an already difficult situation.

Although initially alleviating the symptoms of stress, part of the physiological function of alcohol is to depress functioning on most levels. Hence performance goes down, and, when sober, anxiety goes up. Ann had unwittingly placed herself in the classic trap of allowing an apparent solution to compound the original problem.

Case study 2

Jane is a 21-year-old university graduate. She intended to pursue a career in marketing, but in the meantime was working in telesales. Her supervisor expected her to get good results and to meet all her targets. At evenings and weekends, Jane and her friends regularly went out clubbing, where they often took illegal, recreational drugs, (including ecstasy).

Eventually, Jane began to find her job both demanding and boring, due to excessive pressure and with little or no time for relaxing and 'chilling out'. She discovered (almost by chance) that if she took some 'speed' (amphetamine) before she went to work, she could still meet her targets and have fun as well – tasks that before had seemed so difficult, suddenly became much easier.

This system worked well for her until colleagues observed that sometimes Jane would seem moody, depressed and exhausted. They also began to notice that she was often late and seemed to have lost a lot of weight. They wondered if she had an eating problem, as she never seemed interested in food. Eventually Jane, now down to eight and a half stone in weight, felt unable to leave the house without her 'fix', and when she ran out of supplies would simply stay at home. After two months, she lost her job owing to her poor timekeeping and was obliged to register at the job centre. She is still registered unemployed and draws benefit from her local DSS office every Tuesday at 10 am. A substantial part of this benefit goes in supporting her amphetamine habit which she indulges in together with others who have, unfortunately, also 'dropped out' from the world of work and active employment.

The problems of alcohol and drug abuse for organisations

7.13 In the context of the organisation, the major problem of alcohol abuse is the fact that excess drinking by employees, even out of office hours, will invariably have an adverse effect on their overall work performance, and has the ability to damage both personal and working relationships.

- According to a Health & Safety Executive study, up to 14 million working days are lost each year in the UK as a result of alcohol-related problems. Although the cost to industry is difficult to quantify, it has been estimated at around £3 billion per year. ('People Management' (May 2000))
- Hangovers alone have been estimated to cost between £53 million and £108 million annually. 46% of large companies report alcohol misuse to be a problem at work. (Medical Council on Alcohol (August 2000))
- The cost to industry of drug use is estimated at £800 million per year, and in 1999, 18% of large companies reported illegal drug use among their employees. ('People Management' (May 2000))
- In a survey on drugs and alcohol in the workplace, nearly a third of employers had suffered absenteeism and poor performance due to drug abuse. (*Personnel Today* (August 2001))

These figures speak for themselves. There has been a substantial, general increase in both drug and alcohol abuse, particularly in the under-30's age group as well as in the population at large, and this cannot fail to be reflected in the workplace.

Drugs and alcohol not only affect individual performance, sickness and absenteeism, but also impact on the wider community. Drug abuse has a proven correlation with crime and both impact adversely on the family, across the entire social spectrum.

However, it is widely accepted by some that alcohol is part of the European cultural tradition, with a significant recreational role to play – and that, if consumed in moderation, can have a beneficial effect on health.

Alcohol can be a causative factor in cirrhosis of the liver; a contributory factor in heart disease; places families and relationships under stress; contributes to unemployment, homelessness and suicide, and affects society as a whole in terms of increased violence, social disorder and accidental injury.

The warning signs

7.14 The terms 'drug addict' and 'alcoholic' conjure up stereotypical images, leading most people to think that they could recognise such a person as clearly as if they carried a flashing light. However, dealing with issues of substance abuse is invariably made difficult by the often hidden nature of the problem and the covert actions of the person(s) concerned.

For example, Marie, a sales assistant, walks uncertainly into work, half an hour late and nursing a hangover from her weekend pub-crawl. Michael, an accountant, answers a client call after a liquid lunch, with slightly slurred speech.

In reality, the signs are often far less obvious. Most addicts are adept at concealing their addiction and individuals may have a serious drink problem for years with only their social partner being aware of it – hence the hidden cost to the employer.

Persistent alcohol and drug abuse will inevitably impact on the employer in terms of impaired performance because by definition, employees who are under the influence of drugs or alcohol, or are suffering the effects of the aftermath of having taken them, cannot possibly perform to their full potential. Some of the more obvious warning signs that may help to alert employers to potential drug or alcohol abuse among staff will, therefore, include the following:

- Poor time keeping, eg being consistently late for work in the morning or after lunch.
- Frequent and unauthorised leave.
- Regular absences on Mondays and/ or Fridays.
- Unavailability to take calls during office hours.
- Coming back from lunch smelling of alcohol.
- Being obviously 'distracted' in the afternoons.
- Deterioration in personal appearance and hygiene.
- Giving implausible excuses for poor performance.
- Deteriorating quality and quantity of work, as a result of impaired intellectual functioning.
- A breakdown in relations with colleagues due to mood swings, irritability, poor decision making, over-sensitivity or reaction to criticism.
- Pervasive low mood, indicative of depression.
- Attempts to borrow money.
- Other, more subtle changes of behaviour, that are difficult to identify.

The triangle

7.15 Any problem relating to drugs or alcohol is unlikely to have a single facet. The fact that some people can manage their drug and alcohol intake and others cannot, is a complex issue. However, broadly speaking, it will involve the inter-relationship between three specific factors – the individual, the drug and the specific environment (explained as follows):

- *Concerns affecting the individual*

  For example, how they manage their life, the quality of their relationships, ability to cope with problems and their view of themselves.

- *Effects of drugs*

  Different drugs (including alcohol) have different effects. If an individual wants to feel energised, they are more likely to choose a stimulant, and if that makes them feel good, then they are more likely to want to keep taking it. An example of this is particularly evident in everyday life where many people are unable to get through the day without five or six cups of coffee to give them a 'caffeine lift'.

  Individuals are also becoming more sophisticated in their drug taking, ie using 'uppers' (stimulants) to 'party' with and tranquillisers and other substances to wind down afterwards. Certain drugs, particularly narcotics such as heroin, are physically as well as psychologically addictive and produce extreme withdrawal symptoms.

  It should be borne in mind that any sudden withdrawal from addiction to drugs or alcohol can carry a significant medical risk owing to the body's necessity to adjust to a significant change in the long-term ingestion of a toxic substance and should, therefore, be carried out only under medical supervision. It should also be noted that some people become addicted to over-the-counter drugs, such as analgesics containing codeine that they take to give them a perceived 'lift' in their mood.

- *The environment*

  In the current, prevailing culture (of the early 21st century), substance abuse is commonplace and is too often accepted by a society that perceives it as normal behaviour and not as a problem, unless there is an overt link to crime or ill health. There is little doubt that this attitude will be modified it time, but as with cigarettes, that period can be very many years hence.

Organisational issues

7.16 It should be appreciated by organisations that work-related stress can contribute to the development, or exacerbation, of an existing alcohol or drug problem, and vice versa.

Early identification of the symptoms of stress, followed by effective interventions, can therefore prevent serious problems from developing.

Despite the complications of drug or alcohol abuse, the problem is very often preventable and manageable. The role of a drug and alcohol policy in the workplace is therefore fundamental in tackling these types of problems. Substance misuse should be viewed as a health and safety issue and should be linked to one or more of an organisation's health, safety, personnel, or general management procedures. Where an employer has in place a policy that offers assistance (either in-house or via an external service provider) to any employee suffering from problems such as stress and/ or substance abuse, there is likely to be a clear benefit to the organisation by virtue of the fact that, inter alia, it will help in retaining existing staff rather than having to deal with the costs of dismissal, subsequent recruitment and training of new personnel.

Burnout

The definition of burnout

7.17 Since Freudenberger's pioneering work on the problems of burnout among the caring professions (Freudenberger (1975)), there has been a steady flow of researched information regarding the condition.

The syndrome of 'burnout' has been described as 'to fail, wear out, or become exhausted by reason of excessive demands on energy, strength or resources'. Maslach defined it as 'The loss of concern for the people with whom one is working (including) physical exhaustion (and) characterised by an emotional exhaustion in which the professional no longer has any positive feelings, sympathy or respect for clients or patients' (Maslach (1976)).

As with the other stress-related conditions, the symptoms of burnout are many and varied. Some people become irrationally angry, and may blame any annoyance, large or small, on external factors. Others may become withdrawn, introverted and isolated (possibly leading to serious depression); experience eating disorders or misuse alcohol or other mood-altering (psychotropic) substances.

Burnout can eventually also induce chronic illness and a range of physical symptoms including:

- high blood pressure and frequent headaches;
- whilst in behavioural terms, individuals who are close to burning-out (or are already burnt-out) may become obsessive workaholics, working longer (but unfortunately less productive) hours in an often futile endeavour to complete the job. They may also become more reactive in managing their time, resulting in constant 'fire fighting' that can, in turn, further damage their powers of creativity and productivity.

Reactive physiological factors also play an important role in the development of burnout. Levels of catecholamines, mainly adrenaline and noradrenaline, increase, and the parasympathetic nervous system becomes depleted. An increased dependence on nicotine, caffeine and other dietary stimulants (such as refined sugars, carbonated soft drinks etc), may also exacerbate the problem.

It is often the inability of the individual to effectively manage the recovery of energy, following long-term periods of prolonged and excessive energy expenditure, that leads to the incidence of burnout.

Accompanying these physiological changes are emotional changes in which the individual's mood spirals downwards from anxiety, confusion and anger to depression, hopelessness and helplessness.

The interplay between the organisation and the individual

7.18 The pace of working life has increased dramatically in recent years as a result of globalisation, competitive demands and technological progress – all of which tend to increase the pressure within the workplace. For many workers, schedules rule their lives, whether this is in relation to getting to work on time, hitting targets, attending meetings, writing reports or checking emails every 15 minutes.

James Gleick in his book *Faster* (1999) writes about this phenomenon, and describes how as human beings we are in danger of being driven to our biological, neurological and psychological limits by technological demands. The implication (for people working in organisations) is that there appears to be an urgent need to re-create a culture that will allow adequate rest and recovery in addition to enabling essential energy levels to be renewed.

The organisational culture

7.19 A frequently cited comment from managers and HR professionals who have been involved with burnout cases, is 'why did we not see the signs?'

The difficulty lies in the fact that very often, an individual who is spiralling downwards towards burnout may be unaware, themselves, of the seriousness of the situation. Alternatively, they may become obsessed with concealing that which they do perceive as the problem. Responsibility for burnout, therefore, sits at various levels.

The organisational culture

7.20 As the organisational culture is a major determinant of working practices, the key questions to ask are:

Continuous and prolonged work pressure

=

Headaches/migraine

Frequent minor illnesses

Lowered immune system

Irritability and impatience

Disturbed sleep patterns

Inability to concentrate

Tension, anxiety and confusion

Anger and frustration

Gastric problems

Loss of self-esteem

Fatigue, frequent sick leave

Panic attacks, depression

Feelings of helplessness

'Burnout' syndrome diagnosed

Long-term absence

'Let go' by employer

Sufferers from BURNOUT remain vulnerable. They need patience, considerable help and support and may not be able to resume full-time working in the same capacity as previously.

Figure 1: Development of burnout syndrome

- Is the culture supportive of a healthy work/ life balance?
- Are all employees encouraged to take proper lunch breaks?
- Are the overall work schedules reasonable?
- Are employees expected to take work home at evenings or weekends?
- Do all staff take their annual leave entitlement and is this monitored?

Management

7.21 Managers play a key role in establishing healthy working practices for their teams and also in leading by example. The key questions for them to ask are:

- Am I close enough to my team members to be aware of their working habits and to observe any unusual changes?
- Am I seen as approachable and accessible if problems are encountered?
- Do I lead by example – am I an appropriate model of good working practices and a healthy work/ life balance?

The individual

7.22 Individuals must assume responsibility for being able to recognise that they are heading towards burnout. The key issues are:

- Am I able to communicate with my manager and/ or colleagues about my workload or any other areas in which I may be experiencing difficulties?
- If not, can I develop my communication skills so that I can initiate a meaningful interaction that will assist my position?
- Am I well suited (neither under or over qualified) to my role?
- Do I need to develop and improve my time-management skills?
- Am I able to change and adopt a better work-life balance that will better support me and the work pressures that I am required to withstand?

The interplay between personality factors and the environment in which a person is working is also crucial. The checklist in Table 7.2 below lists some of the factors that may indicate a predisposition to burnout.

Table 7.2: How to assess your risk for burnout

Burnout occurs when you are unable to cope effectively with work-life pressures over a period of time. The following simple test will help to assess predisposition to stress. The more questions answered positively, the greater risk of burnout.		
	Yes	No
1. Are you highly achievement-oriented?		
2. Do you tend to withdraw from offers of support?		
3. Do you have difficulty delegating responsibilities to others?		
4. Do you prefer to work alone?		
5. Do you avoid discussing problems with others?		
6. Do you externalise blame?		
7. Are your relationships asymmetrical, ie are you the person that is always giving?		
8. Is your personal identity bound up with your work role or professional profile?		
9. Do you have a difficult time saying 'No'?		
10. Is there a lack of opportunity to receive positive feedback outside of your professional or work role?		
11. Does your outlook tend to be: 'don't talk, don't trust and don't feel'?		

For predisposed individuals, early symptoms of burnout might also include:

- Excessive time urgency (eg prone to walking, eating or driving quickly).
- Feelings of guilt about relaxing.
- Over-commitment to projects, or failing to meet deadlines.
- Inability to value and protect personal time.

The life skills toolkit

7.23 Essential to the prevention of burnout is an understanding of energy management and how to strike a healthy balance between energy expenditure and essential recovery. The basic strategies for renewing energy at the physical and emotional levels are discussed in more detail in CHAPTER 11. Many of these are so familiar that they have become almost 'background noise' and easy to ignore. For many workers and employees, the absence of these simple practices may, however, be significant in accounting for fatigue, irritability, lack of emotional resilience, difficulties in concentration, and even a weakened sense of purpose.

Rehabilitation back to work

7.24 The key to the successful management of stress-related ill-health lies in appropriate intervention. Fundamental to this (in larger organisations) is the role of personnel and occupational health professionals who should be readily accessible at all levels throughout the organisation.

Interventions will typically include an initial assessment and diagnosis of the problem, referral to an external agency where necessary, worker reassignment where appropriate, training, education on burnout prevention, transitional work programmes and education of managers and their teams on healthy working practices. (See CHAPTER 10).

Suicide

7.25 We have looked at the subject of suicide in CHAPTER 5, in relation to its effects on the organisation together with some of the warning signs. In this chapter, we will examine some of the more detailed aspects of work-related suicide, in terms of the incidence within various occupations, what help can be given and which interventions can be made to help pre-empt such tragic eventualities.

There is no definitive answer as to why people commit suicide. Some people may attempt to end their lives in order to escape from unbearable emotional, or physical, pain or possibly to avoid the inevitability of financial disaster. It is often a cry for help. A person attempting suicide is often so distressed that they are unable to see that there are other options and therefore, a tragedy can sometimes be averted by identifying, realistic alternative choices and ensuring that these are communicated as early as possible to the person in difficulties.

Suicide prevention strategy launched

7.26 The Government's White Paper 'Saving Lives: Our healthier nation' (Department of Health (2002)) sets out a challenging target to reduce the death rate from suicide and underdetermined injury by at least a fifth (a saving of 4,000 lives) by the year 2010 by:

- Reducing access to means.
- Increasing public awareness.
- Focusing on high-risk groups.

- Educating health and social care professionals.
- Audit and investigation of serious incidents.
- Improving primary and secondary health care.

Causes of attempted suicide

7.27 The causes of individuals attempting to commit suicide include:

- Clinical depression.
- Acute loss of self-esteem.
- Feelings of extreme hopelessness.
- Ultimate attention seeking.
- A complete loss of any personal goal.
- To escape the inevitable punishment for criminal behaviour.
- Refusal to believe that death is final.

Factors increasing the risk of a person committing suicide

7.28 The likelihood of a person committing suicide depends on several factors (J Charlton, S Kelly, K Dunnell, B Evans, R Jenkins (1994)). Social problems, especially those related to family stress, separation, divorce, social isolation, death of a loved one and unemployment, as well as mental and physical illness and access to the means of suicide are all potential risk factors.

Suicide and occupation

Suicide and unemployment

7.29 Links between unemployment and the frequency of suicide amongst young adults have been clearly demonstrated (S Platt (1986)). In an international study of male suicides in 22 countries between 1974 and 1986, unemployment was found to be a leading factor (V McDonald 'Suicides in young men on the increase' (1992) Sunday Telegraph.

Men in unskilled occupations are four times more likely to commit suicide than those in professional work (Department of Health (1999)). Certain occupational groups such as doctors, nurses, pharmacists, vets and farmers are at higher risk. This is partly because of access to the means of suicide.

The analysis that follows, (all of which is based on work by the Samaritans), looks at deaths by suicide in people of working age (ie 20 to 59 for women and 20 to 64 for men) between the years 1982 and 1996, analysed by occupational classification. The analysis involves an indicator of risk called the proportional mortality ratio ('PMR'). PMR uses the proportion of deaths due to suicide in a particular group, and compares this to the average for the total population, to give an indication of the risk of suicide in that group. The average figure is 100, meaning that a PMR of 200 indicates twice the average risk, while 50 indicates half the average risk.

It is commonly accepted that high stress, together with, as previously mentioned, with easy access to means of suicide are important factors which put people in certain occupations at greater risk of dying by suicide.

The number of suicides in some occupational groups (particularly among women) are very low, and so this can result in extremely high or low PMRs – meaning that the statistics should be treated with caution. However, from the analysis, it can be seen that those in the medical professions and farming have a higher risk than others. This is commonly accepted to be a result of the fact that high stress, together with easy access to suitable means, are important factors that put people (in certain occupations) at greater risk of dying by their own hand.

High risk professions

The medical profession

7.30 Female nurses and nurse administrators accounted for a substantial proportion (4.8%) of all suicides among women aged 20–59 between 1982 and 1996. There were 240 suicides amongst this group between 1991 and 1996 – an average of 40 per year, or one every nine days. Female nurses were at 1.37 times the average risk for females of working age. Female medical practitioners were also at significant risk (2.85 times the average), although the number of deaths were small in comparison with nurses – there were 25 suicides in this group between 1991 and 1996.

Male dental practitioners were the group with the highest significant risk of suicide over the time period 1991–1996 (2.49 times the average), during which there were 25 suicides in this group. Male medical practitioners also showed higher than average risk (1.47 times), with 71 suicides between 1991 and 1996.

Case study

7.31

> **Suicide in the medical profession**
>
> In February 1998, the widow of a mental health nurse, Richard Pocock, received a £25,000 settlement from North-East Essex Mental Health Trust after claiming that her husband had been driven to suicide. The trade union, UNISON, had argued that he was subjected to a vindictive, oppressive, ruthless and macho style of management, and that although management was made aware that Mr Pocock was suicidal, they failed to do anything about it.

Veterinarians

7.32 From 1982–87, male veterinarians showed the highest PMR from suicide – with more than three times the average male suicide risk. However, this figure came from a total of 17 suicides over the six years. From 1991–1996, the number of suicides dropped to nine, resulting in the second highest PMR (349) of those analysed. Dental practitioners were therefore seen as a higher risk group, as their figure was based on a larger total number of suicides and was therefore the most reliable.

Farmers

7.33 From 1982–87, farmers, horticultural workers and farm managers had the second highest risk of suicide after vets – with a PMR of 202. By 1991–96, the risk had dropped to 1.44 times the average, and they were then the third highest risk group in the occupational suicide table. Nonetheless, numbers of farming suicides were still very high, with farmers accounting for 1.3% of all male suicides (aged 20–64) between 1982 and 1996. Between 1991 and 1996, there were 190 farming suicides, or one every 11 days.

Occupational suicide data tables

7.34

Table 7.3: Suicide data table for men aged 20–64, 1991–96

Occupational	*PMR*	*Suicides*
Veterinarians	324	9
Dental practitioners	249	25
Pharmacists	171	25
Garage proprietors	155	43
Sales representatives: Property and services	151	97
Medical practitioners	147	71
Farmers, horticulturists, farm managers	144	190
Publicans	128	129
Other motor drivers	124	221
Cleaners, window cleaners, road sweepers	122	204
Painters and decorators, French polishers	119	389
Builders	119	332
Shop salesmen and assistants	118	296
Gardeners, groundsmen	117	234
Carpenters and joiners	115	384

[Source: S Kelly, J Bunting (1998)]

Table 7.4: Suicide data table for women aged 20–59, 1991–96

Occupational	*PMR*	*Suicides*
Veterinarians	500	4
Medical practitioners	285	25
Domestic housekeepers	247	16
Waitresses	187	37
Professional and related in education:		
Welfare and heath	183	26
Students	139	132

Occupational	*PMR*	*Suicides*
Cleaners, window cleaners, road sweepers	139	95
Nurse administration, nurses	137	240
Hospital ward orderlies	130	139

[Source: S Kelly, J Bunting (1998)]

Other effects

7.35 The reality of suicide can have a significant effect on anyone having had a connection with the person concerned. The closer the connection, the greater the possible adverse effect and this effect can range from shock to possible guilt at not having anticipated such an extreme action. These feelings can impact on those affected, long after the actual event.

Work colleagues, families and friends of people who have committed suicide may struggle for some time after the event with anger, frustration and the many unanswered questions, eg 'Why did they do it?' . . . 'How could I not have known he/ she was so desperate?' . . . 'How could they leave me?' In addition to the feelings of grief normally associated with a person dying, there may be guilt, resentment, confusion and despair over unresolved issues.

For people who have experienced the suicide of someone they cared deeply about, they can benefit from joining a 'survivor group' eg Survivors of Suicide, where they have the opportunity to relate to people who have been through a similar experience, and know they will be accepted without being judged or condemned.

The act of suicide is still a subject that most people find extremely difficult to comprehend and therefore still regard as a 'taboo' subject, both in and out of the workplace.

There are some suicide 'facts and fables' which, even though they were written many years ago, are just as relevant today.

Suicide facts and fables

7.36 The following are taken from an article 'Suicide', written by Edwin S Schneidman for the *Encyclopaedia Britannica* (1973).

'Fable: People who talk about suicide don't commit suicide.

Fact: Of any 10 persons who kill themselves, 8 have given definite warnings of their suicidal intentions.

Fable: Suicide happens without warning.

Fact: Studies reveal that the suicidal person gives many clues and warnings regarding his "suicidal intentions".

Fable: Suicidal people are intent on dying.

Fact: Most suicidal people are undecided about living or dying, and they "gamble with death" – leaving it to others to save them. Almost no one commits suicide without letting others know how he is feeling.

Fable: Once a person is suicidal, he is suicidal for ever.

Fact: Individuals who wish to kill themselves are suicidal only for a limited period of time.

Fable: Improvement following a suicidal crisis means that the suicidal risk is over.

Fact: Most suicides occur within three months following the beginning of "improvement", when the individual has the energy to put his morbid thoughts and feelings into effect.'

Although these words were written 30 years ago, one point that they make very clear is the extent to which individuals, who are feeling suicidal, give definitive warnings of their intention (quite possibly through the types of warning signs we looked at in CHAPTER 5). If someone is exhibiting some of these warning signs, it is important to enable the person to talk about their feelings, promote an understanding and establish empathy through which it may be possible to alleviate their immediate distress.

It is also necessary to appreciate the type of support that a suicidal person may need (and equally importantly may not need) from their friends, family or colleagues.

Needs of those feeling suicidal

7.37

- Someone to be there for them who has the time to listen.
- A calm, assured voice.
- Confirmation that they are not crazy.
- Not to be judged.
- To feel less frightened.
- To recover respect for themselves.

- To receive undivided attention for their specific problem(s).
- Not to feel a burden.
- To be accepted for whom and what they are.

Attitudes not required

7.38

- Rejection.
- To be told that it is wrong or silly to feel as they do.
- To feel embarrassed for having approached you.
- A lecture, sermon, pep talk or debate.
- Clichés such as being told to 'pull your socks up' or 'think of all those who are worse off than you'.
- False reassurance ('it will be okay').
- To be patronised, criticised, analysed or categorised.
- To be made to feel guilty.

Knowing what to say and do

7.39 It is recognised that there is an inherent difficulty in knowing how to approach someone suspected of feeling suicidal. The following points are offered, as a possible assistance:

- Listen attentively and absorb that which the person is trying to communicate. Encourage the individual to talk and express their feelings. Offer simple, unconditional support even though they may initially reject the help that is offered to them.
- Persons feeling suicidal should be taken seriously and offered empathy, care, trust, space and time. It is important to respond to them on an equal level with unconditional, uncritical acceptance and respect for how they are feeling.
- Listen to, and reflect back, the feelings that are articulated. Focus on the words spoken and ask for expansion. For example, 'I feel down'. Ask them to tell you what they mean by 'down'. 'How far is "down"?'. Stay with the exploration of their feelings.
- Using a low, gentle tone of voice, ask if they are, or have ever been, suicidal. It can sometimes be a relief to the person to be asked the question, even if the answer is 'no'. One sometimes needs to hear the word 'suicidal' verbalised before one dismisses it from one's mind. In the words of a volunteer who worked with the Samaritans for

over 20 years, 'It is important to recognise that merely by asking someone if they are suicidal or are considering suicide, will in not prompt them to do so'.

- Suggest appropriate action(s) and know where and to whom to refer.
- Ask if there is anyone they would like you to call, eg family or friends or they might prefer to call the Samaritans (who are open 24-hours a day and are willing to take third party calls).
- Gently inform them that you may have to break confidentiality and communicate the position to the appropriate person(s) in the organisation, as a matter of duty. It may be also be necessary to make a confidential note in the employee's personnel file. However, it is important to remember that if the person is in a state of crisis, they will probably not hear what is being said to them.
- In the final event, it is imperative to call on external professional support.

In conclusion, there will be individuals who, for one or other reason, are determined to end their life and there will consequently be little that one can do to help. However, there are others who, with help and understanding, can be dissuaded and will, when recovered, be immensely grateful that someone was there who was able to make them see more clearly, how valuable is life – their life – and the available alternatives there were to throwing it away.

Key learning points

7.40 The effects of stress on the individual are many and varied and can extend to cover virtually every aspect of physical and mental well-being. However, with the correct knowledge, individuals can adopt a wide variety of strategies to help themselves 'cope' with stress, although it is important not to employ any action that may exacerbate an already difficult problem.

- Stress comprises three individual responses – emotional, physical and behavioural.
- Unremitting pressure leading to stress can impair individual performance at work and damage health.
- Organisations can minimise the effects of undue stress on individuals, and the impact on the organisation, by educating managers to recognise the early warming signs and symptoms of work-related stress.

(cont'd)

- A stress 'carrier' is a person, (very often a manager), who has the ability to cause stress in others by intentionally raising the anxiety level of all those around them but without suffering any adverse effect on themselves.
- Individuals may attempt to help themselves 'deal' with stress by smoking, or through the misuse of alcohol and drugs.
- Although adult smokers state that cigarettes help them feel relaxed, recent research shows that they actually feel more stressed than non-smokers.
- 'Turning to drink' is a (false) coping mechanism although adopted by many people who are looking for relief from stress.
- Although initially alleviating the symptoms of stress, part of the physiological function of alcohol is to depress functioning on most levels. Hence performance goes down, and when sober, anxiety goes up. Individuals who turn to alcohol, place themselves in the classic trap of allowing an apparent solution to compound the original problem.
- Persistent alcohol and drug abuse will inevitably impact on the employer in terms of impaired performance as an individual's health deteriorates and they become more accident prone and are inclined to take more days off sick.
- Despite the complications of drug or alcohol abuse, the problem is very often preventable and manageable.
- The symptoms of burnout are as varied as the profiles of the sufferers.
- Burnout is often the inability to alternate between energy expenditure and recovery that drives the burnout process.
- The incidence of suicide in the medical professions and in farming is higher than in other fields.
- Any suicidal person who is potentially at risk, should be listened to attentively and taken seriously. It is essential to know what support is needed and is available for them.
- If an employee commits suicide, it can leave long lasting effects within the organisation if not dealt with in an open and sensitive manner.

8 Trauma and Coping with the Aftermath of a Critical Incident

Introduction

8.1

'To conquer fear is the beginning of wisdom'. *Bertrand Russell*

Nothing can adequately prepare organisations or individuals for the experience of a traumatic incident, because by definition it is outside 'normal' experience. This was vividly illustrated by those affected at the collapse of the World Trade Centre, New York (2001) and the disasters of Canary Wharf (1996), the Manchester bombing (1996), the sinking of the *Marchioness* (1989), Hillsborough (1989), Kings Cross (1987), as well as numerous other tragic events which were impossible to have been predicted.

Many victims and witnesses of violence or injury sustained in accidents; criminal activity or natural disasters such as fires or floods, may well require professional, post-trauma support to help deal with the effects of their experience.

Unfortunately, more and more people are the unwitting victims of violent crime both at work and in the street. For example, a young girl who has a Saturday morning job working in a supermarket and is involved in a raid on the store. She may need as much care and support as the young man who has a serious accident in his firm's van as he carries out his delivery schedule. Yet inevitably, some traumatised individuals are unidentified as such, and can slip through the support net.

Defining a traumatic incident

8.2 In 1980, the American Psychiatric Association published the third edition of its 'Diagnostic and statistical manual of mental disorders' (DSM-III) where, for the first time, post-traumatic stress disorder ('PTSD') was defined as a classifiable psychiatric syndrome.

The definition of PTSD has undergone several updates since then and there are now two working definitions available in the following documents published by the World Health Organisation (WHO):

- 'Diagnostic and statistical manual of mental disorders' (DSM-IV) (1994); and
- 'The international classification of mental and behavioural disorders: Clinical descriptions and diagnostic guidelines' (ICD-10).

According to the DSM-IV (1994), the diagnostic criteria for post-traumatic stress disorder (PTSD) require the 'traumatic event' to have been an event (or events) that involved either actual or threatened death or serious injury, or a threat to the physical integrity of the person concerned or others. The ICD-10 defines it in similar terms. However, in the context of working with people who have been traumatised, such definitions are probably too narrow.

Where an incident exerts a traumatic effect on an individual, then it should be recognised as such and should trigger some form of support mechanism, whether or not the event comes within the previously mentioned definitions.

There will be instances where individuals may develop symptoms of anxiety of depression, or some of the symptoms of PTSD, but these may not always be of sufficient range or severity to attract a formal diagnosis.

At-risk groups

8.3 Whilst all organisations should carry out risk assessments to determine whether employees could be at potential risk of psychiatric harm from events that they may have to deal with in the course of their work, there are some organisations whose staff, by virtue of their work, are always potentially at risk. These include:

- the armed forces and emergency services personnel;
- bank staff and certain others in the financial field;
- those working in retail outlets, off-licence liquor stores and petrol stations, where staff may be alone in the premises outside normal shop opening hours or even all night; and
- those who have contact with the general public in circumstances where there is a greater risk of violence.

Companies operating in the travel industry need to be aware of not only the actual incidents that do occur, but also the potential for accidents occurring – and in particular, major disasters for which effective contingency plans should always be in place. Indeed, airlines flying into the USA are required by law to have contingent and workable plans to help families and survivors deal with any airline disaster. [*Aviation Disaster Family Assistance Act 1996* and *Foreign Air Carrier Family Support Act 1997*].

There are also organisations that operate within particular industrial sectors that are inherently hazardous by virtue of the nature of their work; and although good risk management can substantially reduce the incidence of accidents, they may occur at some point in time. In such instances, contingent action plans need to be ready to be implemented, often at short notice, to support those who may be involved.

Those affected, directly and indirectly, following an incident

8.4 It is not only those people who are directly involved in an incident, ie victims and survivors, who may suffer the effects of trauma. It can also affect those who may be indirectly involved, eg witnesses, neighbours, families and work colleagues, or those who may be helping with the setting up of emergency shelters or, in some cases, temporary mortuaries. For example, the people who were inadvertently caught up in an incident in Hungerford, Berkshire in 1987, where passers-by witnessed a heavily armed gunman kill 14 people, (including his mother), before taking his own life.

All of these people have the potential to be traumatised, to a greater or lesser degree, including also the emergency services (police, fire and rescue, ambulance and medical staff), clergy, counsellors, social and voluntary workers.

Although victims will inevitably be emotionally unprepared to deal with a sudden emergency or disaster, trained emergency workers will normally be less vulnerable to emotional overload as a result of extensive mental preparation and training for just such eventualities. Nevertheless, this does not preclude them from being affected and possibly traumatised by being involved.

The following are just some examples of how individuals can become directly or indirectly exposed to traumatic incident and possible post-trauma stress:

- Co-workers who may have to return to work immediately following a disaster will have to come to terms with the injuries and possible death of one or more colleagues, together with possible damage to workplace buildings caused by fire, water, etc. The workplace may have changed dramatically and the effect of this may impact on everyone within the organisation, to a greater or lesser extent.
- There may possibly be feelings of guilt associated with injury and loss of life. Management and staff may feel disorientated and emotional following the harrowing experiences of fellow workers, and will be susceptible to post-trauma stress. Employees may have to be relocated to other premises and be in a position of some turmoil for days, or possibly months, thereafter.

- The designated first-aiders within an organisation, who may have had only limited training and experience, will most certainly be called upon to deal with a major incident before the emergency services arrive. Yet the support that they themselves will require, in the aftermath of the event, is often overlooked.

- The train driver who experiences a person committing suicide under the wheels of his train (known as 'one-under'), and the maintenance team who have to remove the human remains from the track. Some of these people will be required to re-live the situation when they give evidence to an enquiry, which can sometimes be months or even years later. This can trigger again the traumatic reaction to the original incident and the person may be unable to 'close the chapter' until all investigations are complete.

- A traumatic incident may impact on the confidence of other employees performing similar jobs within the organisation.

  For example: Anne, aged 21, works in a retail store on a Saturday morning. Her close friend Jane was sitting at the till when she was attacked by a drunken customer who pulled a knife on her. Anne was badly shaken by the incident, and now when she sits down to work, she looks at every customer very carefully in case they might pose a threat to her safety – as one customer did to Jane.

- Proper consideration also should be given to people involved in potential incidents or 'near misses'. These may include people who believe that they came close to a major accident or incident, even where they sustained no actual damage or physical injury, themselves.

  For example: Bill had booked a ticket on Pan Am flight 103 from London Heathrow to JFK New York, in December 1988, which crashed with no survivors. Bill had cancelled his ticket due to a late business meeting but was left with a feeling of guilt that someone else had taken his place and, for months, experienced nightmares of his close shave with death.

- People who are victims of a malicious hoax that gives harrowing images of danger, may also experience post-trauma stress.

  For example: Gilly, a high-profile public figure, received phone calls and letters saying that her whereabouts were known and that she was going to be murdered. She had no idea who wished to kill or frighten her and was terrified every time she left the safety of her home. Even when the hoaxer was caught she still felt insecure and needed constant reassurance to go about her daily business.

Post-trauma stress

Definition

8.5 Post-trauma stress ('PTS') can be defined as 'the development of characteristic symptoms following a psychologically distressing event outside the range of normal human experience'. Not everyone will suffer the symptoms, but it is necessary to emphasise that to be affected and to experience a reaction, are both natural and normal responses.

PTS cannot only influence an individual's feelings about themselves but can also affect their relationships with others. This can cause difficulties at work, personal relationship difficulties, ill health and sometimes the development of deeper and more disturbing symptoms.

Where the symptoms of PTS persist or intensify, for more than a month or so, the condition of PTSD can emerge (see diagnostic criteria at 8.20 below). Symptoms may develop weeks, months or even years after the event, and these can vary from being mildly disturbing to, in the extreme, completely incapacitating.

The reactions of individuals to traumatic events varies, both in severity and type. Some may react very strongly to what may be regarded by others as a minor incident, whilst certain individuals seem to be able to deal with even major incidents or disasters very much in their stride. The rest of us are spread out on a continuum somewhere between these two points.

However, the majority of people will experience some form of reaction to a traumatic event, although in most cases, such reactions will subside over a short period of time. The emotional equilibrium will eventually be regained, and the incident will become just a part of life experience – not forgotten, but not intrusive.

It is difficult to predict just how any one person might react to an incident but some individuals who may be more vulnerable include:

- Those with existing life difficulties and frustrations, such as relationship problems, divorce, bereavement, chronic illness, etc.
- Those who have an inadequate social support network, such as a partner and/ or family, friends and/ or colleagues with whom they can confide and discuss their feelings.
- Those who are neurotic, anxious, depressed, introverted, unwilling or unable to talk to others.

However, the minority of people who do not recover quickly is sufficiently large to make it essential that they have access to professional help and appropriate treatment – without which their condition could well deteriorate, with possible long-term consequences.

As indicated above, not all those involved in criminal violence or other traumatic incidents will necessarily suffer psychological trauma as a result. In an overview of empirical studies into the outcomes of extreme events, Brom, Kleber and Witztum (1991) found that typically between 18 and 20% of all people who go through extremely distressing events were left with permanent coping disorders associated with PTSD. However, those who did not develop PTSD were not necessarily free of other symptoms caused by the trauma.

Immediate reactions to trauma

8.6 There will inevitably be a number of differing, individual reactions immediately after the experience of a traumatic event. These may include screaming or weeping, silence or disbelief, disorientation or shock and then subsequent tiredness and fatigue, anxiety and a reduced response to stimulus and surroundings.

- Those that survive a trauma may be plagued with the recurrent question, 'Why did I survive?' ... 'Why me?' ... 'I should have done more'. These people will need help to accept that there is no rational explanation or answer to these questions.
- Some people may feel glad to be alive, while others may feel emotionally paralysed. Black humour is frequently used as a protective reaction, and this is a legitimate way of coping, but one that may sometimes be misunderstood by others who may be listening.
- In the absence of any head injury or concussion, amnesia is not uncommon. It is an emotionally protective mechanism and may appear in relation to either just part, or the whole, of the experience, eg 'I can't remember how I got out of the car, I just found myself on the kerb' ... 'I can't remember going to hospital. I have a vague image of blue flashing lights but the first three days were just a blur'.
- The thought by the individual that they themselves could have died, is very difficult to deal with, and an indelible image of their near encounter with death, is often a recurrent thought.
- Some individuals may obtain a sense of satisfaction that they have faced a crisis and handled it well, or of having had the experience of going though a trauma, together with others, and 'coming out the other side'. This often has the effect of subsequently bonding closer together, those concerned.. There may be a feeling that life now has more significance, more meaning, or that family and friends have suddenly become infinitely more important.

Of course, there is an accepted human need, at any time, to feel part of a community, but in times of disaster, we realise how important it is to find comfort and solace in the closeness of others.

Following the catastrophic disaster at the World Trade Centre on 11 September 2001, in New York, when the emblematic twin towers were demolished by terrorists, a spontaneous support group evolved at St Paul's chapel, which adjoins the former World Trade Centre plaza. For eight months after the terrorist attacks, until recovery work stopped at the end of the following May, hundreds of recovery workers, fireman, police, doctors, massage therapists, counsellors and countless other volunteers gave their time. They helped to heal the weary bodies and minds of those who had ceaselessly worked to recover the human remains of over 3,000 people who had been unfortunate enough to be there that day, in the buildings that stood on the site which is now called 'ground zero'. Here, they slept, ate, cried and found essential solace in the human warmth and kindness that filled the chapel – night and day during the weeks and months that followed. Bonds were formed there that gave hope to all who entered that sanctuary – bonds that confirmed their sense of belonging to a giving, caring community.

- It is recognised that we all need, at times, a 'dar-es-salaam' (a haven of peace), a place where feelings of pain, anger and grief can be allowed freedom of expression. Across the front portico at St Paul's chapel, a timeline records the events of the days following 'September 11th' and offers a glimpse of the messages and cards of support that arrived daily from, not only America, but from around the world. Still today, in 2003, many of these cards have been retained to cover the walls of the small chapel, together with the scuff-marks on the pews that were created by the boots and belts of the emergency workers who gathered there. These are being left untouched as a 'sacramental reminder' of those who sought succour and support in this special sanctuary.

'St Paul's, will always be in my mind, Heaven's outpost. By entering through the gates out front, one can leave behind the terror and destruction that leaves you feeling a severe loss and find a place where everyone has a smile'. Robert Senatore, New York Fire Department

Case study

8.7

Post-trauma stress

Paddy was one of the financial traders working in the Twin Towers at the time of the attack. He managed to get to a staircase before the collapse and succeeded in walking down 34 floors. He arrived in the

> ground floor lobby six minutes before the entire structure failed, as floor upon floor collapsed onto the one below, killing all those still within. Paddy suffered from severe PTS and was hospitalised for three months treatment. He now is back at home outside Venice Beach on the East Coast, and spends the bulk of each day just staring out at the ocean. During his weekly sessions he repeatedly tells his therapist that 'God alone knows, it should have been me. What am I going to do and how can I rebuild my life?'

General symptoms of post-trauma stress

8.8 The general symptoms of PTS may include the following:

- There can be a strong reaction of denial which may prevent people from acknowledging that they have a problem and, therefore, from seeking help. To ask an individual who has been involved how they feel after an incident will usually elicit the reply, 'I'm alright, thanks, I can cope'.

- Others will experience increased anxiety, a sense of vulnerability and isolation, and possible survivor guilt. 'Why did I survive when others didn't?' There can also be an induced sympathy for the aggressor(s), eg 'The Stockholm Syndrome', resulting from a hostage situation or being confined together with the hostage takers.

- There may be an inability to concentrate or to make even simple decisions; possible impulsive behaviour including excessive spending, moving home, changing job or lifestyle, or ending or creating new relationships. Some people will want to talk incessantly about the event, while others will feel the need to retreat into isolation.

- Individuals will frequently experience dreams and nightmares, and have a general feeling of either being unwell and/ or of extreme fatigue.

- There are also those who will have experienced changed values or beliefs or re-adjustments in relationships. 'What's the point of marriage or work or even of living when something like this can happen?' Some discover a new faith or lose an existing one. Others may deepen already held beliefs, whether religious or secular.

- It is important to remember that any of these symptoms are very similar to those associated with loss, grief or bereavement.

Specific characteristics of post-trauma stress

8.9 In addition to the above, some of the other specific characteristics of PTS include the following.

Arousal

8.10 This can manifest itself as an increased (over) sensitivity to particular sensory messages, eg sounds or smells, that, in turn, may result in an inability to withstand the normal daily patterns of life and work.

An increased sense of arousal can result in sleeplessness or difficulty in concentrating and some individuals may become over-vigilant. There can be an unwarranted expectancy that something untoward is going to happen at any time and without prior warning.

Re-experiencing (intrusive thoughts and flashbacks)

8.11 The actual trauma of an event can sometimes be re-experienced days, months or even years later with 'intrusive thoughts' that comprise disturbing images of the event that involuntarily come to mind.

A 'flashback', on the other hand, is a feeling that the event, or a part of it, is actually re-occurring. For example, an individual may experience a similar smell or sound that will immediately remind them of the traumatic incident, and act as a trigger to make them feel that they are back at the original scene, experiencing the same sensory impressions that they suffered at the time.

Triggered reactions can come at any time, and may be caused by:

sights: TV, video, photographs, media reports, people

sounds: police sirens, bangs, crashes, voices

smells: petrol, rubber, disinfectant, dampness

tastes: food, water, petrol, alcohol, sweat

touch: rubber, metal, skin, dampness, water

'Out of the blue' reactions can occur randomly and without any warning or obvious 'trigger'. These may happen anywhere, and because there is no apparent cause, can be extremely frightening, both for the individual concerned and others.

Case study

David was inadvertently involved in an altercation in a pub that developed into a violent fight. A few days later, he found himself sitting on the floor in his office, completely confused, disorientated and in a state of panic. David's suppressed feelings suddenly came rushing to the surface of his consciousness with acute physical and psychological effects.

Avoidance

8.12 Those involved in a traumatic event may well seek to avoid anybody or any circumstance (eg flying, horse riding, cycling, going to an airport, driving a car or taking a ferry), that might remind them of the incident and if the incident took place at work, they may experience an inability, or an extreme reluctance, to return to their job. Typically they will try to avoid thoughts, feelings or situations reminiscent of the event. There may also be confusion, a loss of concentration and possibly some feelings of isolation.

Case study

Mary is a clerk who works in the upstairs office of a large supermarket, and handles large amounts of cash. The office is usually quite secure, but as she left to go the toilet one day she was suddenly grabbed from behind by a masked intruder, and bundled into a side office. Threatened with a knife and tied up, she could see that there were three men, all wearing balaclavas. The leader of the gang had deep blue penetrating eyes and threatened her into telling him the entry code for the safe. The thieves got away with £30,000 after leaving the supermarket by forcing open a rear emergency door.

Mary was off work for three days, and upon returning to work for the first time after the robbery, she suddenly visualised the robber's blue, staring eyes piercing into her, and experienced an acute panic attack. After a few minutes, the feeling passed but she left as soon as she could and returned to the safety of her own home.

Mary's doctor gave her a sick note and some medication, and told her not to return to work for at least 14 days. As Mary sat at home, she found it difficult to contemplate returning to work full-time, as any reminder of the incident caused her to feel sick as she started to relive the incident.

Organisational responsibility

8.13 A problem for an organisation that can arise is when an event in the workplace triggers a powerful adverse reaction in an individual that is compounded by the influence of an external factor or previous event. For example, in one particular case of a workplace fire, an employee, who had recently recovered from a serious illness and was under medication, was hit by a falling beam. It may prove difficult to determine the degree of liability the organisation should accept for the treatment of such a vulnerable employee, when part of the cause is not directly connected with his or her work, but is dependent on individual circumstances.

Employers have a duty to consider the robustness of the person attending for work, irrespective of the cause of their stress. For instance, someone who had just suffered a bereavement should probably not be driving a public transport vehicle.

Organisations have a legal duty of care both towards those who they employ and those who may be affected by their operations. There is a necessity to provide an appropriate level of support in order to discharge that duty. A failure to do so may result in legal proceedings and an award, by a court, of substantial compensation in damages. For example, in 2001, the South Yorkshire Police force agreed to pay an out of court settlement of around £330,000 for PTSD, to a policeman to who had attended the Hillsborough football disaster in 1989 that left 96 fans dead. However, should the failure be one of criminal negligence, then criminal proceedings against those responsible could ensue.

Health and safety at work legislation

8.14 The *Health and Safety at Work etc Act 1974* requires employers to do what is reasonably practicable to ensure the health, safety and welfare of employees. It does not differentiate between various forms of harm, and that includes a duty to ensure that safe systems of work are set and followed (see also 2.3).

The provisions of the *Management of Health and Safety at Work Regulations 1999 (SI 1999/3242)* place a statutory duty on employers to conduct risk assessments both of the actual work carried out by employees in addition to that of the workplace and its environment. These assessments are to enable employers to identify any potential hazards to health, who could be harmed, in what way and how often? The assessment of the extent of risk allows the implementation of appropriate preventive or protective measures, or, alternatively, the complete elimination of the identified hazard (see also CHAPTER 3).

Risk assessment

8.15 All employers need to carry out risk assessments to identify any potential risk; whether such risk is significant and what measures should be implemented to prevent or minimise it. This applies to all employers, not just those working in hazardous fields. Risk assessment should focus on the level of risk, the expected type of traumatic incident and the staff roles most likely to be involved.

In many cases when there is a traumatic incident at work, the fact that most people recover naturally could lead the organisation to believe it is unnecessary to provide any form of additional support. However, this is not necessarily the case.

It is therefore essential that organisations design and implement effective systems that will include record keeping and monitoring, both of the causes and the effects of any incident, in addition to the contingent provision of effective support, to employees, following a major incident.

A successful trauma support programme is dependent on the positive attitude of the organisation implementing it and their genuine concern for the welfare of their employees. However, a successful outcome may also be dependent on an employee's perception of the organisation. Does it take seriously the possibility of an accident, or a violent situation occurring, and if so, has the necessary risk assessment taken place and have the appropriate actions been taken?

A post-trauma support programme should include:

- Careful selection and training of staff who are to work in potentially dangerous or aggressive environments.
- Well-designed emergency procedures and action plans.
- An education programme detailing potential hazards.
- Dedicated on-scene support.
- Professional backup following the incident – aimed at providing short and long-term psychological support, as needed.

Post-trauma support strategy

8.16 It is important that senior management 'buy in' and are committed to, a post-trauma support strategy that it is translated into effective systems, procedures and good practice.

The aims and objectives of the policy should be communicated to everyone, throughout all departments. The policy should allow for ongoing support in respect of any individuals affected through the process of debriefing and any other assistance required in order for them to regain their mental and physical health. This may take days, weeks or months, depending on the severity of the effects, and will entail a qualified assessment either by an external or in-house, professional provider.

The methods of access to the organisation's support services should be conveyed to all employees both before and after any incident, and written information should be available for reference at points throughout each department of the company, firm or local authority.

Following any incident, necessary feedback should be conveyed to the line manager(s) of the affected employee(s), but should not break matters of confidentiality. However, opportunity should be given to employees, where possible, for self-referral to post-trauma support services, eg post-trauma counselling that is independent of line management.

Where an organisation decides to set up in-house trauma support teams, then sufficient time must be allocated to their training, and allowance made for 'time-off' from day-to-day activities to receive regular supervision and ongoing training.

Where the organisation is resourcing external professional support, then the nominated providers should be familiar with the company's activities, in advance, and have sufficient professional capacity and expertise to deal with most eventualities.

In some instances, repercussions of an incident may be felt for many months or even years, and any support programme will need to take this into account, in addition to allowing for the differing needs of affected individuals, both in the long and short-term.

Where legal proceedings are instituted in relation to an incident, these may well run for some years, especially after a major disaster.

The role of line management

8.17 In situations where affected employees are not overtly supported by their employer, (including those whose recovery may proceed naturally without any intervention), they may feel angry and disaffected towards the organisation. They may perceive the organisation as being uncaring or even partially responsible for the incident. Such negative feelings could well hinder recovery, affect their work performance and contribute to increased absence from work, as well as adversely affecting other aspects of their lives.

Following an accident, or other event where injury is caused, managers should:

- Make direct contact with the injured or affected employee as soon as is appropriate. This should be done by the immediate line manager, (not the human resources manager), but in an instance where that manager was also involved in the incident, then the next supervisor up the line should take over that role, in respect of those involved.
- Listen, express concern, empathise, acknowledge the individual's feelings and, depending on the circumstances, acknowledge the way in which they handled the situation. This contact is particularly important if the employee is off sick from work.
- Make specific offers of help, eg 'We would like to do the following . . . would that be okay with you or would you rather we left it for later?'
- Keep in touch with the employee and check what form of assistance they need. Some people prefer to be left alone while others like regular contact. In any event it is important to ensure that adequate support is given.
- Be particularly observant in day-to-day interactions with those affected members of staff and take appropriate action in the event of any individual exhibiting obvious symptoms of stress and or trauma.
- 'Return to Work' programmes may need to be offered and encouraged by counselling support.

Business management issues

8.18 Following a major incident an important factor for the organisation will be to contain any damage and disruption to its business activities, whilst ensuring that the working environment is made completely safe and poses no further risk to employees or members of the public. This is particularly relevant in cases of fire, explosion or natural disaster.

The organisation will need to gather detailed evidence from eye witnesses and others, about the incident and to offer support towards all those directly or indirectly involved. This can be a difficult task as it is very easy to be perceived as according first priority to the business, rather than to the employees.

Managers should be aware that a number of different agencies may be involved in any investigation, for example; the Health & Safety Executive ('HSE'), the police, the coroner, as well as the company itself. Individual employees may be interviewed on more than one occasion and it is important not to allow a situation to develop whereby individuals may be made to feel responsible prior to any official report.

Contingency planning

8.19 Bearing in mind that unexpected disasters can happen at any time, organisations need to have a contingency plan in place to enable them to continue to run their business (albeit at a reduced capacity), in the event of a major disaster that may render their premises partially or totally unusable.

Within an organisation, there needs to be a team of people (from different disciplines) who have access to a critical incident management plan that clearly defines the role of each team member and ensures that the following questions can be answered effectively, in order to avoid future problems:

- Do they have a way of accessing immediately all relevant information that is integral to the working of the organisation? Are there arrangements for back-up copies of electronic and other data to be stored externally in an independent location away from the main site of operations?
- Has identification been made of possible alternative premises for emergency use, or alternatively an agency that would be able to find them suitable temporary accommodation?
- How and from where would replacement computer equipment be obtained at short notice and who would be able to re-install essential files onto any new system?
- What contingency arrangements have been made to:
  - (i) publicise temporary telephone numbers and replacement communications systems for customers, suppliers and business associates;
  - (ii) provide practical support for the workforce such as access and travelling to temporary site, replacement of personal belongings, business equipment, tools etc;
  - (iii) provide essential emotional and psychological support for any employee(s) affected by the incident?

The team should review the contingency planning and the control and access of data, at regular intervals and, if necessary, seek expert advice in the field of critical incident management. In specific, high risk areas, such as the City of London, formal training would be appropriate for team leaders who could potentially be caught up in a major incident similar to those that occurred at the Nat West Tower in 1993 or Canary Wharf in 1996.

PTSD diagnostic criteria

8.20 In common with other mental disorders, PTSD is very specifically defined, and a person needs to be found to be exhibiting a number of specific symptoms for a definitive diagnosis of PTSD to be made. According to the 'Diagnostic and statistical manual of mental disorders' (DSM-IV) (1994) (American Psychiatric Association), the diagnostic criteria for PTSD is as follows:

(*a*) The person has been exposed to a traumatic event in which both of the following were present:

(i) The person experienced, witnessed, or was confronted with an event or events that involved actual or threatened death or serious injury, or a threat to the physical integrity of self or others.

(ii) The person's response involved intense fear, helplessness, or horror. (Note: in children, this may be expressed instead by disorganised or agitated behaviour.)

(*b*) The traumatic event is persistently re-experienced in one (or more) of the following ways:

(i) Recurrent and intrusive distressing recollections of the event, including images, thoughts or perceptions. (Note: in young children, repetitive play may occur in which themes or aspects of the trauma are expressed.)

(ii) Recurrent distressing dreams of the event. (Note: in children, there may be frightening dreams without recognisable content.)

(iii) Acting or feeling as if the traumatic event were recurring (which includes a sense of reliving the experience, illusions, hallucinations, and dissociative flashback episodes, including those that occur on awakening or when intoxicated). (Note: in young children, trauma-specific re-enactment may occur.)

(iv) Intense psychological distress at exposure to internal or external cues that symbolise or resemble an aspect of the traumatic event.

(v) Physiological reactivity on exposure to internal or external cues that symbolise or resemble an aspect of the traumatic event.

(*c*) Persistent avoidance of stimuli associated with the trauma and numbing of general responsiveness (not present before the trauma), as indicated by three (or more) of the following:

(i) Efforts to avoid thoughts, feelings, or conversations associated with the trauma.

(ii) Efforts to avoid activities, places, or people that arouse recollections of the trauma.

(iii) Inability to recall an important aspect of the trauma.

(iv) Markedly diminished interest or participation in significant activities.

(v) Feeling of detachment or estrangement from others.

(vi) Restricted range of affections (eg unable to have loving feelings).

(vii) Sense of foreshortened future (eg does not expect to have a career, marriage, children, or a normal life span).

(*d*) Persistent symptoms of increased arousal (not present before the trauma), as indicated by two (or more) of the following:

(i) Difficulty falling or staying asleep.

(ii) Irritability or outbursts of anger.

(iii) Difficulty concentrating.

(iv) Exhibiting hyper-vigilance.

(v) Exaggerated startle response.

(*e*) Duration of the disturbance (symptoms in criteria (b), (c) and (d)) is more than one month.

(*f*) The disturbance causes clinically significant distress or impairment in social, occupational, or other important areas of functioning.

Specify if:

- *Acute* – if duration of symptoms is less than three months.
- *Chronic* – if duration of symptoms is three months or more.
- *With delayed onset* – if onset of symptoms is at least six months after the stressor.

Methodologies in the management of trauma victims

Defusing

8.21 Defusing (a brief informal discussion with an individual or group) should be held within hours of an incident and is most likely to be delivered by a trained manager or supervisor. It will normally last between 30–60 minutes and should be separate from any investigative process. It is often a stand-alone intervention (depending on the nature of the incident) and can also be used for assessment (and sometimes mitigation) of acute symptoms together with acting as a precursor to psychological debriefing to ascertain whether a full debrief is required.

Defusing is a shortened, three-phase version of critical incident stress debriefing (see 8.22 below). It recognises, but does not explore, emotional reactions to the incident (Mitchell and Everly (2001)) and gives people an informal opportunity to say what they want to about the traumatic incident.

Defusing can also be carried out alongside the operational debriefing that addresses practical issues. If so, it should take place in a quiet and comfortable setting, and the participant(s) should be provided with transport home or any other practical support as mentioned above. The environment is similar as required for debriefing and any limits on confidentiality should be enumerated at the start of the session.

Critical incident stress debriefing (CISD)

8.22 A number of debriefing models have been developed since the 1980s, and many individuals have found that talking through traumatic events has been helpful to their recovery (although this has yet to be substantiated by systematic research). The models include those developed by Mitchell (1983), who terms the intervention 'critical incident stress debriefing' ('CISD'); Dyregrov (1989) and Raphael (1986), who both use the term 'psychological debriefing' (PD); and Armstrong, O'Callahan and Marmar (1991), who use the term 'multiple stressor debriefing model'.

The aims of the CISD model are to:

> 'Mitigate the harmful effects of traumatic stress and accelerate recovery processes in individuals who are experiencing normal reactions to abnormal events.'

Formal CISD is recommended to be implemented between 24 and 72 hours subsequent to the incident and be undertaken by trained professionals. This time delay was considered necessary as Mitchell (1983) proposed that emergency workers could suppress psychological reactions for a brief period after an incident as a result of 'training' and would otherwise be too aroused to deal with an in-depth discussion of events.

CISD is run in stages and the debriefer should keep the session open-ended as the different stages can vary in length but an average time would be in the region of 3 hours (including breaks as required). The process enables victims to talk about their experience, normalise their reactions to an event, and receive support and information to reinforce this.

Generally, psychological debriefing has been viewed positively by its participants and this had contributed to anecdotal reports of its effectiveness. However, there has been little empirical evidence to demonstrate the effectiveness of psychological debriefing in accelerating normal recovery processes following trauma (Rose (2000); Kenardy (2000); Orner et al (1999)).

However, CISD has been and is currently used by the emergency services, transport industry, banks, building societies, retail trade, and other organisations, as a standard procedure where people have been exposed to traumatic incident, violent crime or natural disaster.

A debriefing group will usually comprise up to about ten individuals, but where there are greater numbers involved, then the services of a co-debriefer, (whose role should be explained at the beginning of the session), may be required.

The venue should be on neutral ground, eg an office or conference room in the workplace and offer privacy and quiet in order to facilitate open discussion by the participants. Limits on confidentiality should be discussed at the start of the session, and participants in the group should be chosen with care in order to preclude anyone being reluctant to speak openly for fear of recrimination or blame.

Following the debriefing, people whose symptoms are particularly severe may warrant the provision of immediate post-trauma counselling or psychiatric help (see 8.27 below).

A trauma support leaflet setting out the likely consequences of trauma and basic methods of dealing with them, should be distributed, as appropriate, to those involved together with their families. (See APPENDIX A at end of the chapter.)

Trauma support model

8.23 This three-stage model which has been adapted in unpublished research by Suzanna Rose and Gerry Jackson in 2001, as a development of the psychological first-aid model (C Freeman et al (2001)), covers a series of interventions that may commence within the first few hours subsequent to a traumatic event and might well continue over several weeks. Some organisations are using this model as an alternative to CISD. It is less structured, more led by the participants, and is less intense.

The trauma support model has been developed by taking into account available research and the experience of professionals in the field. It is intended to be a programme of support that seeks to look after people in the short, medium and longer term; help them to be heard and feel cared for and ensure that those who are most vulnerable to developing psychiatric illness are provided with appropriate therapeutic help.

Individuals practising the trauma support model must be appropriately trained. However, the training need not be at the same level as that required for CISD – where practitioners need to be able to manage and contain intense emotional reactions, either in individuals or in groups, and conform to a prescriptive, structural procedure.

It should also be noted that those who undertake debriefing services should themselves be afforded appropriate support as it can be distressing to listen to the traumas of others. This may entail discussion with colleagues subsequent to each session, but supervision should also be available from a mental health professional such as a counsellor or psychologist. Their experience can be used to answer questions, develop the experience and skills of persons carrying out the debriefing to ensure that the work is not overwhelming.

Peer support groups

8.24 Peer support is used by some organisations as a means of utilising work-team members rather than staff having a specialist support function.

Supervision is required for any peer support team, to ensure that they can themselves cope with the emotional demands placed upon them.

Points to note:

- It is essential that peer support team members acknowledge their own boundaries and limitations and know when and where to refer on, when necessary.
- They should always appreciate that although they are fulfilling a professional role, they are, nevertheless, not formally trained, trauma professionals.

A 'buddy system' is another model of peer support that is increasingly being used. This would usually comprise individuals who have experienced a particular trauma in the past and are called upon to give support to others. With the empathy they have gained from their own experience, they often have skills and expertise to support others. However, it is essential that they have worked satisfactorily through their own responses and reactions and are not jeopardising their own mental health by helping others. As with any peer support option, supervision, training and support are vital components to ensure that this is a viable alternative,

Post-trauma counselling

8.25 There is a need for organisations to be aware of the type of post-trauma reactions that employees may be experiencing and the symptoms that may be exhibited. It is important to be able to identify changes in behaviour, for example: avoidance or being over-anxious about returning to the place where the incident happened; being unable to talk about the incident or becoming emotionally very detached are often warning signs that after-effects of trauma are being experienced and the individual(s) may need post-trauma counselling support.

It may be that the employee has contacted the personnel or occupational health departments to complain of nightmares or flashbacks and that these symptoms are adversely affecting their relationships.

An effective model when working with traumatised individuals is cognitive behavioural therapy (CBT) (S D Solomon, E T Gerrity and A M Muff (1992)). (See also 1.15.) This is a structured form of therapy that encourages clients to recognise that their thinking processes may be irrational and may be contributing to negative emotional reactions, such as extreme anxiety, that are effecting their lives.

Counsellors or clinical psychologists who are experienced in working with traumatised clients, should always be used wherever possible. Details of specialist consultants may be obtained from the British Association for Counselling and Psychotherapy, the UK Trauma Group or other organisations who specialise in the treatment of trauma (see BIBLIOGRAPHY).

Case study

8.26

The aftermath of a traumatic incident

John and Marcus were part of a 20-strong team of engineering fitters who were welding tank plates on a reserve-feed cooling water tank at a nuclear power station in the north of England. It was late in the afternoon and both men were working from a scaffolding platform about half way up the tank, vertically welding the joints of the steel plates to ensure water-tightness.

Suddenly, a scaffold pipe worked loose from its fixing and the entire scaffold on which the men were working collapsed. John fell over 15 metres and was killed instantly as his head struck an angle iron support at the base of the tank. Marcus fell on top of him, sustaining a fractured pelvis.

The foreman hit the alarm siren, which automatically closed down the electricity supply, and other workmen rushed across the site. Someone had already called an ambulance and Marcus, who was unconscious, was left untouched to await the emergency crew. In the minutes that followed, the entire gang of fitters

(cont'd)

watched as the paramedics arrived, injected Marcus with morphine, and put him gently onto a stretcher then into the ambulance. With blue lights flashing and siren blaring, they sped off to the local hospital, whilst a second ambulance crew gently wrapped up and took away John's body. Official photographs of the accident scene were taken, for record.

Apart from John and Marcus, there were 18 other welders on site that day, in addition to a gang of painters who were working on the adjoining tank. One or two of the men were physically sick, before they and the others gathered up their tools and overalls then slowly made their way off the site, too shocked to talk. The administrative staff's offices also overlooked the scene, and they too saw the incident.

Billy, who had completed the Red Cross training, was the appointed first aider for the company. He had put his certificate up on the wall, but never expected to have to use his training. It was considered quite lightly by him at the time, eg 'I will be here to look after you all, don't worry', he would say. As soon as this terrible accident happened, suddenly people were looking to Billy to give support and advice.

The following day, three of the men who at worked at the site, failed to report for work. A management enquiry was set up to find the cause for the disaster and to allocate responsibility. The scaffolding itself had been supplied and erected by a subcontractor, and the health and safety inspector was making his own official enquiry and report.

In all, over 25 men had either witnessed the accident directly or had arrived in the first few minutes immediately afterwards in response to the alarm. Most of them had seen John's body. However, nobody was prepared for what they had seen, and virtually all were affected to some degree.

Their families were also indirectly affected, as their men-folk tried to come to terms with the image that was now embedded in their minds. In the first few days after the harrowing event, many had difficulty in sleeping and some went to their GP for a sedative. Unfortunately, for many the image was there, immediately upon waking.

The engineering employer was a large company, and fortunately had an occupational health department staffed by three qualified nurses, all of whom had received defusing training. In the days following the accident, they ensured that they spoke to all those employees who had been on the site that day.

A team of debriefers was called in to give a group debriefing to those who wished to receive it. Following this, one-to-one post-trauma counselling was also given.

In the event, two of the men suffered more serious effects and were off work for two months. They have now returned. Marcus recovered quickly and was probably fortunate that he was spared the sight of what had happened to John, as he had been knocked unconscious by the fall.

The scaffolding firm was prosecuted by the HSE for breach of safety regulations, and subsequently fined £25,000 for negligence. John's widow is now taking proceedings to sue his employer in the civil courts for compensation. As John was only 27 years old at the time of the accident, the claim is likely to run well into six figures.

Postscript: It can take a long time to re-establish that feeling of being part of the company again. Mary was on holiday when the above incident occurred. On returning to work, following the incident, she felt a sense of guilt for not having been there, and of isolation because all she could do was 'listen to what people are saying', with the result that she no longer felt included as a member of the group.

Practical critical Incident management – the role of external agencies

8.27 Police in the UK have an overall responsibility for coordinating the work of the other emergency services (but not to control or direct), whilst ensuring that the work is facilitated without impediment or undue restriction.

The various facilities mentioned below may or may not be required in the context of major disasters. (This will depend on the particular circumstances of each disaster.)

Cordons

8.28 Standard police procedure is to set up two cordons: an inner cordon to provide security at the scene itself, and prevent unauthorised access; and an outer cordon to provide a controlled area external to the inner cordon, access for emergency services and a secure area for them to position mobile control vehicles. Depending on the circumstances, a representative from the organisation that is involved in the incident may be needed at the police mobile control centre to give essential information about the company, buildings and/ or staff.

It is usual that between the inner and outer cordons, the police will set up a rendezvous point (RVP) to control access to the scene. Incoming personnel and vehicles will be kept at the RVP until they are required at the scene and then escorted forward.

Survivor reception centre (SRC)

8.29 A survivor reception centre ('SRC') should be set up where there are a number of uninjured (or walking wounded) persons involved. For instance, at the Ladbroke Grove train crash disaster in 1999, the nearby Sainsbury's store was used as the SRC. Typically, this will facilitate care for people, provide hot drinks and/ or food, ensure security and safety from further harm, to have personal details taken down for the record, to ensure nobody has sustained any serious injury to themselves, to establish contact with families, and finally to arrange transport home or to another destination.

These facilities will usually be organised by the police and assisted by the local authority. There should also be a presence of employees of the organisation who were not involved in the incident but who can provide support and assistance with the services mentioned above. These individuals should be identifiable by an identification tag showing their name and position.

Friends and relatives reception centre (FRRC)

8.30 This facility would normally be set up where there is an accident involving passengers travelling by road, rail, airline or ferry – in order to accommodate and answer questions from anxious relatives or friends who are waiting for incoming passengers, or where numbers of friends and relatives are likely to arrive at the scene, looking for those involved.

The rationale is to control the influx of unauthorised persons and to prevent interference or hindrance with the work of the emergency services. Details of people arriving at the centre will be taken and attempts made to reunite them with their relatives or friends. As before, this would normally be a combined effort by the police and the local authority, and a presence by the organisation involved is desirable to assist with the work. The physical act of reuniting people should be carried out somewhere other than at the SRC or FRRC.

Relative liaison officers (RLOs)

8.31 In cases where there are a number of fatalities, the police will appoint relative liaison officers to look after the needs of each bereaved family and assist in the identification processes. In some circumstances the organisation may wish to appoint a 'buddy' (or preferably two) in order to provide care and support and to act as a single link with the employing company.

Casualty bureau

8.32 A casualty bureau is a communications centre set up by the police. (Similar centres are also used by airlines.) Its brief is to collect and collate information from all points and from the scene itself to ensure that those involved receive all the information they require regarding the condition and whereabouts of their colleagues, friends or relatives. Such centres are also responsible for collating information about the identification of any deceased victims and the informing of next of kin.

A liaison representative from the organisation will be required at this centre, to assist with whatever information they have from their internal records regarding the injured and, possibly, also details of close relatives.

Mortuary

8.33 This is the location where the technicalities of the process of identification of the deceased will be carried out. In some circumstances, an existing local mortuary will be used, though often, if there are large numbers of deaths, a temporary mortuary facility may be established near to the point of the disaster. Normally no one from the organisation will be required to work in this facility, though mortuary workers will often require information, most usually from the casualty bureau liaison representative.

Evacuation

8.34 Where it becomes necessary to evacuate buildings, the instructions of the police and/ or the emergency service must be complied with immediately and without query. Modern office buildings often have a small footprint and a high-rise profile, and it is important that an emergency evacuation plan should be developed and regularly rehearsed to ensure that it is practical and effective, prior to it ever being needed. A disorganised evacuation will cause panic, wasted time and further endanger life.

Crowd management

8.35 Where large numbers of people are involved, crowd safety; planning; assessing risks; putting precautions in place; emergency planning and procedures; communication; monitoring crowds and review needs to be taken into consideration (HSE (2000)).

Media

8.36 In any major incident, the media will usually be present in large numbers and will seek to interview anyone who apparently has any knowledge that may be of interest. It is therefore useful to appoint a media liaison person who should be a senior manager, with some appropriate training. Any other person approached by the media for comment about issues relating to either the company or the disaster, should always refer such enquiry to the designated media liaison official.

Key learning points

8.37 The treatment and management of trauma is a difficult and complex subject, and more random controlled studies are needed to establish the outcomes that are being produced by current interventions and which components (of particular interventions) are of most value.

Until further research and substantiated reports are available, professionals involved in post-trauma support will tend to use existing methods. Notwithstanding the fact that we are moving towards more evidence-based practice, it is important that whatever methodology is used, it must identify those who are particularly at risk, so that psychotherapeutic resources can be concentrated on those particular individuals or groups, with the maximum beneficial effect and without undue delay.

CISD is used in many organisations and appears to be beneficial in many cases of trauma support. Nevertheless, many practitioners are reconsidering the efficacy of their approach, and may well adopt something similar to the trauma support system described at 8.23 above.

The most important objective is to provide an ongoing system of support that combines pre-emptive action management and essential reactive treatment, as necessary. The system should include risk assessment, suitable support training prior to any potentially damaging incident, provision of effective management during the event and immediate support after the event

In light of these considerations, it is also important to bear in mind the following.

- The differing reactions of individuals to traumatic events will vary in both severity and type.
- PTSD is the term commonly used to identify the reactions that some people experience in the aftermath of an extreme incident.
- A diagnosis of PTSD is often difficult – only a small proportion of people will be positively diagnosed. However, many more are likely to experience some traumatic symptoms after the event.
- The provisions of the *Health and Safety at Work etc Act 1974* covers the statutory duty of care that applies to the health and safety of all employees and others within the workplace.
- Exposure to traumatic events will vary and employees may suffer from post-trauma symptoms without experiencing PTSD.
- A lack of support and care by the organisation following an incident may cause feelings of anger or frustration in the workforce.
- Managers should be particularly observant of those who have experienced trauma so that any symptoms and reactions are noticed and treated without delay.
- Stress can follow anything from a minor incident to a major disaster and can affect any of those directly or indirectly involved.
- An organisation's crisis management procedure(s) should include all the necessary assistance and support in dealing with an unexpected traumatic incident, including training, identifying those involved, providing all required support and, where applicable, access to professional counselling.
- Where any member of staff is on sick leave absence subsequent to an accident or other distressing event, then managers should keep in regular contact until there has been a return to work.
- Specialist counselling should be provided for those identified as experiencing post-trauma symptoms and they should be given time off work to attend appointments.

APPENDIX A
Example of the Content for a Leaflet for Use with Clients and their Families

8.38

The reactions and effects of involvement in a traumatic incident

You have been involved in a very traumatic incident, and you are likely to have some form of reaction to it. These reactions may happen immediately, or may not occur for weeks, months or occasionally even years after the incident. Not everyone suffers reactions but the majority of us do. These reactions are likely to be worse if:

- Many people died or were injured during the incident, or death or injury was sudden, violent or happened in horrifying circumstances.
- You have a feeling of helplessness or wanting to have done more.
- You do not have good support from family, friends or colleagues.
- The stress resulting from the incident comes on top of other stresses in your life.

What follows has been compiled from the experiences of others who have been involved in similar incidents.

Emotional reactions

Your emotions or feelings are likely to be in chaos after the event, or alternatively you may feel nothing. Some of the more common emotional reactions are listed below.

Guilt:

- For not having done more.
- For having survived when others did not.

Anger at:

- What has happened.
- Whoever caused it or allowed it to happen.
- The injustice or senselessness of it.
- Not being understood by others.
- Those in charge.

Fear:

- Of breaking down or losing control.
- Of a similar event happening again and not being able to cope.

Shame:

- For not having reacted as you might have wanted to.
- For feeling helpless, emotional and wanting others to be with you.

Sadness:

- About the deaths, injuries and the whole circumstances of the incident. You may feel depressed without knowing why.

Mental reactions

You may very likely find that you cannot stop thinking about the incident, dream about it or suffer loss of memory, concentration or motivation. You may experience flashbacks (feeling that part of it is happening again). You may hate to be reminded of it. You could feel always on your guard or easily startled.

Physical reactions

People often experience tiredness, sleeplessness, nightmares, dizziness, palpitations, shakes, difficulty in breathing, tightness in the throat and chest, sickness, diarrhoea, menstrual problems, changes in sexual interest or eating habits, and many other symptoms – frequently without making a connection with the incident.

Other difficulties

You may feel hurt, and your relationships with others, particularly your partner, may feel under additional strain. You may find yourself taking your anger out on your partner or family. You may not be aware that you are doing this, and your partner will probably not understand that it is part of your reaction to the incident. You may find yourself emotionally withdrawing from your close relationships. Just when you need it the most, you may reject the support of those closest to you. Try not to do this.

You may find that the incident has reminded you of some past trauma, at work or in your personal life, and the feelings about that could come back with all their original force. That may also need to be dealt with.

(cont'd)

What can be done to help

Nature often heals by allowing feelings to come out and by making you want to talk about these. Talking about the incident and your feelings about it with your partner, others who were involved, or any sympathetic listener, is very helpful. Take the opportunity if it arises. It will probably help those closest to you to understand and support you more effectively if you show them this leaflet.

Talking to a trained counsellor is often a great relief and can reduce much of the tension and anxiety. Trying to avoid your feelings, or trying to avoid thinking or talking about the incident in the belief that you can cope may be unhelpful and possibly harmful in the long term. This can lead to storing up problems that will come out sooner or later, and possibly in the form of worse physical or nervous difficulties, sometimes a long time afterwards.

When to look for professional help

- If you feel that you cannot handle intense feelings or body sensations, or if you feel that your emotions are not falling into place over a period of time, and you feel chronic tension, emptiness or exhaustion.
- If, after a short time, you continue to feel numb and empty and do not have any feelings.
- If you have to keep active in an attempt to suppress your feelings.
- If you continue to have nightmares or are sleeping badly.
- If you have no one to share your emotions with and you feel the need to do so.
- If your relationships seem to be suffering, or sexual problems develop.
- If you start to have accidents or your work performance suffers.
- If you start to smoke, drink or take drugs to excess.
- If you are suffering from exhaustion or depression.
- If you cannot control your memories of the experience and they are affecting your personal well-being.

DO REMEMBER:

- THAT YOU ARE BASICALLY THE SAME PERSON YOU WERE BEFORE THE INCIDENT.
- THAT TALKING ABOUT YOUR EXPERIENCE AND THE FEELINGS CAN HELP.
- THAT SUPPRESSING YOUR FEELINGS CAN LEAD TO FURTHER PROBLEMS.

APPENDIX B
Brief Screening Questionnaire for Post-Traumatic Stress Disorder

8.39

Your own reactions now to the traumatic event

Please consider the following reactions which sometimes occur after a traumatic event. This questionnaire is concerned with your personal reactions to the traumatic event which happened a few weeks ago. Please indicate whether or not you have experienced any of the following *at least twice in the same week*:

		Yes, at least twice in the past week	*No*
1.	Upsetting thoughts or memories about the event that have come into your mind against your will		
2.	Upsetting dreams about the event		
3.	Acting or feeling as though the event were happening again		
4.	Feeling upset by reminders of the event		
5.	Bodily reactions (such as fast heartbeat, stomach churning, sweatiness, sweating, dizziness or feeling dizzy) when reminded of the event		
6.	Difficulty falling or staying asleep		
7.	Irritability or outbursts of anger		
8.	Difficulty concentrating		
9.	Heightened awareness of potential dangers to yourself and others		
10.	Being jumpy or being startled at something unexpected		

Reproduced with the kind permission of The Royal College of Psychiatrists (2002))

9 Stress and Health

Introduction

9.1

> 'The unfortunate thing about this world is that good habits are much easier to give up than bad ones'. *W Somerset Maughan*

In the World Health Organisation publication 'Classification of mental and behavioural disorders: Clinical descriptions' (ICD-10) (1992), there is no category listed under 'stress' or 'stress-related illness'. There is, however, a section entitled 'Reaction to severe stress and adjustment disorders', indicating that stress is a separate aetiological factor (ie a contributor to, or cause of, other diseases).

Reactions to severe stress are many and varied, but there are seven main areas of disease and health-related issues that are generally acknowledged to have either a work-related stress component, or to be influenced by stress:

- (i) Cardiovascular disease including coronary heart disease (CHD), atherosclerosis (a thickening of the inside walls of the coronary arteries), coronary thrombosis and stroke.
- (ii) Certain cancers, the primary or contributory factors being those related to smoking, diet and lifestyle.
- (iii) Gastro-intestinal problems including irritable bowel syndrome (IBS), colitis and other disorders of the stomach, bowel and colon.
- (iv) Mental illnesses including acute or chronic depression, anxiety, and phobias.
- (v) Musculoskeletal problems including back pain and dysfunction of the neck and upper extremities.
- (vi) Accidents – both those that are work-related and those external to the workplace (such as road traffic accidents).
- (vii) Suicide – which results in approximately 50,000 deaths in the EU each year.

Acute and chronic stress

9.2 There are two stress reactions to excessive pressure – 'acute' (short term) and 'chronic' (long term).

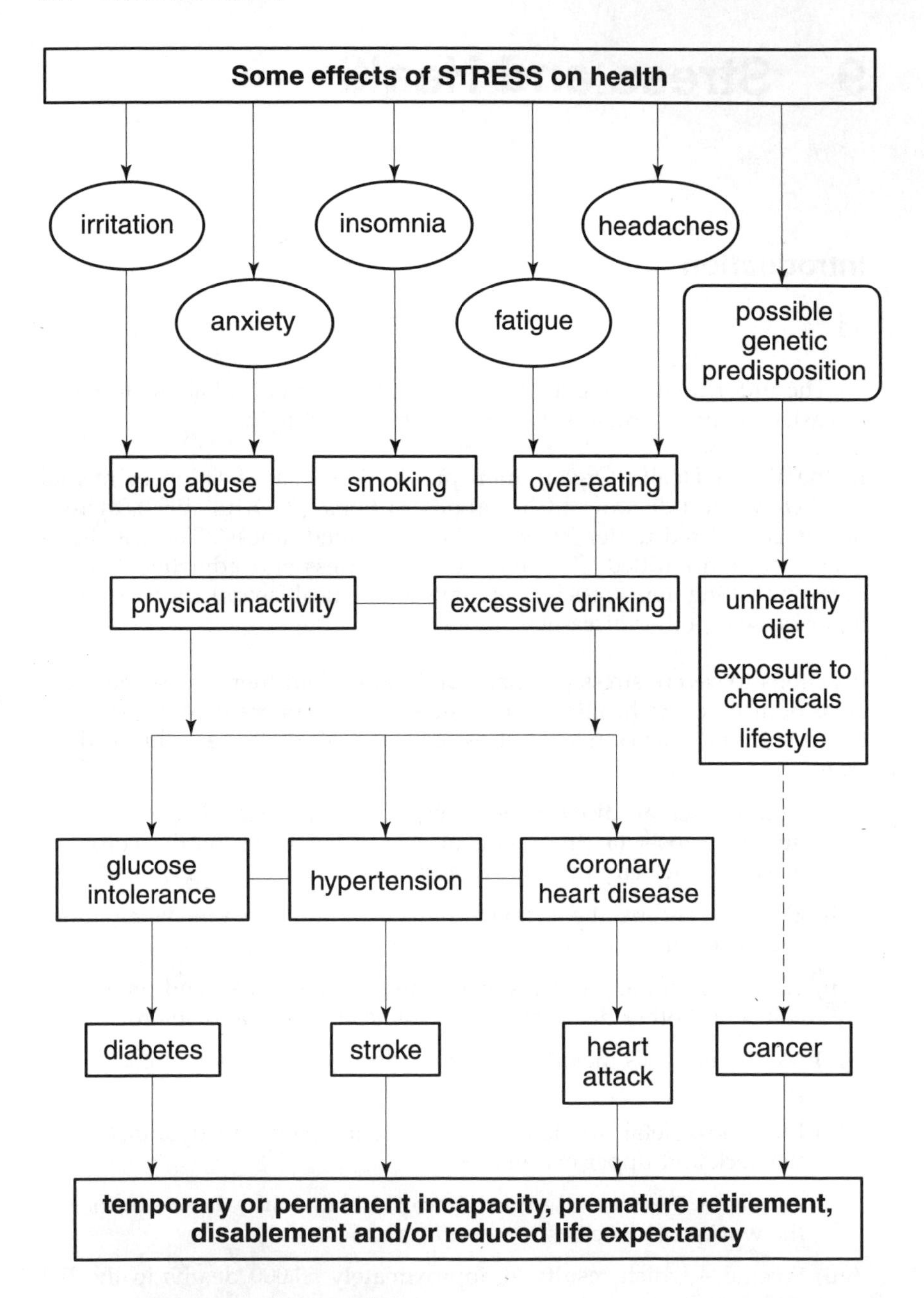

Figure 1: Reactions to stress

Acute stress is an immediate reaction to a threat of danger, either real or perceived. It occurs, for example, when a car driver swerves violently in an urgent attempt to avoid running over a cat or dog, on the road.

Stressors in everyday life that can result in acute pressure being felt by the individual include excessive noise, extremes of temperature, crowds, traffic accidents or other emergencies, fire, violence, isolation or imprisonment, disease and pain.

In medical terms, the physiological effects of acute stress include the following instantaneous responses (see CHAPTER 1, FIGURE 2 at 1.2):

- The endocrine nervous system of the body (the hypothalamic-pituitary-adrenal system ('HPA')) is immediately activated, and releases cortisol (the so-called 'stress hormone') and other corticosteroid hormones.
- The HPA also releases the neurotransmitters adrenaline, noradrenaline and dopamine.
- As a result, blood pressure and heart rate increase immediately, breathing becomes more rapid, and blood circulation increases.
- The immune system automatically changes the body's priorities to meet the sites of anticipated injury.
- The metabolic system slows or shuts down whilst the threat of danger is present.

When the threat (or perceived threat) of danger has passed, the biochemical activity should return to normal over a short period of time, and there will not usually be any long-term effect, other than the stored memory of the event that is filed for future recall.

Chronic stress, on the other hand, is prolonged, with no perceived end in sight. It forces us to repress our in-built survival mechanism of 'fight or flight' – that tries either to remedy the situation, or to escape from it. This can bring about adverse changes in the body's natural chemistry that may lead to physiological dysfunction.

Chronic stressors include continuous work pressures, ongoing financial worries, unhappy personal relationship(s), illness, long-term commitment in caring for infirm dependents, isolation and loneliness.

As a result of chronic stress, high levels of circulating noradrenaline cause the pathways of the vascular system to constrict. The consequent increase in blood pressure entails the heart working harder to force blood through the narrowed arteries. During prolonged periods of stress, there is an increase in blood cholesterol and platelet levels and the excess cholesterol increases the likelihood that the arterial pathways will become partially blocked by atheroma deposits of fatty tissue on the inner walls. In effect, the cross-sectional area of the arteries becomes narrower, offers greater resistance to blood flow, and therefore increases the workload on the heart.

Another effect of prolonged stress is that blood sugar levels increase, making the individual more susceptible to glucose intolerance and diabetes. Chronic stress may also lead to lower than normal levels of serotonin – a neurotransmitter now acknowledged to be implicated in clinical depression. In addition, stress can lead to chronic anxiety, nervous breakdown or burnout, as well as other mental dysfunction.

In the case of individuals already suffering from an illness or disease, stress can contribute to an increased awareness of their symptoms, thereby exacerbating them. This can lead to a tendency to interpret symptoms as more distressing or even life threatening, than they actually are and the stress-induced, psycho-physiological effects may, over time, actually reinforce the existing symptoms.

There is increasing evidence that confirms the long-held suspicion that stress and hypertension (high blood pressure) are associated and that prolonged stress may increase the incidence of coronary heart disease. There is also the contention that chronic stress, in females, may cause depleted levels of the hormone oestrogen, while in highly stressed males, the elevated blood pressure often linked to stress can lead to strokes.

When evaluating the possible damage to the cardiovascular system, it should be borne in mind that people under stress frequently resort to unhealthy lifestyles in an effort to mitigate or ameliorate the problem. Alcohol and drug abuse (including smoking) are common within this category, as is the consumption of processed food (leading to a diet high in fat and sodium but lacking in fresh fruit and vegetables). All too often, this diet will be combined with insufficient exercise.

Chronic stress also appears to have an adverse effect on the immune system and may well increase an individual's susceptibility to both viral and bacterial infection.

A compromised immune system can obviously lead to longer recovery times from any infection or disease. Unfortunately, work-related stress may also induce workers to suppress or deny symptoms of sickness or disease, thereby delaying necessary medical treatment. Admitting to being unwell is frequently avoided by employees because of concerns they may have regarding the possible repercussions.

Finally, whilst there is currently no concrete evidence that prolonged stress either causes or predisposes to cancer, (the primary causative factors being smoking, diet and lifestyle, plus a genetic influence) – studies do suggest that an impaired immune system can be a contributory factor to tumour establishment and growth.

Systems of the body that can be affected by stress

9.3

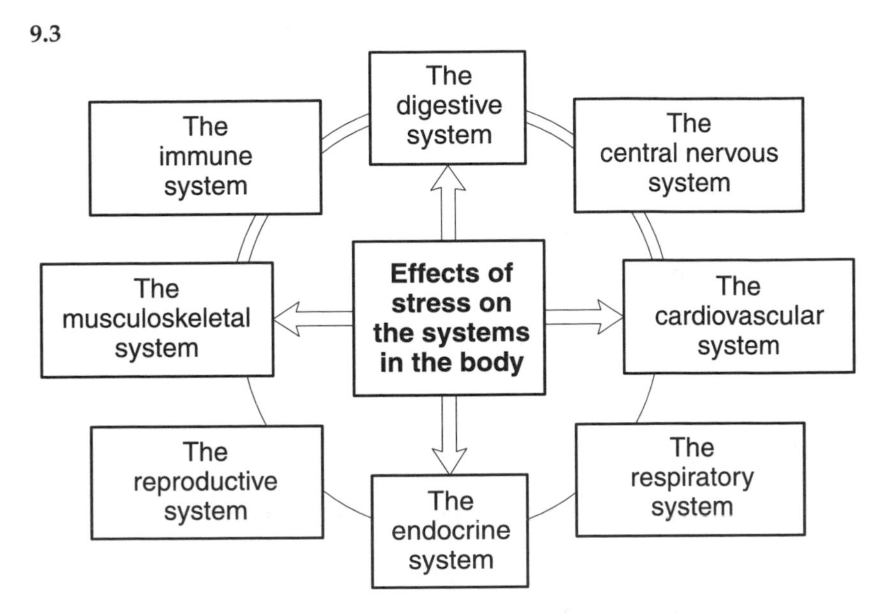

Figure 2: Effects of stress on the systems in the body

Prolonged and continuous stress can be debilitating, both physically and mentally and can be an influencing factor in the cause or course of various dysfunctions. Some of these are listed alphabetically, as follows:

- Angina pectoris.
- Asthma.
- Anxiety.
- Back, neck and shoulder tension.
- Bulimia.
- Coronary heart disease.
- Depression.
- Dermatitis.
- Digestive disorders.
- Eating disorders.
- Eczema.
- Fatigue.
- Headaches and migraine.

- Hypertension.
- Impotence.
- Insomnia.
- Irritable bladder.
- Irritable bowel syndrome.
- Menstrual problems.
- Mouth and peptic ulcers.
- Muscular pain.
- Palpitations.
- Panic attacks.
- Psoriasis.
- Reduced libido.

These ailments can be allocated to specific body systems:

The central nervous system

- Anxiety, depression and fatigue.

The cardiovascular system

- Impaired heart function can cause angina in those with underlying disease.
- Constriction of the peripheral blood vessels, thereby raising blood pressure.

The digestive system

- Stomach upsets.
- Diarrhoea.
- Gastritis.
- Ulcers.
- Irritable bowel syndrome.
- Colitis.

The respiratory system

- Asthma.

The musculoskeletal system

- Tension in skeletal muscles and joints, leading to backache and muscular pain.

The immune system

- Weakened body defences, with lowered resistance to infections.
- Viral illnesses (often due to a depleted immune defence system).
- Allergies.
- Malignant cell changes, and possibly cancer.

The endocrine system

- Menstrual disorders.
- Thyroid disorders (under-active, overactive, and thyroiditis).

The reproductive system

- Infertility.
- Premature ejaculation.
- Impotence.

Skin

- Eczema.
- Psoriasis.

General

- Tissue degeneration.
- Acceleration of the ageing process.

Some of the possible physical conditions that may be influenced by stress

9.4 It is clear that there is virtually no aspect of the human condition that is not, in some way, affected by prolonged stress. In this section (and the two that follow at 9.26 and 9.33), we will look in more detail at the physical, emotional and behavioural disorders that can influenced by stress.

Physical disorders that come into this category include:

- Susceptibility to infection.
- Immune disorders.
- Pain.
- Gastrointestinal disorders.
- Diabetes mellitus.

- Cardiovascular disease.
- Stroke.
- Cancer.
- Sexual and reproductive dysfunction.
- Other disorders.

Susceptibility to infections

9.5 Chronic stress appears to blunt the immune response and increase the risk of infections, and may even impair an individual's response to immunisation.

Individuals suffering chronic stress may have low white blood cell counts, making them possibly vulnerable to colds and other infections. Stress may also exacerbate illnesses and prolong recovery time.

People who harbour the herpes or HIV viruses may also be more susceptible to viral activation following exposure to stress. A serious factor, highlighted by research, shows that HIV-infected men with high stress levels, progress more rapidly to AIDS when compared to those with lower stress levels (*American Journal of Psychiatry* (August 2000)).

Immune disorders

9.6 According to Reuters Health Information Inc (www.reutershealth-.com), the effects of stress on the immune system can also have mixed effects on auto-immune diseases (those that are caused by inflammation and damage from immune attacks on the our own body).

Pain

Back pain

9.7 In 1998, 40% of adults (or more than 16 million people), had back pain lasting more than one day. For more than half of these people, the pain lasted more than four weeks, while 2.5 million people suffered back pain every day of the year. Back problems affect all kinds of people, men as much as women, and young as well as old. Among 16–24 year olds, one in three had back pain in the past year (Department of Health (1998)).

The incidence of back pain is spread fairly evenly across the community in terms of age, sex and geography, but occupation makes a difference, with over one million people having work-related back pain or upper limb disorders. Manual workers are, not surprisingly, more likely to have back problems than white-collar workers, since their work is more likely to involve heavy lifting. Around 10% of all major employee injuries take place while handling, lifting or carrying. (HSE (1999)).

Employees are more likely to suffer a serious back injury at work (involving a fracture or hospital admission for more than 24 hours), if they work in:

- agriculture;
- construction;
- mining;
- transport and communications.

[HSE (1997)]

Some occupations can adversely affect the body and musculoskeletal problems may develop without necessarily involving specific injury.

Among these are:

- Driving a motor vehicle

  A study found that people driving over 25,000 miles a year average have just over 22 days off work with a bad back, compared with just over 3 days for low mileage drivers. ('Driving and Musculoskeletal Health' (1999)). Train drivers are twice as likely as HGV drivers to report back pain (P S Helliwell and J R Smeathers (1998)).

- Work involving intensive use of the telephone, without a headset

  A survey of London office workers showed that half of office workers who use a telephone for at least two hours a day and who also use a computer, report neck pain, and 31% suffered from lower back pain (University of Surrey, quoted in the *Safety and Health Practitioner* (May 1999)).

- Being a supermarket cashier

  A study found that 57% of supermarket cashiers experience lower back pain, in a year ('Musculoskeletal disorders in supermarket cashiers', HSE (1998)).

Research by Manchester University (reported by www.holistic-online.com) has found that when an individual feels underpaid and dissatisfied, they become a prime candidate for backache. The study, based on interviews with 1,600 people, half of whom were employed, found that although there was little difference in the risk of back ache between those in work and those who were not; those who were unhappy about their job status (whether employed or unemployed) were twice as likely to suffer from backache.

Similarly, the study found that workers unhappy about their salaries were three times more likely to go for medical help, or simply take time off work, than those satisfied with the money they were earning. The researchers concluded that the study refutes the myth that those who suffer from pain in the lower back are simply work-shy.

Many other studies from around the world have also linked back pain to stress (see BIBLIOGRAPHY).

Muscular and joint pain

9.8 Chronic pain caused by arthritis and other conditions may be exacerbated by stress. Psychological distress also plays a significant role in the severity of back pain. Some studies (such as the Manchester University study mentioned at 9.7 above) have clearly associated job dissatisfaction and depression to back problems, although it is still unclear if stress is a direct cause of the condition.

Headaches

9.9 Tension-type headaches are frequently associated with stress and stressful events. However, individuals often experience headaches or migraine attacks at the weekend or on holiday, ie when the particular pressure has been lifted and the body then exhibits a (delayed) reaction. Some research suggests that sufferers from tension-type headaches may actually have some biological predisposition for translating stress into muscle contraction. Emotional strain is among the wide range of possible migraine triggers (together with dietary factors), although the attack will classically commence after the stress has eased.

Gastrointestinal problems

9.10 The interactions of the brain and the intestine are strongly related, being linked by hormones, neurotransmitters and other biochemical components of the nervous system. Prolonged stress can disrupt the natural operation of the digestive system, irritating the colon and causing diarrhoea, constipation, cramps and bloating. Excessive production of gastric acid, in the stomach, can lead to ulceration and potentially serious illness, if not swiftly treated.

Peptic ulcers

9.11 It is now well established that most peptic ulcers are either caused by the *H pylori* bacteria or by the use of non-steroidal anti-inflammatory drugs ('NSAIDs'). Nevertheless, studies still suggest that stress may either predispose an individual to ulcers, or sustain existing ones. Some experts estimate that social and psychological factors play some contributing role in 30–60% of peptic ulcer cases, whether they are caused by *H pylori* or NSAIDs. Experts also believe that the anecdotal relationship between stress and gastric ulcers is so strong that attention to psychological factors is still warranted.

Irritable bowel syndrome (IBS)

9.12 Irritable bowel syndrome (or spastic colon) is strongly related to stress. With this condition, the large intestine becomes irritated, and its muscular contractions become spastic rather than smooth. The abdomen becomes bloated, and the person experiences cramping with alternating periods of constipation and diarrhoea. Sleep disturbance can further exacerbate the condition.

Inflammatory bowel disease

9.13 Although stress is not a cause of inflammatory bowel disease (Crohn's disease or ulcerative colitis), there are reports of an association between stress and symptom flare-ups. One study, for example, found that while short-term (past month) stress did not significantly exacerbate ulcerative colitis symptoms, long-term perceived stress tripled the rate of flare-ups compared to individuals who did not report feelings of stress (*American Journal of Gastroenterology* (May 2000)).

Diabetes mellitus

9.14 Chronic stress has been associated with the development of insulin resistance, a condition in which the body is unable to use insulin effectively to regulate glucose (blood sugar). Insulin resistance is a primary factor in diabetes, and stress can also exacerbate existing diabetes by impairing the patient's ability to manage the disease effectively.

Cardiovascular disease

9.15 According to the British Heart Foundation (2002), heart and circulatory disease kills more people in the UK, than any other disease. In total it accounts for over 235,000 deaths per year (or around 40% of all deaths); and in 2000, for example, 120,000 women died from heart and circulatory disease – 50,000 more than died from cancer.

The main forms of cardiovascular disease are coronary heart disease ('CHD') and stroke. About half of all deaths from cardiovascular disease are from CHD, and about a quarter are from stroke. Cardiovascular disease is also one of the main causes of premature death: A total of 36% of premature deaths in men and 28% of premature deaths in women are from cardiovascular disease.

CHD by itself is the most common cause of death in the UK, accounting for around 125,000 deaths per year (approximately one in four deaths in men and one in six deaths in women). Nearly all deaths from CHD are because of a heart attack. Over 270,000 people in the UK suffer a heart attack each year (around one every two minutes), and in about 30% of heart attacks the patient dies before reaching hospital. In addition, over 1.5 million people have angina, half a million have heart failure, and these numbers are rising.

Regular exercise, good nutrition and giving up smoking are some of the keys to controlling the risk factors for coronary heart disease, in addition to a stress-reduced lifestyle.

Stress and cardiovascular disease

9.16 Researchers have found a link between work-related stress and heart disease. Doctors in Finland have concluded that work stress needs to be tackled as a way of improving employees' health. The stresses and strains of the workplace and an imbalance in the amount of effort workers put in, against their rewards, were both associated with a doubling of the risk of cardiovascular death.

Stress was also a factor in other workplace problems, such as why people ate too much, continued to smoke or did not take enough exercise. The researchers studied 812 healthy men and women who worked in factories in Finland over a period of 25 years. During that period, 73 died from heart disease or stroke. After taking age and sex into account, those who faced high demands or had little control over the way the way worked were 2.2 times more likely to die from cardiovascular disease. Workers in jobs with high demands, low salaries and a lack of social approval were 2.4 times more at risk. After five years, there was an increased risk of having higher blood cholesterol levels, associated with high job strain (*British Medical Journal* (2002)).

The combination of high psychological demand at work and low control over decision-making, can increase the risk of cardiovascular disease. As described above at 9.2, stress constricts the blood vessels, and at the same time raises blood pressure and heart rate, increasing the heart's demand for oxygen. The result is that the heart may be starved of blood – a condition known as ischaemia.

Stress is also a major factor in angina pectoris (severe left side chest and/ or shoulder pain); and incidents of acute stress have been associated with a higher risk of serious cardiac events, such as heart rhythm abnormalities and heart attacks, and even death from such events in people with existing heart disease.

- Sudden stress influences the heart's pumping action, rhythm and rate as well as increasing platelet activity causing arterial constriction and posing a risk of thrombosis, stroke or potentially serious arrhythmias.
- Stress may also signal the body to release fat into the bloodstream, raising blood cholesterol levels (at least temporarily).
- Stressful events may cause individuals who have relatively low levels of the neurotransmitter serotonin (and therefore a higher risk of depression or anger) to produce more immune system proteins (called *cytokines*), which in high amounts can cause inflammation and damage to cells, possibly including heart cells.
- Cardiologists often find it difficult to get patients back to work after diagnosing and treating heart disease. Three factors seem to determine the likelihood of the patient returning to work: prognosis of the condition, functional capacity and the individual's psychological status.

A study by the University of Tsukuba in Japan (2002) involving 73,000 people over an eight-year period, found that men and women who report high levels of perceived stress were at much greater risk of heart attacks than people reporting low levels of such stress. Highly stressed women also showed increased rates of fatal stokes. Women who reported high stress levels were more than twice as likely to have a fatal stroke than women reporting low stress, and about three times as likely to have a deadly heart attack. Men who reported high stress levels were about 1.5 times more likely to die from a heart attack, although they did not show significantly increased deaths overall from strokes. (The researchers said that one possibility for the difference in results between women and men, may be that men are less likely to admit to high stress levels.) Simply because there is an association between the two factors does not indicate that stress causes heart attacks. The researchers pointed out, for example, that the women who reported high stress were more sedentary, more likely to have a history of hypertension or diabetes, smoked more and were more likely to work full time. Even after adjusting for these factors, however, the researchers found a significant association between stress and heart attacks in both men and women, and an association between stress and strokes in women (*Circulation* (Journal of the American Heart Association)).

Stroke

9.17 The Tsukuba study (see 9.16 above) is not alone in finding a link between stress and strokes. Other studies have also shown that men whose blood pressure rises in response to stress have an increased chance of developing a stroke compared to men whose blood pressure does not react in stressful situations.

Strokes occur when part of the brain is deprived of oxygen and sustains damage as a result, leading to a wide range of possible neurological symptoms including paralysis, difficulties with speaking and memory, vision disturbances etc. Strokes affecting vital centres in the brain (those that control life-sustaining functions such as breathing) can also be fatal.

Researchers from the Department of Epidemiology at the University of Michigan, along with colleagues in Finland, studied the blood pressure response to a stressful situation in over 2,300 men in eastern Finland. The men had a mean age of 52.8 years and were being observed as part of a long-term study of risk factors for heart disease. Investigators measured resting blood pressure in each participant; and one week later the blood pressure was measured again during a waiting period for a treadmill test to diagnose heart disease. The waiting period for the test was considered a stressful situation, and the differences in blood pressures between the resting and stressed values were determined. The participants were followed for 11 years, during which time 113 strokes occurred in the population being studied. Results published in the June 2001 issue of *Stroke* (S A Everson, J W Lynch, G A Kaplan, T A Lakka, J Sivenius, J T Salonen) showed that men who had exaggerated blood pressure rises in response to stress (defined as a difference of 20mmHg or more in the systolic blood pressure) had a 72% greater risk of developing a stroke than did men whose blood pressure reacted less strongly to stress. Amongst those men whose blood pressure rose in response to stress, poorly educated men were found to be at even more risk of a stroke than better-educated men.

Cancer

9.18 Cancer is a disease that affects cells in the body. It arises when there is an incidence of uncontrolled cell reproduction, with cells continuously dividing and replicating and eventually developing into a tumour. Tumours can be either malignant or benign. Benign tumours, however, are harmless, or are only a problem where they put pressure on surrounding organs. However, in some cases, the tumour can spread beyond its original site and begin to invade or destroy the surrounding tissues.

Although it has not been shown that stress-induced changes in the immune system are a direct cause of cancer, the complex relationship between physical and psychological health is not, currently, completely understood. Scientists know that many types of stress activate the body's endocrine (hormone) system, which in turn can cause changes in the immune system – the body's defence against infection and disease, including cancer. However, the immune system is a highly complex network whose activity is affected not only by stress but by a number of other, unrelated factors.

The relationship between breast cancer and stress has received particular attention, in recent years. Some studies of women with breast cancer have shown significantly higher rates of this disease among those who experienced a traumatic life event within several years prior to the diagnosis. Although studies have shown that stress factors (such as death of a spouse, social isolation, or even examinations) alter the way the immune system functions, they have not provided scientific evidence of a direct cause-and-effect relationship between these immune system changes and the development of cancer. One study sponsored by the US National Cancer Institute (B L Andersen, W B Farrar, D Golden-Kreutz D et al (1998)) suggests that there is no important association between stressful life events, such as the death of a loved one or divorce, and breast cancer risk. However, more research is needed to establish whether or not there is a relationship between psychological stress and the transformation of normal cells into cancerous cells.

One area that is currently being studied is the effect of stress on women already diagnosed with breast cancer. These studies are looking at whether stress reduction can improve the immune response and possibly slow cancer progression. Researchers are doing this by determining whether women with breast cancer who are in support groups have better survival rates than those not in support groups.

Many factors come into play when determining the relationship between stress and cancer. At present, the relationship between psychological stress and cancer occurrence or progression has not been scientifically proven. However, stress reduction is of benefit for many other health reasons.

Sexual or reproduction dysfunction

Sexual function

9.19 Stress can lead to decreased sexual desire (libido) and an inability to achieve orgasm in women, whilst the stress response can also cause temporary impotence in men by preventing a sustainable erection and intercourse.

Fertility

9.20 Stress may affect fertility. Stress hormones have an impact on the hypothalamus gland, that produces reproductive hormones and continuously elevated cortisol levels can temporarily shut down the menstrual cycle.

Effects on pregnancy

9.21 The old wives' tales regarding a woman's emotional state during pregnancy, affecting her baby, may have some credence, as maternal stress during pregnancy has been linked to a higher risk of miscarriage. A study by the University of California in 1997 (reported in the *Journal of Occupational and Environmental Medicine*) found that pregnant women lawyers who work long hours (45 hours per week or more) were five times as likely to feel high levels of job-related stress, and also three times as likely to have a miscarriage than as those who worked 35 hours or less.

Stress is also associated with lower birth weights and increased incidence of premature birth, both of which are risk factors for infant mortality. In addition, stress may cause physiological alterations such as increased adrenal hormone levels or resistance in the arteries, which may interfere with normal blood flow to the placenta.

Other disorders

Allergies

9.22 Research suggests that stress may contribute to so-called 'sick-building syndrome', that produces allergy-like symptoms, eg eczema, headaches, asthma and sinus problems in office staff who work in air-conditioned, high-rise buildings with permanently sealed, double glazed windows.

Skin disorders

9.23 Stress can play a role in exacerbating a number of skin conditions, including hives, psoriasis, acne, rosacea and eczema. Unexplained itching may also be influenced by stress.

Unexplained hair loss (alopecia areata)

9.24 Alopecia areata is hair loss that occurs in localised (or discrete) patches. The cause is unknown but stress is suspected as a factor in this condition. General hair loss often occurs during or after periods of intense stress, such as mourning or serious illness, though this effect may be reversible.

Teeth and gums

9.25 Stress has now been implicated in increasing the risk for periodontal disease, which is a gum disease that can cause tooth loss.

Some of the possible emotional conditions that may be influenced by stress

9.26 The effects of prolonged stress will often be initially manifest in lowered mood and altered behaviour.

Emotional disorders influenced by stress include:

- Rapid and frequent mood change.
- Depression.
- Anxiety.
- Panic attacks.
- Phobias.
- Pain.

Depression

9.27 Clinical depression is the most common mental health problem in Britain. As a medical term, depression covers a broad range of psychological distress, ranging from a mild lowering of mood to a suicidal feeling of hopelessness. People who are depressed often experience anxiety, and depression is frequently related to wider aspects of life such as bereavement, illness and childbirth (including antenatal and postnatal depression).

It is possible that such sufferers will have difficulty in sleeping and eating regular meals, and may more easily become addicted to alcohol and drugs, as a way of seeking relief.

Social isolation is a major risk factor in depression, and men are particularly at risk because they are reluctant to ask for help from friends, partners, family or health professionals. The perceived need to exhibit a 'macho' image is a factor in causing some men to keep their emotions 'bottled-up'. This can have serious consequences, as revealed for example in the national figures for suicide. Of the 5,986 suicides in the UK in 2000, 75% were by males (source: Samaritans (April 2002)).

According to a survey by the UK charity, Depression Alliance for National Depression Week (April 2002), the UK has one of the highest attempted suicide rates in Europe, with a massive 50% increase since 1990. At least three-quarters of suicides are the result of depression, and the survey clearly showed that many people with depression who had sought help from their employers, continued to be made ill through unrealistic workloads and bad management. Over a quarter of the people interviewed (29%) felt their depression was caused, or made worse, by unrealistic workloads, with 35% of people blaming bad management.

Most people will recover from bouts of depression and use them as a learning experience to reflect on, and take stock of, their life style and priorities. The survey's results are therefore particularly concerning, especially given the importance of providing support to those who are feeling depressed and of removing the stigma and taboo surrounding the condition. It is also generally acknowledged that the most important initial step in treating depression is acknowledging the problem and the need for help, so that treatment, usually comprising a combination of counselling, cognitive behavioural therapy and possibly antidepressants, can begin.

Anxiety

9.28 Anxiety is one form of the body's response to fear that plays a vital role in self-preservation, as fear often prevents us from taking unnecessary risks.

In certain situations, such as taking examinations and competing in sporting events, a 'fight or flight' response can be highly beneficial, as the increased adrenaline and body stimulation heightens perception and concentration. However, if the feelings continue for longer periods, then they can be detrimental. Prolonged muscle tension can cause headaches, tightness in the chest and pain in the neck, shoulders and back.

Anxiety results from the fear response being out of proportion to the actual risk. These anxious feelings often occur involuntarily, despite the best endeavours of the person concerned, to avoid or repress them.

Excessive anxiety can lead to over-stimulation of the sympathetic nervous system and be manifested through physical symptoms such as a higher than normal pulse, sweating and trembling, as well as psychological symptoms such as restlessness, insomnia and a difficulty in concentration. Digestion may also suffer, and many people experience faintness and dizzy spells, particularly in restaurants or other crowded places.

Mentally, anxiety may create the feeling of being unable to switch off the mind, making it difficult to concentrate.

If left untreated, anxiety is likely to produce a feeling of exhaustion. Sufferers often experience panic attacks where they feel faint and physically sick or believe that are suffering a heart attack, even though none of these are likely to happen. Long-term anxiety may also lead to high blood pressure, ulcers as well as an inability to develop and maintain meaningful social relationships.

Treatment for anxiety often relates to controlling the 'flight or fight' response through relaxation techniques, counselling or therapy. Often a change in lifestyle, and reducing the intake of alcohol or stress-boosting chemicals such as caffeine and nicotine, can help the situation, while in extreme cases, medication and professional counselling or therapy can be useful.

Panic attacks

9.29 Panic attacks are usually associated with rapid, heavy breathing (hyperventilation) that entails the body swiftly absorbing oxygen to deal with a perceived emergency.

Some researchers believe that during a panic attack, the brain falsely signals a shortage of oxygen (or an increase in carbon dioxide) that sets off a suffocation alert in a predisposed individual. However, because both an excess and a shortage of carbon dioxide are associated with panic, experts suggest that a panic attack may occur in a susceptible person who develops any imbalance of gases in the blood sufficient to cause intense physical sensations.

Case study

9.30

Anxiety and panic attacks

For many years, Mary was an Assistant Buyer in the soft furnishing section of a large department store in London. Late one August, her 72 year old mother, to whom she was very close to, was diagnosed with advanced terminal cancer. Over the next five weeks Mary's mother deteriorated rapidly, and died at the Royal Free Hospital in mid-October. It all happened so swiftly that Mary was unable to immediately absorb what had happened, or to accept that her one remaining parent had gone. A great cloud seemed to descend on her as she was forced to accept that she now no longer had any parents living. With her mother's passing, the last tenuous connection with her childhood had finally been broken.

During this period, and to add to her grief and isolation, her live-in partner decided that he was 'bored', and had moved out of the house in Fulham that had been their home for the past three years. Mary was devastated, but continued with her job, having taken only a few days off work for the funeral and to attend to her mother's last wishes. She gave no indication, to her colleagues, of her internal turmoil that would eventually tear her apart.

The first clear indication that all was not well happened soon after her return to work. Mary had been persuaded to join a colleague for

(cont'd)

lunch, but before she had even touched her food she experienced her first panic attack. Not realising what was happening, she suddenly felt a wave of nausea sweep over her and fell to the floor in a faint. Coming round after only a few seconds, her friend helped her to a chair as a waiter ran over to give her water. Mary's face was ashen and she remained sitting for some minutes before she felt well enough to be driven home in a taxi. Once there, she immediately went to bed.

From then on, Mary avoided restaurants or any place where there were crowds or where she could not 'escape' easily. She ceased to use the lift at work, preferring to walk the two floors up to her office. Meanwhile, to help her sleep, her GP prescribed medication and advised her to seek professional counselling.

It took Mary over a year to recover her composure, during which time she worked with her counsellor through the many unresolved issues in her life that had contributed to her stressed state of anxiety and fear of the future. Two years on, she still avoids crowds, and cinemas, but has learned to realise that she is able to control her situational fear of places and people. With the support of a new boyfriend, she has successfully built up her self-esteem and, most nights, sleeps soundly. She now knows who she is and where she is going, and is very much more in control.

Phobias

9.31 A phobia is a hyper-intensive fear of a situation or object of a type that would not normally cause any particular anxiety in others. Its effect on an individual's life depends upon the nature of the phobia experienced and the frequency of its occurrence or the effort of avoidance.

The mental health charity, MIND, estimates that 10 million people in the UK have phobias of varying degrees (*Understanding phobias*, MIND Publications). In the main, these are mostly simple and fairly harmless phobias that do not seriously affect the life of the person concerned. However, in extreme cases, even common phobias such as a fear of spiders (arachnophobia) or enclosed space (claustrophobia) can seriously disrupt an otherwise normal existence.

In medical terms, phobias are categorised as 'anxiety disorders' as they centre around the same reactions as those of fear.

The major difference is that a phobia is a fear of the feelings that are produced by being frightened, or in other words, a fear of fear itself.

Phobias can provoke the same feelings associated with anxiety and panic, eg rapid heart beat, profuse sweating, shortness of breath and trembling.

This if often coupled with an overwhelming desire to escape when confronted with the perceived source of the fear.

Treatment usually starts with a GP, who will generally refer individuals to an expert source of advice and/ or help. There is no general cure for phobias, although it is possible to learn, over a period of time, how to confront irrational fears with professional help.

Pain

9.32 There has been much research into the relationships between pain and emotion, but this is complicated by many factors including the effects of personality types, fear of pain, and stress itself.

Some changes in behaviour that may be influenced by stress

9.33 Stress will not normally be identified as a causal factor in behavioural changes but patterns of behaviour that can be influenced or exacerbated by continuous stress include:

- Memory loss and lack of concentration.
- Obsessive compulsive disorder.
- Eating problems.
- Disrupted sleep patterns.

Memory loss and lack of concentration

Loss of concentration

9.34 Stress has significant effects on the brain, particularly on recall and memory. The typical victim of severe stress will suffer from loss of concentration (at work or home) and as a result of a reduced efficiency and alertness may become accident-prone. In children, the physiological responses to stress can clearly inhibit learning. As we become older, although some memory loss occurs with age, stress may play an even more important role than simple ageing, in that process.

Memory

9.35 The immediate effect of acute stress is to impair short-term memory, particularly verbal recall – although fortunately in such cases, memory is usually restored after a period of relaxation. There is also a strong association between excess cortisol (a major stress hormone) and shrinkage in the hippocampus (the centre of memory). However, it is not yet known if this shrinkage is reversible.

Obsessive Compulsive Disorder (OCD)

9.36 Obsessive compulsive disorder ('OCD') is a mental illness characterised by the repetitive employment of ritualistic behaviour in order to satisfy the uncontrollable urges or impulses (obsessions) that arrive, uninvited, in the mind of the sufferer. Such rituals, eg the repeated washing of hands, may be harmless whilst others, enacted in the pursuit of the satisfaction of an extreme obsession, may well become more serious.

However, under the influence of stress, perfectly normal actions taken in the everyday course of living, eg the weekly cleaning of carpets to remove dirt, may turn into obsessive compulsive behaviour and the sufferer may believe that there is a necessity to cover the floor with newspaper in order that his/ her feet are not contaminated by germs. This idea may be controlled by the compulsive cleaning of the carpets, repeatedly throughout each day.

All OCD behaviour need not necessarily be classified as an illness. However, should the symptoms become so extreme or persistent that they interfere with normal functioning, then professional medical help will be required.

Eating problems

9.37 Stress can influence eating disorders and weight loss/ gain.

Weight gain

9.38 Weight gain and obesity may be linked to increased levels of stress. Many people develop cravings for salt, fat and sugar to counteract tension, and thus gain weight. However, in some people exposed to stress, weight gain can occur even with a healthy diet, and the weight gained is often abdominal fat, a predictor of diabetes and heart problems. The release of cortisol, a major stress hormone, appears to promote abdominal fat, and may be the primary connection between stress and weight gain in people affected in this way.

Weight loss

9.39 On the other hand, some people suffer a loss of appetite and lose weight as a result of stress. In rare cases, stress may trigger hyperactivity of the thyroid gland, stimulating appetite but causing the body to burn up calories at a faster than normal rate (Reuters Health Information Inc).

Eating disorders

9.40 Anorexia nervosa and bulimia nervosa are eating disorders that are associated with adjustment problems in response to stress and emotional issues. According to the Mental Health Foundation ('Eating Disorders: Information from the Mental Health Foundation', available via the Foundation's website at: www.mentalhealth.org.uk/), they affect up to 2% of women, and some men. As many cases of eating disorders are unreported or undiagnosed, the actual figures are likely to be much higher, particularly among men.

Eating patterns will vary from person to person, and almost all of us will have experienced some difficulty in eating or have food cravings, at some stage during our lives. It is when food becomes the centre of an individual's life that problems begin to occur, particularly if there is a refusal to eat when hungry, or a compulsion to eat all the time.

In this way food can become an addiction just like anything else and such abuse is particularly problematic. As we all have to eat to live, an individual with an eating disorder has little choice but to face their particular problem many times each day. Symptoms of anorexia include sudden weight loss, and the taking of drastic measures to avoid gaining weight such as drugs, excessive exercising and induced vomiting after meals. This can lead to weakness, depression and a lack of concentration and because anorexia involves self-starvation, it can be, and often is, fatal.

Bulimia nervosa involves binge eating of large amounts, coupled with guilt and the taking of drastic measures to lose weight, immediately afterwards. Often this involves vomiting and the use of laxatives.

Treatment of eating disorders usually involves counselling, that can be arranged through a GP or self-help group.

Disrupted sleep patterns

9.41 The tensions of unresolved stress frequently cause insomnia, generally keeping the stressed person awake at night and consequently fatigued during the day.

Sleep is the body's natural way of replenishing energy. Adequate sleep increases the body's ability to handle stress and without this ability, coping with everyday demands can be either difficult or impossible.

If we spread our physical reserves of energy too thinly by insufficient sleep and poor nutrition, then irritability, frustration and stress, will result. If we fail to meet our body's natural demands, then we may become depressed. Fatigue and stress feed off each other, creating a vicious circle.

Serotonin and melatonin are neurotransmitters and essential biochemicals that affects, amongst other things, the quality of our sleep. However, serotonin production measurably decreases during times of stress and anxiety.

Mental health at work

9.42 Howard Davies, Director General of the Confederation of British Industry, is quoted as saying:

> 'We at the CBI are convinced that the mental health of a company's employees can have an important impact on business performance in the same way as do the industrial relations climate or inadequate training. That is why the CBI continues to add its voice to the campaign to raise the profile of mental health as a workplace issue.'

Mental illnesses such as anorexia, bulimia, schizophrenia, depression, bipolar mood disorder, anxiety and drug/ alcohol addiction, are much more discriminated against both in society and the workplace, compared to illnesses that have a purely physiological basis.

Unfortunately, a majority of people believe that depression (like cancer or a heart attack) will never happen to them. Regrettably, the facts simply don't bear this out.

Facts and figures

9.43

- According to the World Health Organisation ('WHO'), one in four people will be affected by mental or neurological disorders at some point in their lives ('World Health Report' (2001)).
- By 2020, the WHO says depressive disorders will be second only to heart disease as a contributor to the 'global disease burden' ('World Health Report' (2001)).
- 'As many as one in four of us will experience a mental health problem at some time in our lives, which will in turn affect family and friends' (Department of Health (2001)).
- 'It is estimated that about one adult in six in the UK aged between 16 and 64 suffers from some form of mental illness. This suggests that about six million adults between the ages of 16 and 64 in the UK

suffer from some form of mental illness, mainly anxiety and depressive episodes. It is also estimated that about one child in ten between the ages of 5 and 15 suffers from a mental disorder – about 800,000.' (Department of Health Minister John Hutton, written answer (13 June 2000)).

- Evidence suggests that one in four people with a mental health problem have not contacted any professional about it (A Mann et al (1998)).
- Nevertheless, an estimated 10–20% of a GP's time is taken up with mental health-related problems, and this figure is thought to be rising (A Kendrick (1991)).
- A survey found that 16% of the people interviewed used minor tranquillisers on an occasional basis; 14% had taken them in the past; and a quarter of the sample had found them to be damaging (A Faulkner A (1997)).

The implications of mental health problems for employment

9.44 Although these figures are serious enough in themselves, the implications of mental health problems on employment, make even more disturbing reading:

- Lost employment accounts for 37% of the cost of mental health problems in England, which in total amounts to £11.8 billion (A Patel, M Knapp (1998)).
- People who experience mental distress have the highest rate of unemployment among people with disabilities. The Government's 'Labour Force Survey' shows that 71% of those giving mental health problems as their main disability receive state benefits and are out of work, and only 19% are employed. Out of 16 classifications of different types of disability, those with mental health problems were the most likely to be unemployed (Office for National Statistics (1999)).
- Research shows that stress in the workplace can cause psychological problems such as depression and anxiety (HSE (1998)). Research by the HSE in 1998 also found that 182,000 cases of stress or depression are either caused, or made worse, by work each year.
- About half a million UK employees believe that they have some form of work-related stress, including anxiety and depression (HSE (1998)). Of adults who have worked, 2.2 million people suffer from some form of ill health which they believe was caused or aggravated

by work. The most commonly reported conditions are musculoskeletal disorders, followed by stress and depression (Health Education Authority (1997)).

- In 1995, it was estimated that 91 million working days were lost every year due to mental health problems. These accounted for 15% and 26% of days of certified incapacity in the early 1990s in men and women respectively (Department of Health (1998)). In spite of this, one study found that only 12% of organisations had a policy on mental health, although 95% had said that mental health was a concern to their organisation (Confederation of British Industry and Department of Health (1997)).

Case study

9.45

Mental health at work

For many people, coping alone with the responsibility of a managerial position, can sometimes prove to be too stressful.

When the manager of the power station where John worked resigned, John was persuaded to put his name forward for promotion. He was subsequently offered the position and initially all went well. John found the work to be challenging and everyone had high expectations of him. He had always been a committed and driven person, and thought that he could 'bring this one out of the bag'. However, the combined responsibility and stress of the workload that was involved in his new job, soon became too much for him to manage successfully.

After six months, the department, of which he was head, was increased in size by 25% – adding further to his overall responsibilities. However, he felt disinclined to share his concerns with anyone, as he felt he 'should be able to cope' and so struggled on alone.

John found himself working longer and longer hours in order to keep up with the demands of the job, that even extended into weekends. His wife and family were neglected, and his social life became non-existent. Toward the end of the first 12 months, John was having sleepless nights and had started to suffer with severe migraine. After some persuasion by the occupational health doctor, he saw his GP. The advice was that the stresses and demands of his position were making him unwell, and that he should modify his lifestyle. However, this would have meant him resigning from his job, and as he had taken on further financial commitments, this alternative did not appear, to him, to be viable.

Unfortunately, John eventually started to exhibit irrational behaviour, and frequently became agitated by small issues that caused him to over-react. Instances occurred where he started to lose touch with real issues, and often talked about concepts that were incomprehensible to others.

One day John was found crying at his desk, and was sent home for medical attention. He was diagnosed as suffering from stress-related psychiatric illness and was given paid leave of absence. Eventually, a severance package was negotiated, as John was found to have suffered irreparable damage. He was just 30 years old and had struggled to handle increased responsibility for just over twelve months.

During the period that followed, John experienced bouts of depression. With help from his GP and support from his wife, he recovered and was ready to find re-employment. However, he was faced with the dilemma of dealing with the gap in his CV, or having to be 'economical with the truth' regarding his illness. He found it to be extremely difficult to even be selected for an interview, and came to realise very quickly that it was a tough world where understanding and compassion were not easy to find. Companies were apprehensive of employing anyone with a history of mental illness or breakdown, believing that they would be taking on a liability.

Eventually John was offered work at a local business where, coincidentally, the Managing Director's wife also suffered from depression, and so had some empathy with him. Today he is again working successfully, and the Managing Director has said that it was one of the company's best decisions to employ him, as he is a credit to the organisation. John also works for a charity that runs courses for industry, educating employers on mental illness.

However, John's case raises as many questions as it answers:

- Was it, in fact, John's work that caused the problem?
- Was he really the wrong person in the wrong job?
- Were his pressures internal or external?
- How could he best have been rehabilitated back to work?

The difficulty of returning to work

9.46 Dostoevesky said 'you can tell what sort of society you live in by the way they treat mental illness'. In the UK, this is demonstrated by the need to be more open and to disseminate more widely accurate facts regarding mental illness.

For example, it is not generally understood that people can eventually recover from schizophrenia and go on to live healthy, active and productive lives – a case study illustrated the life of the mathematician John F Nash. Nash won the 1994 Nobel Prize in Economic Science for his pioneering analysis of equilibria in the theory of non-cooperative systems – yet his career could have been ruined by paranoid schizophrenia. Born in 1928, he was struck by the disease in 1959, and suffered partially delusional thinking for the next 25 years, before his return to rational thought.

There are fundamental barriers to the employment or re-employment of individuals who have experienced mental illness:

- Stigma, discrimination and misinformation.
- Ignorance of modern advances in medication/ therapy.
- Few work opportunities.
- Complexity of existing work incentive systems.
- Financial disadvantages of working part-time.
- Loss of health benefits, including disability/ social security.

A report ('Out at work') published by the Mental Health Foundation in 2002, based on a survey of 400 people with personal experience of mental health problems, highlighted a number of these issues.

- Only one in three people with experience of mental health problems felt confident in disclosing this on job application forms. However, many of those who had succeeded in finding employment were 'pleasantly surprised' with the support and attitudes of employers and colleagues when they did 'come out' about their mental health problem.
- Less than half of those with psychosis, schizophrenia or bipolar depression, who took part in the survey, were in full-time or part-time employment.
- People with anxiety or depression were more likely to be employed, but still less than six out of ten were employed on a full-time or part-time basis.
- Despite the fact that only one in three people felt confident in disclosing their experience of mental health problems on an application form, nine out of ten of those who were currently working, whether in a paid or voluntary capacity, had told somebody at work about their experience of mental health problems.
- Within the workplace, those who had been open about their experiences generally felt supported and accepted. Over half reported that they always or often had support when they needed it, with a

further one in five sometimes getting support. Around two out of three said that people at work were always or often accepting towards them.

- However, the report also painted a picture of pressures at work causing or exacerbating mental health problems. Nearly two out of three believed that unrealistic workloads, too high expectations or long hours were a major contributor to their mental health problems; while one in three believed that bullying at work had caused or added to their mental health problems.

Managing mental health issues in the workplace

9.47 The problem of how to manage mental health issues in the workplace, raises some difficult questions for human resources practitioners. A Recruitment Manager might well perceive that taking any unnecessary risk, when employing new staff, could jeopardise their position, should things go wrong.

Prospective employees, with a history of clinical depression (or other mental health problems), may feel it necessary to conceal those details in order to increase their chance of re-employment.

Mental illness is covered by the *Disability Discrimination Act 1995*, the provisions of which require organisations not to discriminate and obliges them to make 'reasonable adjustments' where necessary.

A Court of Appeal case (*Kapadia v Lambeth Borough Council [2000] IRLR 699*) supported the application of an accountant who was forcibly retired by Lambeth Council after he suffered depression resulting from an increased workload. Pravin Kapadia said that he had been discriminated against on the grounds of disability.

There is, however, little willingness among employers to take mental health issues seriously. Although the HSE recommends that mental health should come within an occupational health strategy, only one in ten employers have such a policy in place.

Some organisations include mental health problems within their employee assistance programmes (EAPs), but for organisations without this provision in place, dealing with mental health issues rests on education, good management and the support and help that is available from the Employer's Forum on Disability and Mental Health charities. This may include the need to restructure certain jobs.

Some larger companies have developed policies to address mental health issues in the workplace and have defined key elements of good practice in relation to the promotion of well-being at work.

These include the following:

- A fundamental step is for management to recognise and accept that mental health is an important issue.
- It is essential to evaluate information for identifying existing stress levels and the incidence of mental ill health problems within the organisation, and to elaborate on ways in which organisational structures and functions may be contributory factors.
- All employees must be made aware of the route to take if they wish to seek help or advice and this should be enabled in complete confidence.
- The process of analysing the current situation helps to identify areas and goals for intervention via a mental health policy, and to target the specific needs of the individual organisation.
- A mental health policy in the workplace can reduce the stigma associated with mental ill health, and provide assistance to employees suffering from stress and other serious mental health problems.

The economic logic of employers seeking to protect their workers' mental wellbeing is justified as it compliments their compliance procedures to ensure they meet their legal duty of care.

The introduction of a mental health policy, backed by senior management, embodies the organisation's commitment to mental health.

Key learning points

9.48 In this chapter we have looked in detail, at both acute and chronic stress, and their effects on a wide range of physical, emotional and behavioural conditions.

People under chronic stress frequently seek relief through drug or alcohol abuse, smoking and abnormal eating patterns.

A sedentary lifestyle, an unhealthy diet, alcohol abuse and smoking are all conditions that promote heart disease, interfere with sleep patterns, lead to increased, rather than reduced, tension levels and develop into a self-perpetuating cycle of inappropriate lifestyle choices.

- There are two stress reactions to excessive pressure –'acute' (short term) and 'chronic' (long term).
- Stressors in everyday life that can result in acute or chronic stress include conflict, work pressures, financial and relationship problems, noise, crowds, accidents, violence, illness, isolation and/ or loneliness.
- Prolonged exposure to stress can be debilitating both physically and mentally and can impact adversely on all the major systems of the body.
- Physical conditions that may be influenced by stress include increased susceptibility to infection, immune system dysfunction, back pain, gastrointestinal problems, cardiovascular disease, stroke and certain cancers.
- Emotional conditions that can be influenced by stress include mental health problems such as depression, anxiety and panic attacks and also sexual and reproductive dysfunction.
- Behavioural changes that can influenced by stress include, memory loss, eating problems and sleep deprivation.
- Mental illness is covered by the *Disability Discrimination Act 1995*, the provisions of which require organisations not to discriminate and obliges them to make 'reasonable adjustments' where necessary.
- A fundamental step is for management to recognise and accept that mental health is an important issue.

10 Developing and Maintaining a Healthy Organisation

Introduction

10.1

> 'A policy of good human relations at work is not about [the] job, it is about people'. *Marcus Sieff*

Individuals respond to stress in many different ways. They may show an acute or immediate reaction, or the symptoms may appear over time (and may possibly be mistaken for other health problems). In light of the health implications of stress discussed in CHAPTER 9, it is important for everyone in the workplace to understand the causes of stress, how to recognise it and what actions should be taken to manage it. Through co-operation, consultation and effective management techniques, a healthy work environment can be created, and maintained, in which excessive pressure is discouraged, stress is either eliminated or minimised, and individuals are motivated to work to their optimum levels of performance.

The right amount of pressure can stimulate us to succeed, and success gives us satisfaction. However, problems arise when pressure is too great, goes on for too long or comes from too many directions at once. It is when we feel we may be losing control that the symptoms of stress are most likely to occur.

In order to avoid the occurrence of work-related stress, stress management initiatives should ideally emanate from the boardroom, in order to bring about positive change. Where stress is detected, its cause should be identified and the underlying problem(s) addressed through a comprehensive risk assessment.

Stress factors at work

10.2 Being aware of the likely causes of stress within an organisation and who may be affected by it, will clearly help in determining the type of intervention appropriate in order to create and maintain a healthy working environment. For example, individual employees may feel unhappy about:

- Work that is too boring, or difficult;
- Work that confers too much responsibility or, perhaps, too little;
- Feeling isolated;
- Constantly having to deal with grievances or complaints from colleagues or customers;
- An overbearing management or an unhealthy departmental culture;
- Poor remuneration, lack of parity or other conditions of employment;
- Lack of support or poor communications;
- Having to fulfil an ill-defined role or to deal with conflicting priorities;
- An inflexible or inconsistent system of control;
- Too little job satisfaction;
- Damaged career prospects, frustration, insecurity, lack of recognition or training;
- Bullying, harassment or intimidation;
- Extremes of temperature, bad ventilation, noise or odours;
- A management disregard for the ergonomic design of office furniture or equipment;
- Work-related ill-health.

Interventions that can increase job satisfaction and reduce stress include:

- Improved working conditions and environment;
- Work recognition and reward that are at least equal to those general to the industry;
- More consultation regarding conditions of work;
- Greater responsibility and control including involvement in any process of change;
- Developing a supportive culture;
- Improving accessibility to management;
- Establishing clear roles, objectives and priorities;
- Providing promotional paths and prospects, at all levels.

Table 10.1 (below) summarises some of the possible causes and effects of work-related stress, together with possible interventions.

Table 10.1: Cause, effect and possible intervention

Cause	Effect	Intervention
Poor communication	Low morale	Listening skills training
Long hours	Fatigue	Job restructuring/ Time management
Poor conditions	Sickness	Remedial action and joint consultation
Bullying	Low self-esteem/ Performance	Performance review and Dignity at Work training
Work overload	Exhaustion	Job appraisal

Management strategies to minimise work-related stress

10.3 Strategies to manage stress at work include:

- Prevention, preferably by risk reduction and/ or employee education and training.
- Timely and appropriate action to identify, eliminate or mitigate reported problems, as soon as they arise.
- Provision of contingent treatment facilities for any employee complaining of stress or stress-related illness.
- Supporting rehabilitation back to work after sickness absence.

Risk assessment and the use of surveys can provide management with valuable information on which to base appropriate interventions. Some surveys tested against benchmarked data from the 'occupational stress indicator' (*Occupational Health Review* (September/October 2000)), have shown 'relationships with other people' to be reported as a major stressor. Perhaps unsurprisingly, 'factors intrinsic to the job' and the 'home-work interface' are two other significant areas.

Good communication channels are important within any organisation. It is essential that managers and supervisors feel they can communicate easily and effectively with employees at all levels, enabling the early identification of any problems that might be causing stress at home or at work. An 'open door' culture is most likely to enable difficulties with pressure at work to be identified and dealt with speedily.

Organisational policies and procedures

10.4 One of the first steps in helping to minimise stress at work is to ensure that all employees understand what is required of them and the level of co-operation they can expect from others within their own or other departments.

In a large organisation, policies and procedures determine how it manages its people, their workload, and many of the other factors that might otherwise contribute to work-related stress. For example, a health and safety policy and safe systems of working are a requirement under the *Health and Safety at Work etc Act 1974.*

Within smaller organisations, written procedures tend to be very basic, with more emphasis on custom and practice, but in any organisation, a written policy provides a useful framework for how business should be conducted and managed. It is therefore important that policies relating to terms and conditions of service, employee management and welfare, should be clear and concise in setting out best practice, and should be communicated to all staff.

The effectiveness of any policy will, in part, be dependent on the managers who are required to understand it and invoke the necessary procedures. Adequate management training in this area is most important if policies are to effectively support the elimination and reduction of work-related stress. Organisations need to evaluate, on a regular basis, the effectiveness of their policies and procedures through the process of risk assessment and it is important that reviews are ongoing and take account of any major changes within the organisation.

Stress policy

10.5 In CHAPTER 3 we discussed how a stress policy may be part of an organisation's health and safety policy, or how it may be a document that stands alone. Whichever approach is taken, stress must be viewed in the same way as any other workplace hazard.

A stress policy should be implemented in conjunction with staff liaison groups or trade union representatives, where appropriate. Commitment to it should begin at the most senior level and be cascaded through the organisation. There is little point in introducing stress management training at line management level if senior management have little or no commitment to minimising or eliminating excessive pressure, at all levels, within the organisation. The policy must 'live and breathe' within the organisation, and take its place alongside other legal obligations that employers have to their employees.

Every organisation should know how to deal with the consequences of stress and be committed to preventative support. Managers therefore have a vital role to play in recognising the signs and symptoms of stress, and taking appropriate action. There will be times, however, when they may feel out of their depth and require the services of occupational health or personnel professionals to guide them. It is always necessary to appreciate where individual boundaries are, and naturally these will be different for everyone. Whereas one manager may feel comfortable in talking to a member of their team who may have a home-related problem, another may not, and this could flag up a need for the individual to consult with the human resources department, who may refer him or her on to others.

With many employees perceiving themselves to be simply 'a number on a page', it is becoming increasingly important for managers to support their teams and play their part in ensuring that individuals feel that they have a presence and a value within the organisation.

Organisations need to look at stress management interventions as a holistic way of moving forward. No longer is stress management a soft option of 'touchy-feely support' – it is a hard reality issue where the organisation must comply with its legal responsibility in demonstrating a duty of care towards all persons that it employs.

The benefit that can be gained from timely interventions can give an organisation a competitive advantage that will, in turn, hopefully be reflected in its 'bottom line'.

Recruitment and selection policies

10.6 It is important, when recruiting and selecting people for a specific job, that both the organisation and the applicant understand the requirements of the post and the potential pressures involved. One of the conclusions of the Court of Appeal case in February 2002 was that 'there are no occupations that should be regarded as intrinsically dangerous to mental health' (*Hatton v Sutherland and Others [2002] ECWA Civ 76*). It is therefore essential to employ an appropriate selection policy, combined with sufficient job-specific and practical training, to enable the individual to discharge the responsibilities and duties of the job within his/ her capabilities and with the minimum of stress.

Absence management policy and procedures

10.7 The management of absence, and in particular the early identification of any factor that appears to be a stress-related issue, is of the utmost importance. Sickness absence is estimated to cost the UK economy as a whole £11.8 billion, and organisations an average of £476 per employee (CBI (2002)). It is essential, therefore, that organisations have a formal structure within which to monitor and manage sickness absence and support the rehabilitation to work of employees as quickly and efficiently as possible. Effective rehabilitation works to help the individual regain their maximum potential within the workplace, at the earliest reasonable opportunity.

An absence management policy should set out the parameters and lay down procedures for managers and staff to follow when reporting and monitoring absence. The management of attendance is firmly within the remit of managers who have employee responsibility, and a clear understanding of an organisation's absence policy and procedures is essential, as is the need to communicate this information to all employees.

'Return to work' interviews are an integral part of managing sickness absence and present an excellent opportunity for managers to explore, on a one-to-one basis, the reasons for absence, and to offer employees support, where appropriate. It may well be that absence is not actually caused through ill health, but is merely a mechanism for taking time off to deal with pressing personal problems. A 'listening ear' and some flexible management can possibly reduce the risk of unscheduled absence, and give employees the opportunity to take more responsibility for their attendance.

Where it is apparent that there is an ongoing health problem, there may be the opportunity for the organisation to offer support to employees, either via the occupational health department or through private medical care. Finally, for 'rehabilitation back to work' following a long period of illness, the procedure and the people involved will again depend upon the size of the organisation and the nature and extent of the absence. (See section on rehabilitation at 10.52 below.)

Disciplinary and grievance procedures

10.8 Organisations need to have clearly defined disciplinary and grievance procedures, and there is a wealth of employment case law to support this. In the context of stress management, such procedures are clearly important in dealing with the perpetrators of bullying, harassment or victimisation at work – all being situations that are likely to be responsible for stress in the workplace.

Grievance procedures need to lay down an agreed set of rules both for raising and resolving problems between employee and employer. There will usually be a requirement for two procedures: one for individual grievances and the other to deal with collective issues. In general, a grievance exists when an employee wishes to complain about job demands or treatment that is perceived to be either unreasonable or unfair. While it is sometimes the case that groups of employees might raise a 'collective' grievance, procedures will tend to be directed towards providing individuals with a right of redress.

Disciplinary procedures are designed to encourage all employees to achieve and maintain standards of conduct, attendance and job performance. Procedures should apply to all employees, and aim to ensure consistent and fair treatment for all. For further information on this important subject, see the Code of Practice on Disciplinary and Grievance procedures, published by ACAS.

Alcohol and drug policy

10.9 Although alcohol and drug-related problems are amongst the most common causes of sickness absence, a study by the Institute of Personnel Development shows that a large number of organisations (43%) do not have a drug and alcohol policy in place, nor do the vast majority (84%) run any awareness programmes for staff (TUC (2001)).

When formulating an alcohol/ drug policy, it is important to take into consideration the prevailing attitudes towards alcohol and drug use that are currently acceptable, together with any problems with the use of these substances that the organisation may have encountered in the past.

Under the *Management of Health and Safety at Work Regulations 1999 (SI 1999/3242)*, employers are required to assess the risks to health and safety to which their employees are exposed while at work. Knowingly allowing an employee who is under the influence of substance misuse to continue working (if that employee's behaviour or negligence puts the employee or others at risk), could amount to a criminal offence. Were the risk to materialise, then the employer might well also incur a civil liability to anyone sustaining injury as a consequence.

The aim of an alcohol and drugs policy should be to:

- Promote greater awareness of how alcohol and drugs dependency can be prevented.
- Encourage and support self-referral or intervention at an early stage of dependency.
- Manage dependency so that there is an efficient and supportive working environment.
- Comply with the organisation's legal obligations to discharge its duty of care both to its employees and to any third party.

The use of alcohol and drugs by individuals who are under pressure and suffering from stress is well documented. It is therefore important that organisations have a mechanism for dealing with such issues promptly and fairly. It is worth noting that some organisations are introducing random testing for substance abuse within the workforce.

Culture and management of the organisation

Workplace culture

10.10 The culture of the organisation, as well as the way in which people are managed, can be just as important in determining an employee's level of stress, as the demands of the job itself. A supportive working culture needs to be established in order that employees feel integral to the organisation and are motivated to bring loyalty and commitment to their work. This can be assisted by a good understanding and acceptance of the organisation's objectives and corporate goals, in order to help employees feel personally involved in the achievement of them.

On the other hand, an organisation that implicitly condones a culture of fear (and the consequent stress that it brings), will find that in addition to breaching employment law, such a method of management can eventually lead to serious problems of stress-related illness, and that the maintenance of such a culture is therefore not a valid option.

Managers need to accept that stress is not necessarily due to factors of competence, weakness or even fear, and that if existing, it should be investigated thoroughly to identify any apparent symptoms that might be work-related. It may then be necessary to bring in changes to the existing methods of working, including staff responsibilities and staffing levels.

Management style and support

10.11 Clearly, there are some fundamental differences in the causes and prevalence of stress, between larger and smaller organisations. In a smaller organisation, the management style or personality of the chief executive can make a significant difference to the levels of stress experienced within the organisation, whereas in a larger company or firm, the various layers of management tend to diffuse any idiosyncratic style (or prejudice) of the 'top man'. However, it is still possible in all organisations to encounter a clash of personality and, in such instances, the individuals concerned will have to endeavour to resolve differences within established procedures.

Management skills

10.12 All managers will benefit from bearing in mind the following:

- Expecting the best from people improves the odds of achieving it. Encourage a 'can do' culture, in a realistic way, in order that individuals contribute new ideas on how to achieve given tasks – as opposed to reasons why they cannot be achieved.
- Encourage personal feedback. People appreciate the chance to comment on how they perceive management input or supervision, and such insight is beneficial.
- Be approachable. Listen whenever there is an opportunity, and develop a reputation for being a good listener.
- Reduce uncertainty to a minimum and endeavour to ensure that employees receive timely information about proposed changes within the organisation. Receiving adequate notice of news, be it good or bad, is better than being kept in ignorance, as it enables individuals to make informed decisions and to plan for the future.
- Be aware of scheduled working hours – both those of your staff and management. It is unrealistic to expect 100% performance for 14 hours a day, from anyone. Ensure that everyone takes regular, short breaks away from their job, however brief.
- Foster an atmosphere in which everyone can enjoy both their own work and each other's company and build relationships between manager/ employee. Think about team building opportunities and the occasional, external, social activity to engender goodwill and team spirit. Enquire about employees' families and life outside work, as well as encouraging general discussion of any problems they may be experiencing, eg moving house, obtaining a mortgage, finding a school or dealing with illness. (Expressing interest and making suggestions in the finding of solutions can establish or reinforce a non-intimate bonding that can give satisfaction to both parties).
- Staff should know exactly what their work entails and have a clear, written job description that details their specific duties and responsibilities.
- When delegating tasks, the requirements and expectations should be made clear, either verbally or in writing. At any one time, staff should know exactly what is expected of them, when things need to be done by, and how often they should keep management informed of the progress of their work.

Targets given to employees should always be realistic and achievable within available resources and, wherever possible, individuals should have some input into the setting of them. Objectives should be reviewed regularly. Remember the acronym 'SMART': specific, measurable, achievable, realistic and time-limited.

Managers need be both assertive and honest, both with themselves and with others. They need to be open and direct in informing an individual when they may have taken an incorrect action whilst ensuring that issue is taken with the action (or inaction) complained of, without making it a personal attack. A decision to say nothing may result in a repression of feelings that can lead to resentment and, in time, an aggressive response.

Managers should:

- Remember to give praise, when appropriate. It is worth questioning whether it is correct to always complain when things go wrong, if there is neglect to give praise when work is done well.
- Consult over deadlines and targets whenever possible and ensure that they are realistic and achievable.
- Endeavour to ensure that the environment is appropriate for the job. Working conditions should be compatible with the results required. For example, the work area must usually be clean, warm and dry in order to produce acceptable results.
- Where possible, try to avoid employees being bored by continuously repetitive work, by changing their routines at regular intervals.
- Always endeavour to behave calmly and speak slowly and distinctly when giving instructions. Make an effort not to become unduly annoyed or irritated by events.
- Be aware of the importance of confidentiality in certain areas when dealing with staff, together with the limits of that confidentiality.
- If presented with a problem that may be beyond their own skills and resources to deal with, make sure there is an awareness of where to get specialist help. In particular the personnel and occupational health departments, or an external counselling service, are likely to be excellent sources of assistance.

Above all, a manager needs to retain their sense of humour. A smile can benefit both themselves and others in the workplace.

Appraisal and performance review

10.13 Carrying out regular performance review interviews with staff is one of the most important tasks for managers and supervisors and it is essential that these are conducted effectively.

Interviews need to take place, optimally, within an environment conducive to a discussion in relation to the job and future prospects. Both manager and employee need to be aware of how the appraisal system operates and how it should be conducted. It is often of benefit to link the discussion with the organisation's business plan and overall objectives. This can assist the individual to understand more clearly where they fit into the organisation and in which way their contribution is of value.

- Appraisal interviews should be conducted in a quiet and private setting and should be regarded as confidential by both parties. It is important to give the matter undivided attention and telephone calls or other interruptions should be confined to matters of urgency, in order to not disrupt the flow of discussion and thought.
- It is recommended that appraisal interviews should ideally start and end on a positive note. Any negative issues should be dealt with during the middle of the discussion. Attentive listening skills should be employed in order to properly appreciate both points of view; to convey genuine interest and to encourage participation in any decisions such as the setting of goals or objectives.
- Any concerns should be raised clearly and concisely and agreement reached for the extent and timetable of any proposed improvements.
- Obviously, the interview should be conducted with restraint and mutual respect, and any inclination to raise voices should be avoided. The purpose of the appraisal is for both parties to learn something of the aims and objectives of the other.

Career development

10.14 For most employees it is important to have a viable, planned career path, and it is equally important for them to have the necessary support and information in order to achieve this.

Individual career aspirations should, wherever possible, be recognised and supported by offering training, and possibly mentoring. This is likely to be of benefit by increasing motivation and reducing the potential for frustration and consequent stress.

A career promotion scheme, whereby individuals are sponsored by their manager, can be helpful – whereas a lack of opportunity for promotion (or being over-promoted) may often be a cause of frustration and stress. In this context, it should be borne in mind that managers sometimes assume that all employees are seeking promotion, whereas in fact, many are quite content to be 'foot soldiers', and the idea of promotion, with the responsibility that entails, could well be anathema to them.

Of course, status and recognition will be more important for some than for others and most people do benefit from their work being recognised. However, for a majority of workers, job security will be of particular importance, implying that conversely, job insecurity can be a significant cause of stress. In many instances, there may be no actual foundation for a fear of job loss and it is important for line managers to address this problem, where it exists.

Managers role and responsibility

Participation in decision making

10.15 Involving employees in decision making can be a very effective intervention. Unfortunately, there may be occasions when some staff feel uncomfortable participating in this process and, in this case, joint consultation may need to be used in setting formal guidelines.

It is obviously important to find the right balance between, on the one hand, individual employees having complete control over their work, and on the other hand, having their work fully controlled for them with no consultation process that can offer any opportunity for input. Obviously, the organisation and the individual will tend to have varying perceptions as to the degree of control that is best suited to a specific job.

Induction and job training

10.16 Induction training is an important part of the introduction of new employees into the established workplace ethos and culture. When carried out effectively, it can help to lead to a clearer understanding of how a specific job role and function fits into the overall objectives of the organisation. The training should include an introduction to the organisation's policy and procedures, in particular those relating to sickness absence and disciplinary/ grievance issues. Stress awareness training should also be included at this stage, as this is an ideal point to deliver the consistent message that stress is unhealthy and should be avoided.

When planning an induction package, it is useful to ask the following questions:

- Does it provide the right information?
- Is it still relevant, ie has it been updated to reflect any recent changes?
- Is it focused exclusively on the needs of the organisation or also on the needs of the individual?
- Does it convey the overall organisational aim(s) as well as the role of the individual within the realisation of those objectives?
- Does it provide practical advice on where to receive further information, support and training and does it fully explain organisational policies and procedures?
- Does it clearly define terms and conditions of service as well as employee benefits such as leisure facilities and the employee assistance programme provision?

Role conflict and role ambiguity

10.17 While policies and procedures can be used to specify standards of behaviour for the organisation as a whole, it is equally important for individual employees to fully understand that which is expected of them in their day-to-day work by having their roles and responsibilities clearly defined.

Staff should be encouraged to discuss any work-based problems, within a supportive organisational structure.

However, it is recommended that job descriptions should be reviewed at regular intervals and not just at times of appraisal. It is equally important to review job descriptions where the organisation is going through a process of change or where work patterns are being modified as a result of a risk assessment procedure.

Regular team meetings to discuss and evaluate departmental efficiency and productivity, are beneficial in eliciting ideas for improvement in work flow, and staff should be encouraged to contribute individual ideas for the improvement of organisational efficiency.

In instances where an organisation operates by a system of matrix management, it is important to be aware of the pitfalls, as well as the benefits. Recommendations include:

- The establishment, where necessary, of a process to clarify procedures and priorities.
- The involvement of staff in these discussions.
- The setting up of clear and accessible channels of communication.
- The dissemination throughout the organisation of the details of all benefits perceived to be integral to the particular system of working.
- Discussion and prompt action regarding any valid concerns expressed by the workforce.
- The recognition and reward of both individual and departmental success in achieving group goals and/ or targets.

It is important that commitment is given to clear reporting lines in order to facilitate progress of allocated workloads. Wherever possible, agreement by consultation should be obtained to given deadlines including reasons why particular tasks need to be completed by specific dates and why particular deadlines need to be imposed. Similarly, if a task is required to be completed in advance of the time originally allocated, eg escalation of contract completion, managers should explain the reason for the accelerated deadline, and support the efforts of their staff in attempting to achieve it.

Individual employees may be concerned at the level of responsibility that they hold, whilst in other cases this may be a motivating factor. Obviously, the actual extent of any responsibility should always be made clear, while taking due account of both the ability of the particular individual and whether he, or she, feels that responsibility to be unduly onerous.

Managing 'presenteeism'

10.18 It is the responsibility of management to ensure that, in the ordinary course of events, scheduled hours of work are maintained as a norm. Managers should avoid setting an example of regularly working extended office hours, as this will tend to set an unhealthy precedent that staff may feel obliged to emulate.

Employees, (especially those who fully subscribe to the mission and vision of the organisation), usually realise that in order to 'make a deadline' or to secure an opportunity for the company, they will be expected, on occasion, to put in extra time. Few would object to such times of 'doing whatever it takes', not only as a mark of loyalty, but also because they know that when the company succeeds, so can they. However, instances where employees regularly spend excessive time at work, for no apparently valid reason, should provoke enquiry by line managers.

Work-life balance culture

10.19 It is essential for organisations to facilitate a culture that enables a correct work-life balance to be enjoyed by all its employees. Whilst policies to allow more flexible working have been introduced into some businesses (and there have been recent legislative changes to ensure the provision of this, in specific circumstances), the prevailing culture in most organisations, of rigid working hours, remains unchanged. Ideally, there needs to be a greater knowledge and understanding of the benefits of flexible working in addition to other successful work-life balance initiatives that are available and that have already been introduced by some organisations.

Tackling human resource problems is the increasing challenge of the future. The changing demographic profile means that many firms are now tapping into a reducing pool of younger talent. Furthermore, the so-called 'X generation' have different expectations of work from their 'baby boomer' forebears, and tend to attach a much greater importance to work-life issues.

Using technology to better effect can help to enable some employees to work from home, thereby reducing the stress of commuting.

Flexible working arrangements

10.20 Employees who are eligible have the right to request flexible working hours and employers have the duty to consider the request seriously. The right applies to employees with at least six months' service and who have children under the age of six (or aged eighteen in respect of disabled children). The right is outlined in the *Flexible Working (Eligibility, Complaints and Remedies) Regulations 2002 (SI 2002/3236)* and the *Flexible Working (Procedural Requirements) Regulations 2002 (SI 2002/3207).*

Flexible work schedules have the potential to improve employee satisfaction and to reduce stress. Employees must feel able, and confident, to discuss their work schedules with their line manager and to know that they will receive a considered response. It may be necessary, for example, to modify shift working hours in one particular week for an individual employee (because of a home-related problem), and this can be dealt with effectively to pre-empt the employee becoming an absence statistic.

As the DTI's recommendations suggest, this can be especially important for employees with a young family. Some degree of flexibility in hours of work can dramatically help to reduce work-related pressures and assist in improving and maintaining employee satisfaction, and loyalty, to the organisation.

Nevertheless, endeavouring to maintain a reasonable work-life balance is, for many, a common area of stress. Domestic pressures can exacerbate pressure(s) at work (and vice-versa), and managers need to be aware, as far as they reasonably can, of this; and be cognisant of the fact that there are ways in which greater work flexibility, such as working from home or a variation in working hours, can assist the situation.

Although it is well known that scheduled leisure time away from work is essential for optimum performance and health, many employees still fail to take their full holiday entitlement – men being less likely than women to take their entire annual leave.

The *Working Time Regulations 1998 (SI 1998/1833)* entitle employees to a statutory annual holiday entitlement of four weeks' paid leave each year, which they must be allowed, and even encouraged, to take. It would seem, however, that for some it can prove difficult, for whatever reason, to be away from the office. Therefore, good discipline in holiday planning should be encouraged in order that full leave entitlement is actually taken up.

Organisational change

Managing change

10.21 Change is acknowledged to be an area that can have a profound effect on both morale and the incidence of stress, and because its effects are so well known, most organisations pay at least lip service to the human dimension of the change (although, in reality, this is often too little, too late). The time to start working on the effects of the change process is at the feasibility stage, in order that employees do not feel that they are being presented with a *fait accompli*, whereby the change has been decided without any consultation with staff representatives or before they have been given any opportunity to contribute.

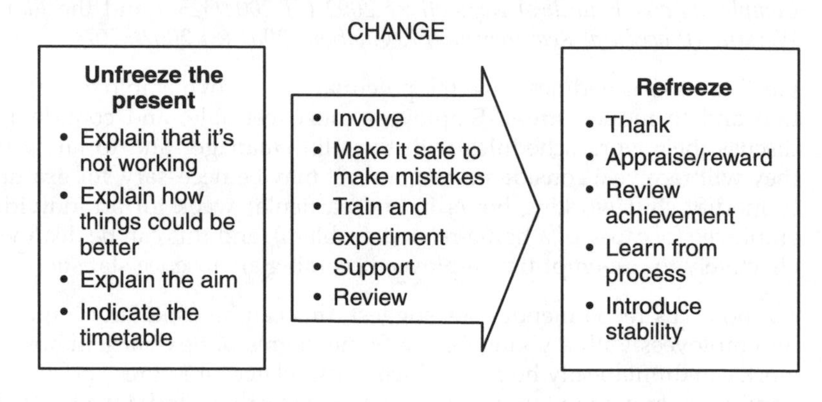

Figure 1: Managing the change process

As illustrated by Figure 1, the change process should start by explaining the need for change, ie that current operations, procedures, equipment, premises or staffing numbers can be improved to produce discernible benefits. If employees can accept this as a starting point, then they will not only accept the need for change but will feel involved in the process.

During the change process itself, the key requirements are to:

- Facilitate communication between all those affected, and preferably directly, ie person-to-person, as opposed to electronically.
- Identify and meet training needs.
- Involve staff in problem solving at an early stage.
- Clarify roles and objectives, as they are established.
- Be aware of the people dimension – moving individuals either to a new location/ into different teams or changing responsibilities, can cause anxiety and should be carried out sympathetically and with sufficient time allowed to adjust to the change(s)
- Assess the risks of stress in the new (or changed) processes and the positions affected
- Create a safe, blame-free environment so that errors, as they occur, can be used to refine the process, rather than being hidden and subsequently becoming the cause of future problems.

Change and stress management

10.22 The introduction of improved work procedures, as a result of the risk assessment process, can be the most direct way that an organisation can reduce stress in the workplace. The identification of any stress-related aspects of work, such as excessive workloads or conflicting expectations, and the implementation of strategies to reduce or eliminate them, are essential to a healthy working environment. However, the process itself needs to be efficiently managed, particularly in relation to changes in work routines, production schedules or the organisational structure.

Typically, there are three distinct steps in a stress prevention programme:

- Identification of any potential problem area (ie stressor).
- Considering the most effective and appropriate management intervention.
- The implementation of the proposed remedial action together with ongoing monitoring of its results.

For the overall process of stress management to be effective, organisations need to:

- Build general awareness regarding job stress (ie causes, costs and control).
- Secure commitment and support for the programme from board level.
- Where possible, consult with staff representatives in all phases of the programme.
- Provide, or arrange, adequate technical capacity/ expertise to implement the programme (eg specialised training by in-house staff, or alternatively the use of an external, professional stress consultants).

It is advisable to gather information and invite observations, criticisms and suggestions from all relevant departments and individuals, in order to help to identify the cause of any specific work-related stress problem.

Problems such as excessive workloads may only exist in a particular department but other stressors, such as inflexible or unhealthy working conditions, may exist throughout the entire company. In these latter circumstances, it may require specific solutions such as a comprehensive review of the organisational structure or possibly a complete (or partial) refurbishment of premises, in order to eliminate all identified causes of stress in order to comply with health and safety legislation.

Managing conflict at work

10.23 Herbert Kindler, in his book *Managing disagreement constructively* (1988), proposed a systematic approach to disagreement. Kindler conceived a four-point process for managing conflict:

- Diagnose, or in other words, anticipate disagreement before it boils over into heated conflict. Identify stakeholders – those people who have a vested interest in the outcome of the conflict.
- Plan an appropriate strategy.
- Prepare the strategy by problem solving and practising it.
- Implement and put the plan into action and evaluate the results.

Such an approach is, in many respects, commonsense. Yet, because of the unpleasantness often associated with conflict, many managers avoid addressing the issue until it is too late and damage has been sustained, either by the organisation or by the individual(s) concerned, or possibly both.

Dealing with conflict – the role of the manager

10.24 Julie and Brian have been working together in the IT department for the last five years but they refuse to co-operate fully with one another. Julie prefers not to speak to Brian because she considers him to be lazy and incompetent, whilst he believes her to be arrogant and rude towards people who are subordinate to her. Over the last three months the deteriorating atmosphere between them has been variously described as 'difficult' and 'impossible'. Meetings that they both attend are punctuated by mutual criticism. John, their manager, doesn't approve of this 'bad atmosphere' in his department, but prefers to ignore the antipathy between these two members of staff, because it happens to be the easiest option for him.

However, because John allows the current situation to persist, this has consequences for other members of staff who are on the periphery of this interpersonal conflict and who may feel:

- Pressured to take sides.
- Continually wary of both parties ('feeling like I'm walking on eggshells').
- Resentment of the bad feelings engendered.
- Disappointment with John's management of the situation in failing to deal with the problem.
- Feeling frustrated with the situation and the ensuing disruption to co-operative working.

Although inter-personal conflict is viewed by many as an (occasional) occupational hazard, the method of addressing it will usually depend upon the individual manager. Managers should have a vested interest in addressing workplace conflict, not least because:

- Promotion of a harmonious workplace atmosphere is not only good practice but can have significant influence on productivity, reduced incidence of work errors and improved quality control.
- There is a mandatory 'duty of care' to all those on work premises. Failure to intervene in a dispute that impacts adversely on employees can have serious consequences, not least of which is the possibility of either prosecution or litigation (or both), should they develop a stress-related illness as a result. Taking this into consideration, and the fact that an employer can also be taken to an employment tribunal, means that a decision to allow any disruptive or damaging situation to persist without intervention, would be ill advised, to say the least.

Therefore, in the circumstances given, what should John do to address the ongoing problem between Julie and Brian?

As the departmental manager, he should arrange separate meetings with the two parties with the aim of establishing and understanding the views and grievances of both, and eventually propose a solution that will hopefully be acceptable.

However, when mediating between members of staff, managers should not expect the problem to be always 'black and white'. Each party is invariably likely to believe that they are in the 'right' in regard to the matter in contention and will behave accordingly. Rarely will one party accept that they may have been wrong and, more often than not, they will maintain that they are the victim. It is worth noting that whilst trying to find an equitable solution to the dispute, in reality, it is more likely to be about finding common ground whereby both parties can give way to some extent whilst retaining dignity and maintaining 'face'.

Having obtained a list of grievances and complaints from both parties, John should then meet again with Julie and Brian individually. This time he should put Julie's complaints to Brian (and vice-versa), and allow him to refute, agree with or merely acknowledge them. Presented in this way, the position of each side can be discussed in an emotion-free atmosphere. John will then need to record Brian's replies with a view to feeding them back to Julie. He will also need to ask Brian what changes he believes are required for them to be able to work harmoniously together.

Parties to the conflict will frequently lay down what they feel to be cast iron conditions that *must* be in place before they will consider co-operating again with their opposition.

A further joint meeting should then be arranged with both sides, at which John provides a brief summary of the events leading up to it. The emphasis should be on finding a way for Julie and Brian to work together productively and professionally in the future. However, in such situations, the contentious parties are invariably much less likely to want to be seen as the 'difficult one' and this can make the role of the manager easier. Each side must be seen to be reasonable, and more often than not, this will result in a 'contract' being agreed between all three parties, establishing the standards of behaviour for each person that will be acceptable – in this case to John, as the manager responsible for the smooth running of the department.

It should be borne in mind that this is an informal attempt at conflict resolution without including, for example, union representation or other external agency. It is in everybody's interest to try to resolve conflict as quickly and as informally as possible – yet very few organisations have such mechanisms in place.

Good managers who are able to quickly recognise conflict and take remedial action, can also act as a catalyst in developing pre-emptive solutions to work-related people problems: for example by regular discussion during group meetings, suggestion boxes, bi-annual appraisals, an open-door policy etc.

An understanding of the relevant psychology, ie the basic mental processes that trigger and escalate conflict, will be of value to managers who intend to apply effective interventions in its resolution.

Mediation and negotiation

10.25 Mediation is a process of intervention whereby an independent specialist assists all parties to try to express succinctly their views on the particular matters in contention, to try to establish some common ground and, utilising this, to move towards a solution that is arrived at by negotiation and is acceptable to all. The mediator's role in this process is to collect information from all parties, set ground rules, facilitate a joint discussion whilst enforcing those rules, encourage creative problem solving, and to subsequently ensure that the agreed solution is adhered to by all parties. It is important to emphasise that the process is not about seeking to apportion blame or ascertaining who is right, but to find a solution acceptable to all sides.

A mediator has no part in the decision making process but instead focuses on the process by which the parties interact. He, or she, must pass the test of neutrality in order to be totally indifferent to the settlement in respect of whether the outcome favours one party or the other. In this respect, the mediator's role is simply to help the parties negotiate a settlement to which they will both commit. Mediation is often effective because although the parties are too angry to talk to each other directly, they have little trouble discussing issues and settlements with the mediator, who can avoid escalation by using 'shuttle diplomacy' to avoid face-to-face contact between the parties during those times when emotions are running high.

As an empathic listener, the mediator also facilitates the discharge of emotion, which can reduce the level of anger that is fuelling the conflict, as well as assisting in conflict de-escalation. A mediator who is accepted and respected by both sides, will also tend to reduce the severity of the arguments used, because the parties will be continuing the conflict in front of an audience whose opinion of them they deem important, and this dynamic can prevent further retaliatory escalation.

The choice of mediator is clearly of consequence, and the personal qualities ideally required include that they be:

- Seen by both parties as independent, impartial, trustworthy, credible and competent.
- Appointed from outside the direct ambit of any party to the dispute.
- Experienced in the fields of mediation, conciliation or arbitration.
- A good facilitator and communicator.
- Able to generate creative ideas.
- Able to manage group discussions and the emotional responses of individual parties.
- Able to maintain an equitable balance and to not appear judgemental.

The use of a mediator should be seriously considered in the event that:

- The parties wish to retain a mutual control over the eventual solution, whereas a more formal approach, such as an employment tribunal, would impose a solution upon the parties.
- Communication between the parties has broken down.
- Either one or all parties to the dispute lack the necessary skills of good communication and the ability to entertain an equitable compromise.
- There is little or no trust between the parties.
- A negotiated solution without the services of an intermediary appears unlikely.
- One or other of the parties fear some loss, were they to negotiate directly.
- A speedy resolution of the conflict is required in order to limit ongoing damage and costs.

- The maintenance of existing relationships is of substantial importance.
- Both sides have, apparently, an equally good case.
- There is a need to express strong feelings within a safe forum.

There may be circumstances under which mediation should correctly be preceded or accompanied by additional training, counselling or disciplinary action.

Mediation should not be used when:

- Either side appears to be unwilling to consider any resolution to the dispute.
- One side seeks an explicit punishment for the other.
- A legally imposed decision, that cannot be ignored, is sought.

There may also be individual issues that cannot be mediated, and occasionally during a mediation session, points may arise that cannot be dealt with there and then, for example:

- When the issue is in respect of a situation that patently cannot be changed.
- The issue is regarding a person or persons who are not present.
- Those present do not have the authority to make changes.

Typical work situations where mediation is particularly helpful include:

- Disputes between colleagues that hinder their co-operation.
- Inter-departmental quarrels.
- 'Personality clashes' or poor working relationships.
- Communication breakdowns.
- Planning deadlocks.
- Change of roles.
- Conflicts over task allocation.
- Disputes over aims and objectives.
- A long-standing history of animosity.
- Where there is a need to broker a return to work programme.

Where mediation is successful, decisions should also be taken as to how the agreed solution will be implemented within an agreed timescale. It is also important that any negotiated agreement is carefully monitored so that the conflict does not re-emerge at a later date. Further meetings may be arranged if necessary, and the parties involved should be encouraged to use the communication skills learned during the mediation process.

Negotiation

10.26 Direct negotiation is the most frequent conflict intervention strategy that is used by managers – and is often viewed as 'lock them in a room together and don't let them out until it's sorted!' type of strategy.

Two-party intervention strategies

10.27 Negotiation involves engagement, by both parties, in a focused discussion to endeavour to reach agreement on a course of action that satisfies at least some of the claims of both sides to the dispute. It is not an argumentative discussion that has the objective of scoring points or of one side to be seen as emerging victorious. That scenario invariably means success for the winner and resentment by the loser, and is rarely an effective means of conflict resolution.

Occasionally, the outcome of a dispute will be determined by one party persuading the other to alter their position by the employment of a highly convincing argument. However, this is unlikely to be successful as it is likely to mean a loss of face for one side, as well as a possibly humiliating climb-down in having to concede that the former position taken was erroneous. There is always also the danger, inherent in most acrimonious disputes, that the losing party may allow their resentment to escalate into anger or even violence.

Third-party intervention strategies

10.28 The foregoing discussion is the reason that a decision is often made to delegate the process of negotiation to an external agency. This can be done constructively, by turning the issue(s) over to unbiased, professional negotiators or conciliators who are skilled in problem solving and creative strategy options.

However, such interventions can sometimes fail, and an alternative is to submit to arbitration, whereby a professional arbitrator hears both sides of the case and then adjudicates between the parties.

The arbitrator will settle the dispute on the basis of his, or her, interpretation of the facts of the case. The difference is that the parties to the dispute must agree, prior to submitting to arbitration, that the decision reached by the arbitrator will be accepted as binding upon all parties. This agreement then has legal force. However, the arbitrator is not constrained by the same rules of law that are observed by a court in a judicial process. This latter process is an even more formal route whereby the decision, having legal force, is binding and final, unless successfully appealed.

Work practices/ systems

10.29 Clearly defined work practices will help to reduce the incidence of ambiguity and uncertainty, that if prolonged, can cause stress. Another factor is that, for many, it is important to have a wide variety of work, while for others, variation may cause problems. It is therefore important to find the right work-balance through discussion with the individual(s) concerned.

Work systems and processes should be designed and formulated to take account of any potential stress factors, and should include suitable safeguards and procedures to either eliminate or minimise them. This means incorporating many of the personnel controls that we have already discussed, such as regular and robust reviews of job specifications and skill assessments. The important point here is for the organisation to keep its practices and systems under continuous review in order to identify and control any work-related stress that may emerge, at any level, throughout the organisation. Stress has no bias and is therefore not confined to any particular type or methodology of working practice or environment.

Workload and work pace

10.30 Work overload and work underload can both be sources of frustration and stress. Finding the right balance for the individual is the best way to pre-empt such workplace problems, as well as optimising performance and improving productivity.

There may be instances where particular individuals may feel underutilised, and therefore experience stress when skills that they possess, and want to use, are then not required within their role. Clearly, being given the opportunity to utilise such skills will not only improve the well-being of the individuals concerned but may also be of wider benefit to the organisation, as a whole.

For many individuals, time pressures and deadlines are a continuous source of stress. If this is the case, as well as reviewing the constraints under which a person is working, it may help the individual to develop better coping strategies through training in the use of time-management techniques.

The extent of control over the pace of work and the setting of deadlines can also be beneficially adjusted to minimise the possibility of stress. There will always be a variance depending on the individuals concerned, and greater control will be more important for some than for others.

Equally, meaningful work is invariably more satisfying, and therefore less stressful to those employed in its completion.

Environmental issues

10.31 There are a number of environmental factors that can contribute to stress and may need to be evaluated and examined. Many of these are requirements that are basic, or fundamental, to health and safety, such as good lighting, ventilation, temperature control, adequate workspace, ergonomically designed equipment etc. However, exceeding the minimum legal requirement by improving the work environment through, for example, the use of greater natural lighting; increased frequency of air changes; good temperature control; light, airy atria; and increased communal space to encourage communications (and cross fertilisation of ideas between departments) – can all help to maximise employee potential and productivity.

Ergonomics

10.32 There is a close correlation between stress and the use, or lack of, ergonomic design. A body made tense by using badly designed furniture or equipment may become more susceptible to injury, such as that caused by repetitive movement. The ability to be able to rest and relax tight muscles and joints through regular (if only brief) breaks during work, can therefore be very important. It is recommended, for instance, that computer users take a short break away from their workstation every 20–30 minutes or so.

Case studies

10.33

Case study 1

Sarah's work on an assembly line involved her having to do frequent, rapid and repetitive movements with her soldering-iron tool, in the course of assembling printed circuit boards (PCB's) for electronic control units.

Sarah found the work physically and mentally tiring. An ergonomic assessment of her workstation showed that she was being required to reach over too far for comfort and needed to use excessive energy to position the work tool.

(cont'd)

The company adjusted the height of the workstation and trained Sarah on how to hold the tool more easily and effectively without gripping it too tightly. As a result, although the work speed was unchanged, Sarah felt less stressed by time pressure on the line, and more able to work in a relaxed manner.

Case study 2

Peter had never learnt to touch type. He would sit hunched over his desk at the office for long periods, and would use the keyboard in a position whereby his fingers were at an acute angle to his wrist, instead of his hand and lower arm correctly being kept level. He would also tend to use his mouse with the same hand position. As a result, he started to experience pain in his wrist that refused go away, except at weekends. An ergonomic assessment showed that Peter's working position at his desk was incorrect – the height of his chair relative to the desk being too high. This also resulted in neck strain, due to the screen height not being correctly at eye-level. The solution to the problem entailed taking the time to readjust his sitting and typing position and to relax his body at the keyboard, which helped to relieve the discomfort and minimise the stress that Peter had felt after using the computer. In addition, the pain in his right wrist also eventually disappeared, as a result of his improved working posture.

Process design/ redesign

10.34 Process design and redesign are frequently allied to ergonomic changes in operating machinery or equipment and the following case study illustrates this point.

Case study

A car manufacturing company introduced a suspended rotating table system on its assembly line. This meant that the cars being assembled were suspended and could be rotated through 180°, enabling the fitters who were installing fuel and brake lines on the underside of the chassis to work on them while standing. Previously these operations had involved assembly workers crouching beneath the vehicle or stretching up and being required to twist the body into awkward positions in order to assemble the parts. The line height was also modified so that it was possible to raise or lower certain sections relative to the operations that had to take place. These ergonomic changes significantly reduced the number of cases of work-related, upper limb disorders that had been reported on the earlier line setting, thereby reducing the stress factor experienced by working in awkward postural positions, and increasing staff morale and job satisfaction. The changes were introduced at the planning stage when the line was being redesigned – an example of system re-engineering that served to improve both employee health and productivity.

Skills training and Continuing Professional Development (CPD)

Training

10.35 Lack of 'suitable and sufficient' training has been identified by the Health & Safety Executive ('HSE') as a key factor in the evaluation of risks related to stress in the workplace. Furthermore, there is a legal duty to ensure that staff are provided with adequate training to carry out their work.

Insufficient training has a significant effect on the amount of job-related pressure felt by the individual. In instances where staff may be justified in feeling that a job is beyond their ability or resources, work-related stress may result. Employees in this position may feel that a reluctance by the employer to offer adequate training is an indication of the general lack of support being offered by management. It is not surprising, therefore, that employees in this situation often become stressed. Eventually they may well take time off, rather than continuing with a job that they feel they cannot perform to the required standard. It should also be borne in mind that 'one size training' will not necessarily fit all. There are frequently differing requirements for different employees, such as varying numeracy or literacy skills or allowing perhaps for a slight dyslexic impediment etc.

In some cases, there may be specific training needs related to the post itself, ie technical instruction on new hardware/ software, or possibly new interpersonal skills that may be necessary, such as those required for dealing with complaints from the general public. It is this area of individual need that is often overlooked. Each employee will approach a job in their own unique way, bringing with them their own individual skills. Managers should always discuss the requirements of a position with the person concerned, carry out a training needs analysis, and devise a training plan as part of the individual's personal development. It is also important to monitor and review such a plan once the member of staff is in their post, in the event that other skill gaps might be identified that need to be met.

Time management training

10.36 The planning and pacing of a personal workload is a critical factor in managing stress. Unfortunately, time management training is occasionally seen as a panacea for reducing stress, due to the unreasonable expectation that it should enable the individual to successfully control all types of pressure. The reality is that it can form one of several useful coping strategies, but is limited by the fact that not all pressures are necessarily related to time or under the control of the individual.

Communications training

10.37 Effective communication is often a neglected factor in management training – yet it is essential to good management by reducing the incidence of misunderstanding and consequent errors. Good communications also enable individual employees to be more readily aligned to the vision and leadership of the organisation, and this in turn serves to reduce the opportunity for disharmony, discontent or dissatisfaction.

In basic terms, clear and concise communication improves understanding and helps with the accurate dissemination of information throughout the organisation, enabling both management and staff to work efficiently. It goes without saying that effective communication also includes active listening skills, engaging with the person you are listening to and responding appropriately to them. Good communication on all levels will help to ensure that everyone within the organisation is enabled to work with knowledge and confidence.

Effective communication can also assist managers to be better aware of external pressures on employees. This may well enable the provision of suitable interventions in the instance of problems such as sickness or bereavement that might benefit from temporary, flexible working arrangements.

Assertiveness training

10.38 Assertiveness training can be a useful tool to both the employee and the manager. It can lead to a better ability to deal with both workload and work capacity, insofar as the individual is enabled to say whether they can take on more work, and just as importantly, when they cannot.

Stress management training

10.39 To ensure that stress management becomes integral to corporate culture and company philosophy, serious consideration needs to be given to careful planning and training in order to raise awareness in identifying work-related stress, where it exists, whilst taking due account of the culture and nature of the organisation.

It is necessary to be able to recognise at an early stage the warning signs and symptoms of stress, and this should be an integral part of the management strategy. By monitoring sickness absence (especially short-term absence), carrying out confidential staff surveys, observing working relationships (especially in terms of team dynamics), and questioning changes in attitude and behaviour, it is possible to raise awareness. This will invariably lead to a greater understanding and recognition of the causes and symptoms of work-related stress, and the interventions that can be used to minimise it.

Stress management training programmes typically teach employees about the nature and sources of stress, its effects on health, and the personal skills necessary to reduce it, including the achievement of a correct work/life balance. Training may also help reduce stress symptoms such as anxiety and sleep disturbances, and has the added advantage of being relatively inexpensive to provide.

A typical stress awareness programme would be likely to include the following modules:

- The relationship between pressure and stress;
- Understanding our biological response to pressure;
- Examining the individual nature of stress;
- Key causes and effects of stress in the workforce;
- Recognising the difference between internal and external pressure;
- Handling the pressures of time demand;
- Effective stress management techniques;
- Commitment to a stress management contract.

A stress management programme designed for all grades of supervisory and management staff might include the above topics plus an overview of the legal implications of stress, together with the option of interpersonal skills development training.

Leadership training

10.40 Effective leadership is an important component of good teamwork. It may include an employee being required to progress a group task – a responsibility not only a pre-requisite of senior management. Leadership training should ideally be combined with communication skills training, since the ability to communicate clearly is also an important component of good leadership. A team that has a clear sense of direction, with individual members understanding their role and encouraged to be part of the group, is more likely to be successful. It is also more likely that members of such a team will feel less stress than those in which one person simply assumes a dominant position, without any consent of the group, and insists on demands being met, whether they be reasonable or not.

It is advantageous for an organisation to provide adequate support and encouragement in order to enable individuals to realise their full potential within the company. Occasionally, a more senior manager (not normally the employee's direct line manager) will act as mentor to an aspiring individual with drive and ambition. Alternatively, external coaching or mentoring can be sought outside the organisation for prospective leaders, supervisors and managers.

Cultural awareness training

10.41 This is usually understood to be the raising of awareness of the cultural factors emanating from differing ethnic or religious backgrounds within the workforce. Such awareness can help to reduce the incidence of any ethnic-related problems or tensions within the workplace and minimise the likelihood of bullying or harassment. It can also serve to reduce discrimination, whether it be intentional or unintentional.

Evaluation, monitoring and review of interventions

10.42 Subsequent evaluation of any remedial actions taken, is an essential step in the intervention process, being necessary to determine whether or not a specific intervention is producing the desired effects. It is important to plan the timescale, as interventions involving organisational change should, ideally, receive both short and long-term scrutiny. Short-term evaluations might well be carried out monthly in order to provide an early indication of programme effectiveness or, alternatively, a possible need for redirection. Too many interventions tend to produce initial, beneficial effects that are, unfortunately, not ongoing. Long-term evaluations are often conducted bi-annually as a means of determining the effectiveness of interventions that are already in place.

Evaluations need to take into account information collected during the problem identification phase, and should include feedback from employees regarding working conditions, levels of perceived stress, health problems and job satisfaction. Employee perceptions are usually the most sensitive measure of stressful working conditions, and often provide the first indication of the effectiveness, or otherwise, of subsequent interventions. Examining attendance records in order to measure the incidence and cause of absenteeism prior, and subsequent, to any intervention, can be indicative of the effectiveness of change.

Evaluation is not the final step. It should be seen as part of an ongoing process of continuous improvement. Monitoring the effectiveness of organisational strategy in relation to employee health and safety, reviewing policy and procedures and re-appraising risks, should be a continuing process. It is important to take special care during any critical periods for the company, such as redundancy, restructuring, change of ownership or relocation, when staff are more likely to feel anxious or sensitive regarding their role or prospects within the organisation.

The results of the review, whether they be of the organisation's policy and procedures, or individual employee performance, should highlight the need for any further measures to be put in place. Such reactive measures will typically include further training for individuals and a possible re-evaluation of their job design or process.

Where an organisation has been unsuccessful in completely eliminating all identified work-related stress, it has a duty of care to mitigate the effects of any excessive pressure, in the workplace, as far as is possible whilst at the same time implementing necessary remedial support.

From the employee's perspective, this is likely to be a process of rebuilding any diminished self confidence or self-esteem, as well as possibly gaining one or more of the new skills previously mentioned.

Workplace employee support systems

Active listening

10.43 The ability to listen actively enables the person who is speaking to talk without interruption or contradiction and, by virtue of having a 'sounding board', facilitate a clarification of his or her particular opinion or circumstance; improve self-confidence and encourage a better assessment of any proposed action, prior to it being taken.

However, some employees still feel that it is a sign of weakness to voice their concerns, in the event that they are experiencing stress. They may not feel confident enough to talk to their line manager, knowing that there is an implied expectation from management that they can adequately cope. There is also the reality that any such admission may, indeed, compromise their position by it being noted on their personnel records. Many individuals fear that this could adversely affect their chance of future promotion, and it is for this reason that they often endeavour to cope with any such difficulty and to cover up their problems and the fact that they are in need of support.

Unfortunately, it is often the very people who require help most, who will deny that they are in need of such support. They will frequently assure colleagues that they can manage and are in control of a situation, when the reality is not only that they cannot cope, but that they have never been actively encouraged by the organisational culture to seek help.

Active listening should therefore be seen as an essential managerial tool and part of effective people management. It should be within the skills portfolio of all managers and available to be used in the maintenance of a stress-free environment and the avoidance of any disruption or discontent within the workforce.

Although such interpersonal skills are an important aspect of management training, and vital for team leaders in order that they can communicate effectively, it is also essential that they acknowledge their own limitations and the importance of accepting the need to refer on to a professional counselling service, where necessary.

The role of external services and providers

10.44 Employers increasingly recognise the fact that both personal and work-related problems can, unfortunately, preoccupy employees sufficiently to affect their work capability, motivation and/ or performance. Of course, many organisations (especially if they are small to medium-sized) may not have an occupational health or personnel department in-house. In such event, a variety of external agencies are available to offer support to employees.

Counselling

10.45 In February 2002, the Court of Appeal ruled, *inter alia,* that 'any employer who offered a confidential counselling service was unlikely to be found in breach of duty of care, by the courts'.

Counselling should be regarded as a necessary intervention to be included alongside other supportive services available to employees. Counselling is about facilitation: allowing people the opportunity to recognise their own feelings and reactions and the causes thereof, and to develop their own potential strengths in dealing with them. It involves the utilisation of caring, listening and prompting skills, and is based upon a trusting relationship that is built up between client and counsellor, where fears and concerns can be articulated within a safe context. It is not the same as an advice-giving service, such as that provided by a medical practitioner or the Citizens Advice Bureau.

A counsellor will be supportive, but will offer little or no direct advice, since the aim is to help individuals to find a solution, themselves, through the development of an insight into their own problems. This is done by enabling individuals to draw on their own personal strengths and resources with a new approach of self-empowerment. It is a sensible and cost-effective intervention to provide counselling support to any individual who may require it.

In addition to the services provided by the British Association for Counselling and Psychotherapy (BACP), many GPs now employ counsellors within their health centre practices. Other UK organisations that provide counselling and support services include Age Concern, Andrea Adams Trust (on bullying), Alcoholics Anonymous, Cancer BACUP, Cruse Bereavement Care, the International Stress Management Association^{UK}, MIND, Relate, Samaritans, Victim Support and Westminster Pastoral Foundation, amongst others.

Employees who work in larger organisations may have the opportunity to contact a company occupational health practitioner (OHP), who in turn may make a referral to their GP. (It should be noted that such a step may be seen as a valid defence for an employer in the event of subsequent litigation or prosecution under health and safety legislation). It is important to note that once a company becomes aware of the vulnerability of a particular employee, they should take proactive measures to offer support. The effect, otherwise (in any legal proceedings), would be to lose the defence that the danger of injury to health was not possible to foresee.

Clearly, the maintenance of confidentiality within counselling is paramount, but a promise of absolute confidentiality may be a restriction, meaning that ground rules need to be established at the commencement of the initial session. There may be instances where counsellors are required to disclose certain information in order to comply with their own professional code of ethics, and employees need to be made aware of this obligation, prior to the start of any session. This could include situations where there is a threat to life or physical safety (either of the client or anyone else), or where there is a serious threat to the interests of the organisation.

In cases where an employee is demonstrating signs of psychological disorder such as anxiety or depression, the aim of counselling is to try to ensure that those disorders are not exacerbated or do not develop into chronic states of dysfunction, thereby making rehabilitation difficult or impossible. A suitably qualified counselling psychologist will be best placed to distinguish between psychological difficulties arising either from work or from personal or social problems, and it is advisable that only professionally qualified counselling services are used.

It should go without saying that accurate and up-to date records should be kept of all actions taken, including details of names, dates, roles, responsibilities and outcomes. This should be done routinely and as a matter of 'good practice', and the records kept confidential and in a secure location. Proposed actions and recommended interventions should also be recorded. Although confidential, it must be remembered that such records could be called for evidence in the event of any possible future legal proceedings.

First contact counselling teams (buddying teams)

10.46 These teams comprise of volunteers (from the organisation) who are trained in basic counselling skills, and receive ongoing training and supervision. It should be borne in mind, however, that they are not professional counsellors, and are typically intended to fulfil a different, although complementary, role. Often they are used within an organisation as a 'first contact' for employees experiencing personal problems. When an employee has a problem, these teams can provide an 'active listening' service – with the proviso that they are always aware of their boundaries and limitations, and when necessary will refer the employee on to a professional counsellor, or to internal personnel or occupational health departments.

First contact counselling teams ('FCC') can be used to help deal with work-related problems such as stress, bullying, organisational change and mediation. They should provide an immediate and alternative initial route for the employee, and one advantage of such teams is that they should also be able to provide a (third party) buffer between the employee and line management where there is a perceived need for mediation. This service should usually offer confidentiality, but this may have to be broken if there is an awareness of a life-threatening situation or a major organisational issue.

First contact teams have been run very successfully for many years, and form an important facility within the conflict and grievance resolution procedures of a number of UK organisations.

Case study

10.47

An example of a first contact counselling team

'The 'Listening Ears' group was formed (by Carol Baker and Anne Cooper of Quest International, Ashford, Kent) early in 2000, when the number of people in the organisation who were trained in listening/counselling skills increased to seven. The purpose was to raise the profile of the group and to advertise the services offered (with confidentiality being the most important aspect of the work). Posters were put up on all notice boards around the site giving photographs and telephone extension numbers of the group, and in addition a leaflet was produced and given to all new employees in their induction packs. The group is also mentioned in the company handbook.

The counselling team has now grown to 10 members and they offer first-contact counselling skills to everyone on site (in excess of 600 people). They are kept busy, and in 2002 alone, they saw in excess of 100 people on a vast range of issues, both work-related and personal. At a time of change within the company and projects taking key people out of their normal roles, long working hours, time pressures and learning new skills, put a great number of people under increased pressure.

The mutual support within the group is vital to its function, with openness and honesty between members forming a strong bond. They meet on a quarterly basis to share experiences, off-load and benefit from each other's experience and knowledge. If needed, any one of the group can call an extra meeting for support or, of course, talk to another member of the group at any time.

The group's six-monthly ongoing training and supervision is of the utmost importance, keeping them up-to-date on legal developments, and working through case studies to gain more knowledge from their trainer, Carole Spiers, who nurtured the group from the very beginning. The Listening Ears group supports National Stress Awareness Day each year with its own activities on-site, such as visiting every department during the morning handing out leaflets and goodies. Invitations are extended to everyone within the organisation to join in during an extended lunchtime period in the sports and social club to try out some stress-busting techniques such as yoga, tai chi, reflexology, aromatherapy, indian head massage, neck and back massage, shiatsu and even juggling!

The group also held a 'Stress Buster' day in Spring 2002 along similar lines.'

Employee assistance programmes

10.48 Employee assistance programmes ('EAP') offer employees access to a professional counselling service on a confidential basis.

An EAP is usually supplied by an external provider, and organisations may use their services to support employees who find themselves to be unable to work effectively because they are experiencing either work-related or personal problems that are affecting their performance.

EAP counselling and information services are available through well-publicised telephone lines and/ or websites, which typically operate 24 hours per day. Helplines are usually backed up by face-to-face counselling on a short-term, solution-focused intervention. Where necessary, employees may be referred on to further specialist agencies.

For an EAP to be effective, it must have the backing and commitment of senior management. However, although they can play an important role in helping to deal with employees' stress-related problems, this service should not detract from the importance of line managers actively listening to their staff. Furthermore, an application to the EAP service must not be misinterpreted by managers as a lack of confidence in their own ability to deal with stress-related issues.

Clearly, employees must be assured that the service is confidential (subject to the proviso as mentioned at 10.45 above) otherwise they will be reluctant to utilise its services.

Some EAP providers offer additional services such as legal, financial and medical advice, lifestyle issues, childcare and elderly care services, and guidance on issues such as health, diet, nutrition and exercise. Some even include concierge services as an option.

The role of occupational health practitioners

10.49 Employees may arrange to see their occupational health nurse ('OHN') because they do not feel able to cope by themselves with either home or work-related stress problems. The OHN is there to listen to the employee, understand the problem(s) and make a referral to an external support agency, should this be necessary. Employees often view the OHN as a neutral person with whom to talk, and one who is, in the context of neutrality, independent of management.

Where occupational health departments have been reduced, as part of overall cost cutting exercises, it may be necessary to consider buying-in external professional resources such as early intervention remedial therapy, psychological assessment and group therapy provision. Occupational health employees have a unique role as they have both the trust of the employee and the ear of management, but this can present a dilemma when dealing with illnesses where stress at work is possibly a contributing factor.

Occupational health can hold a critical, neutral role in the rehabilitation of the employee. This may be through the intervention of the OHN or the company doctor. They may well be in a position to more readily interpret the individual's need for a change to either their work role or the work environment. They will also be able to recommend, with the agreement of the individual concerned and their line manager, the most appropriate, phased return to work as part of a rehabilitation programme.

It should be noted that where the OHN is aware of cases of work-related stress, they have a duty to inform the organisation so that the necessary steps can be taken to remove the causes of stress.

Medical treatment

10.50 Obviously, employers are not qualified to prescribe any form of clinical treatment for work-related illness. Where there is a suggestion that an employee might be suffering from an illness that has been or is influenced by his or her work or the work environment, and he or she is technically still at work, then the employer should arrange for an examination by a registered physician.

An occupational physician is well placed to communicate between employee, employer, GP and any other specialist, as well as being able to report when the employee is fit for work again.

The employee's consent will be necessary to allow medical details to be divulged to other health professionals. In extreme cases, an occupational physician may advise a suspension from work on medical grounds.

Small to medium enterprises, which are less likely to retain the services of an occupational physician, are advised to communicate personally with the employee's GP in instances where they suspect that stress may be work-related. However, it is important to note that they must first obtain the employee's consent before taking this action. Some GPs may not be that well informed regarding the effects of work-related stress and possible remedial interventions. They will also be less familiar with the needs of the organisation compared to an OHP. Opening up any such channels of communication may however help to facilitate recommended courses of action. Employers should be seen to be willing to assist, particularly in the return to work process.

It should be recognised that employees are under no obligation to give consent to an employer to contact their GP. Where an employee does not give consent or declines an examination by an occupational adviser, then any intervention is limited to ensuring that everything possible has been done, by the employer, to identify, remove or reduce the particular cause of stress. Employers should make it clear that in such a case, their decisions and actions will have been made on the basis of the facts that are within their knowledge – although in most cases a medical report is to the benefit of the employee.

Key accidents/ illnesses requiring early intervention

10.51 Currently, musculoskeletal injuries and psychological illness top the league in respect of the number of days lost through specific illness. However, it is prudent to bear in mind that any accident or injury sustained (either at home or at work) requires early assessment and appropriate treatment to ensure that any such injury does not develop into a more serious, long-term problem.

Back pain is the leading cause of sickness absence from work in the UK and it is estimated that in 1995, 10 million working days were lost because of this complaint. The cost to the NHS of treating back pain and injury has been estimated at £481 million and the overall cost to industry has been estimated at £5 billion (as reported by the HSE in 2002).

Psychological illness is likely to be extremely difficult to positively identify in the early stages. Stress itself is not a recognised medical condition and its role may not be immediately identifiable from medical certificates. There is a need to examine the data available within the existing record(s) of absence and attendance to ensure that patterns of sickness absence are highlighted, and identified, at the earliest opportunity, and actions initiated to explore the root cause(s).

Rehabilitation

10.52 Rehabilitation is not currently a legal requirement (although it may yet become so), even in those instances where it has been established that the illness or injury was work-related. However, if primary interventions (those designed to remove the sources of stress) have been ineffective, for whatever reason, and as a result have causes stress-related illness, then there may ultimately be a need for a 'rehabilitation back to work' programme.

Case study

10.53

Rehabilitation

Simon was off work and was in hospital recovering from a severe breakdown for a month. His employer kept in regular contact with him through his wife. When he was eventually discharged from hospital, he was not confident enough to return to work. However, the personnel officer met with him to discuss how best to phase his return. They both agreed that he would first attend briefly one morning, just for the opportunity to meet his colleagues again. This he was happy to do, having also met with his line manager. Agreed arrangements were then put in place for him to be phased back into work, initially for a few hours only per day (coming in slightly later in the morning since his medication also made him sleepy) and leaving earlier in the afternoon. This was gradually increased from two to three and four days a week until he was able, and sufficiently confident, to return to work full time. During the entire process, there prevailed a good, two-way communication with his manager and personnel officer. Additionally, throughout the process, there were regular checks made to ensure that Simon was not finding the workload either too high or too low.

The employee's perspective

10.54 Returning to work following a long period of absence is daunting in itself, but returning following a stress-related absence is often so daunting that some individuals never successfully make the transition back to full time employment. There are many reasons why an individual may feel unable to take that first step back into the workplace. If the illness was brought about by stress at work or there are unresolved bullying or harassment issues, it is likely that fear of a relapse, along with lack of confidence and low self-esteem, will inhibit rehabilitation. When work pressures only partially contributed to the illness, there may be a feeling of guilt on the part of the employee that he/ her had let fellow workers down and put unnecessary pressure on others in the run-up to their illness. Such anxieties may well be groundless, but individuals feel very fragile following stress-related illnesses – anxiety, depression and panic attacks being common symptoms of breakdown or burnout. A considerable amount of support and encouragement is required if a full recovery is to be both achieved and sustained.

The employer's perspective

10.55 The absence of an employee for a long period of time naturally puts pressure on an organisation, both in terms of the costs of covering the absence and also in maintaining the morale of team workers. Achieving the smooth return to work of an employee who has been absent for some time following an accident at work or period of sickness, requires early steps to be taken to establish a non-threatening rapport with the individual. Approaching an employee who is absent through ill health must be undertaken with care and sensitivity. It should be the responsibility of the line manager or departmental personnel officer to keep in contact with the employee, either by telephone, or possibly a home visit. It is important that the absent employee feels valued but not pressured into returning to work before they are completely recovered.

Once it has been established that an employee is well enough to return to work, it is imperative that a phased return is planned. This process should involve the employee's medical support team in the form of GP, counsellor/ coach, a member of personnel or the line manager; together with an external mediator if there are unresolved workplace issues that still need to be addressed. Trade unions will often take on the role of mediation on behalf of their members, and trained mediators can be found working both with solicitors and within organisations specialising in absence management. Initially, a phased return should be planned with a short induction programme and any retraining which it may be agreed is necessary. Workloads at this stage should be carefully monitored – too much too soon will result in a crisis of confidence and a relapse. Too small a workload, on the other hand, can have the effect of making the employee feel superfluous.

Serious stress-related illness, and in particular 'burnout', are conditions from which it is difficult to make a complete recovery. There is, unfortunately, a high risk of relapse should either the employer or the employee not have learnt the necessary lessons. It is for this reason that the employer must monitor the employee's return to work and be alert to the early warning signs of reoccurring personal stress or any inability to cope with given tasks. Regular appraisal will identify further training needs and provide a discussion forum to enable both parties to raise issues that may be inhibiting a full return to work, but it is only reasonable to accept that whilst the employer can provide the structure and support mechanism for an employee to return to work, they cannot guarantee that the employee will slot back, successfully, into the original position that they formerly occupied.

Benefits to the organisation of rehabilitating employees to work

10.56 The cost to employers of absenteeism has already been highlighted. The need for a strategy to manage the rehabilitation to work of employees with either long-term sickness absences or repeated short-term absences, is paramount. Supporting employees in their return to work by means of mentoring, coaching, counselling or mediation is imperative, as is a review of their job descriptions and working environment. The support and understanding of managers is crucial if a sustained return to work is to be achieved. Training needs to be given to management with employee responsibility in order to effectively manage absence policies and actively support members of their team to fulfil this role.

'Britain could learn a lot about rehabilitation from the example of other countries' said TUC General Secretary, John Monks, at a conference in May 2000. For example, 'In Sweden, people suffering serious injuries have a one in two chance of getting back to work. In Britain, the proportion is one in ten. In London alone, the TUC has calculated that about 5,000 working people leave the labour market permanently each year as a result of work-related injury or illness . . . the challenge of rehabilitation is a moral and humanitarian one . . . but there is also an economic imperative at work here'.

The longer an employee is absent from work, the harder it is for them to psychologically adjust to a return to the workplace. Self-esteem and confidence are invariably severely dented by long-term sickness absence, and rehabilitation to the workplace needs to be a gradual process that is managed in a sensitive way by both management and colleagues.

Learning from athletes and 'over-training' syndrome

10.57 Some useful parallels can be drawn from comparing the over-training syndrome in athletes, with burnout in the workplace. These include:

- *The duration of the rest period* – if an athlete returns to training too soon the likelihood of injury recurrence is high; conversely, if they are off for too long, confidence and self-esteem start to become eroded.
- *Appropriate diagnosis* – this is a critical element in determining the duration of rest and the treatment programme, and must be made by the appropriate authority.
- *Appropriate treatment* – in the treatment of athletic injuries there is a role for the sports physician, the osteopath, the physiotherapist and the coach. Similarly, for those who have been 'injured' in the work environment, treatment and rehabilitation should involve a multi-disciplinary team which may include the manager, the OHP, the GP, the independent medical expert and the treatment specialist (eg counsellor, psychologist, cognitive behavioural therapist).
- *Understanding how the problem occurred is essential* – the coach might reassess the training programme, and the athlete might assess his/her strengths and weaknesses.
- *Rehabilitation* – the coach and individual will eventually develop a training programme which will rehabilitate the injury and strengthen the supporting muscles and joints.
- *Reassessment* – the athlete will work with the coach, sports physician and/or physiotherapist to assess the effectiveness of the treatment and any ongoing changes to the programme.
- *Motivation* – this is critical to the recovery and continuing success of the athlete – do they want to continue in the sport?

With the appropriate diagnosis, recovery period and rehabilitation programme, the athlete can return to their sport far more successful than ever before. The following case study also illustrates how all these elements, when working together, can ensure a full and productive return to work.

Case study

10.58

This case study involves an individual who has recently been treated by a rehabilitation centre. The sooner that an employee with problems such as these can be assessed by a centre, the sooner they can return to work. Referrals are taken directly from employers, GPs or from the individual themselves. Where appropriate, all work and medical agencies involved are consulted and regular progress reports are written.

PATIENT: John Parker, a 33-year-old male.

JOB: A 'medium physical demand level' job as a process optimiser in the water industry. This involves driving a light van which John has to enter and exit about 35 times a day, using manual mechanical handling equipment, and lifting weights of up to 30kgs. His desk work totals up to two hours a day.

PROBLEM: John had been off work for five weeks because of back and leg pain, and an inability to squat. His sleep reduced to two hours a night because of pain. Prior to attending the centre he had been seen by his GP and had undergone some medical treatment for about a month including traction and manipulation. This provided some relief but not a solution.

ATTENDANCE: After an initial assessment by the centre, he attended a back rehabilitation course one day a week for six weeks, complemented with twice-weekly local exercise sessions.

OUTCOME: When reassessed after two months, he had no back pain, full trunk flexibility and strength, good aerobic fitness and lifting techniques. Six months since completion of functional rehabilitation, he has had no reoccurrence of pain and has not had any days off work. This was confirmed by a routine questionnaire at six months.

FUTURE: A physical reassessment will be given 12 months after discharge. John's problem will be monitored by questionnaire again at 24 months. He agreed to an exercise plan to maintain general fitness.

Success with rehabilitation is a partnership between employer and employee, and both must want the partnership to work. It is an employer's duty to support an employee back to work wherever possible. It is the employee's duty to do everything they can to keep themselves in good health and fit for work. Old habits die hard, and recovering from burnout is a long-term process. It is all too easy to revert to old habits and succumb to pressure whether it be external or self-induced.

A strict regime that recognises the need for a work/ life balance is essential. Exercise and diet will play an important part in the recovery process and it is essential that once rehabilitated to work, care is taken not to neglect these areas. Burnout changes people – permanently. They may never be the person they were and they may not function again at the level they once did. It is for employers (in conjunction with occupational health doctors) to assess an employee's ability to perform their duties, and it may be that future employment will be offered in another area of the organisation. It is advocated that employees consider such redeployment carefully, as an inability to carry out their job once passed as fit may be regarded as due to capability rather than stress-related illness.

Guidelines for rehabilitation back to work

10.59 The following identifies suggested levels of intervention after an episode of stress-related illness. Each case has to be judged on its merits in order to assess the appropriate level of response to the individual, and in cases of return to work after several months of illness it will be important to work in conjunction with an occupational physician

The importance of this section cannot be overstated. Once an employee has highlighted that they have had a stress-related illness, positive action must be taken to remove the stressors and/ or give adequate training and support to the individual to enable them to cope with the demands of their job

The normal work-related pressure factors should be removed, as far as is possible, for the initial return to work (see rehabilitation below). Appropriate levels of pressure will then gradually be reapplied as the individual becomes fit enough to accommodate them as part of their normal, everyday work.

In some cases re-training may be appropriate. This will depend on discussions with the individual, the job evaluation and skills analysis.

Ongoing appraisal will effectively be a guided conversation, with the manager helping to review aspects of the job that may be difficult, and identifying areas where the individual is happy to initially return to begin work.

Key questions for consideration in ensuring successful rehabilitation of an employee back to work include the following:

- Has the individual been off work for the optimum recovery period? Individuals may return to work too soon and without having had enough time to rest and recover and to rebuild their self-esteem and confidence.
- What are the factors that caused the original problem, and who needs to assume responsibility?
- Is it appropriate that the individual returns to exactly the same role?
- Where the individual's role needs to be changed, has this been communicated clearly, ie are the manager and individual clear about roles, responsibilities and expectations?
- What working practices need to be in place to support the returned worker and what can management do to facilitate this?
- Has the individual been given any advice on burnout prevention? (The value of appropriate education of the individual in burnout prevention cannot be overstated. Particular key areas of education are sleep, hygiene, energy management through nutrition, and graded physical exercise programmes.)
- Have procedures been established for the regular review of the situation? Progress should be continually monitored with regular communication between occupational health, the manager and the individual.
- Motivation – is the individual suited to the job? What do they want from their job?

Summary of effective interventions

10.60 Table 10.2 below outlines positive interventions (control measures) that organisations can put in place to control and manage stress. These measures are in line with good management practice, and the table may therefore be used as a checklist to help assess current practice in stress management, and as a guide to framing future strategies. Compare the existing interventions used by the organisation now, with the points listed in the table.

Table 10.2: Effective stress management interventions

Building corporate awareness and managerial understanding	
Awareness and understanding	Improve the level of awareness of stress in respect of its: • symptoms; and • causes. Ensure that supervisors and managers: • are able to recognise problems of stress in the workplace; • take some responsibility for dealing with it. To help them, use: • training seminars; • booklets, etc. Consider using stress indicators[1] (such as the risk assessment questionnaire) to see whether there are organisational symptoms of stress present such as: • high absenteeism; • staff turnover; • poor time keeping, productivity and performance; • low motivation; • low morale; • an increase in employee complaints, accidents, incidents and ill health reports.
Avoiding and minimising stress	

Structure of organisation and resources	Methods for planning and directing work:
	• Ensure that the performance and production targets are:
	  ○ reasonable;
	  ○ stretching;
	  ○ 'SMART'.[2]
	• Avoid overload or lack of challenge.
	
	Resourcing:

	If you now have fewer employees doing the same amount of work, or increased demand without additional workers, consider: • changing work practices; • enabling people to work more 'smartly'. This may be done through introducing new technology, or re-appraising with everyone what work they do and how they do it.
Management style	Employee involvement: • Ensure good communications with employees. • Involve employees in planning and organising their own job. Teamwork: • Operate in teams. • Vary tasks. • Let individuals undertake tasks from beginning to end. Training: • Use careful selection procedures for new or significantly restructured jobs. • Ensure that staff have the skills/ knowledge and confidence required. • Provide training for managers, for example on: ○ stress awareness and recognition; ○ managing pressure; ○ counselling skills; ○ change management; ○ assertiveness; ○ time management; ○ managing conflict.

<table>
<tr><th colspan="2">Helping the individual to cope</th></tr>
<tr><td>Individual capability</td><td>Remember that:
• individual capability will vary;
• a positive management style will support coping strategies;
• people and groups who may be at risk from stress may benefit from learning coping skills.*
*Coping skills will involve:
• appropriate assertiveness;
• time management;
• knowing when to ask for help;
• avoiding blaming or negative behaviour;
• social support from spouse/ partner, friends, colleagues/ boss;
• appropriate lifestyle: exercise; sensible eating; relaxation/ hobbies.</td></tr>
<tr><td>Access point for help</td><td>• Provide contacts for employees both inside and outside the company in times of crisis.
• Provide access to expert and confidential help through an independent agency, eg confidential help and advice lines covering:
○ legal issues;
○ domestic issues;
○ financial issues;
○ health issues.
With these provisions in place, it is important to have some form of anonymous feedback to act as a barometer of employee welfare.
At least provide a readily available list of local voluntary sources of advice and help, eg:
– local counselling services;
– Benefits Agency;
– Citizens Advice Bureau;
– Samaritans;
– Victim support etc.</td></tr>
</table>

[1] *Occupational stress indicators are questionnaires or systematic interviews which provide a structured analysis of a job and ask about problems outside work. They are useful to provide a baseline once an organisation has become aware of stress issues. However, they can be of limited value and may not be useful as a 'first step' when developing a strategy for stress at work.*
[2] *'SMART' specific, measurable, achievable, realistic and time based.*

Remember: for success, what you decide to do you must *do*!

Key learning points

10.61

- Risk assessment and the use of surveys can provide management with valuable information on which to base appropriate interventions.
- An 'open door' culture is most likely to enable difficulties with pressure at work to be identified and dealt with speedily.
- One of the first steps in helping to minimise stress at work is to ensure that all employees understand what is required of them and the level of co-operation they can expect from others within their own or other departments.
- Commitment to a stress policy should begin at the most senior level and be cascaded through the organisation. The policy must 'live and breathe' within the organisation and take its place alongside other legal obligations that employers have in relation to their employees.
- It is important when recruiting and selecting people for a specific job that both the organisation and the applicant understand the requirements of the post and the potential pressures involved.
- The management of absence and in particular the early identification of any factor that appears to be a stress-related issue is of the utmost importance.
- 'Return to work' interviews are an integral part of managing sickness absence and present an excellent opportunity for managers to explore, on a one to one basis, the reasons for absence and to offer employees support, where appropriate.
- The culture of the organisation, as well as the way in which people are managed, can be just as important in determining levels of stress as the demands of the job itself.

- Carrying out regular performance review interviews with staff is one of the most important tasks for managers and supervisors.
- Involving employees in decision-making can be a very effective intervention.
- Managers should avoid setting an example of regularly working extended office hours, as this will tend to set an unhealthy precedent that staff may feel obliged to emulate.
- It is essential for an organisation to facilitate a culture that enables a correct work/ life balance to be enjoyed by all its employees.
- Flexible work schedules have the potential to improve employee satisfaction and to reduce stress.
- The introduction of improved work procedures, as a result of the risk assessment process, can be the most direct way that an organisation can reduce stress in the workplace.
- Good managers, who are able to quickly recognise conflict and to take remedial action, can also act as a catalyst in developing pre-emptive solutions to work-related people problems.
- Clearly defined work practices will help to reduce ambiguity and uncertainty, which if prolonged, can cause stress.
- Work systems and processes should be designed and formulated to take account of any inherent stress factors and should include suitable safeguards and procedures either to eliminate or minimise them.
- Work overload and work underload can both be sources of frustration and stress. Finding the right balance for the individual is the best way to pre-empt such workplace problems, as well as optimising performance and improving productivity.
- To ensure that stress management becomes integral to corporate culture and company philosophy, serious consideration needs to be given to careful planning and training in order to raise awareness in identifying work-related stress, where it exists, whilst taking due account of the culture and nature of the organisation.
- Active listening should be seen as an essential managerial tool and part of effective people management. It should be within the skills portfolio of all managers and available to be used in the maintenance of a stress-free environment and the avoidance of any disruption or discontent within the workforce.

(cont'd)

- Counselling should be regarded as a necessary intervention to be included alongside other supportive services available to employees.
- Employee assistance programmes (EAP) offer employees access to a professional counselling service on a confidential basis.
- If primary interventions (those designed to remove the sources of stress) have been ineffective, for whatever reason, then there may ultimately be a need for a 'rehabilitation back to work' programme.
- Serious stress-related illness, and in particular 'burnout', are conditions from which it is difficult to make a complete recovery.
- The need for a strategy to manage the rehabilitation to work of employees with either long-term sickness absences or repeated short-term absences, is paramount. Supporting employees in their return to work by means of mentoring, coaching, counselling or mediation is imperative, as is a review of their job descriptions and working environment.
- Success with rehabilitation is a partnership between employer and employee, and both must want the partnership to work.

APPENDIX
Proactive and Reactive Management Interventions

10.62 The primary, proactive, stress management intervention, (that is also mandatory), is the carrying out of a risk assessment, in accordance with the *Management of Health and Safety at Work Regulations 1999 (SI 1999/ 3242).*

Following the initial risk assessment are the secondary, proactive interventions that are focused on eliminating and or reducing any stressors and their impact on employees.

The tertiary, reactive interventions, are those primarily of counselling and employee assistance programmes, used in the treatment and rehabilitation of affected employees.

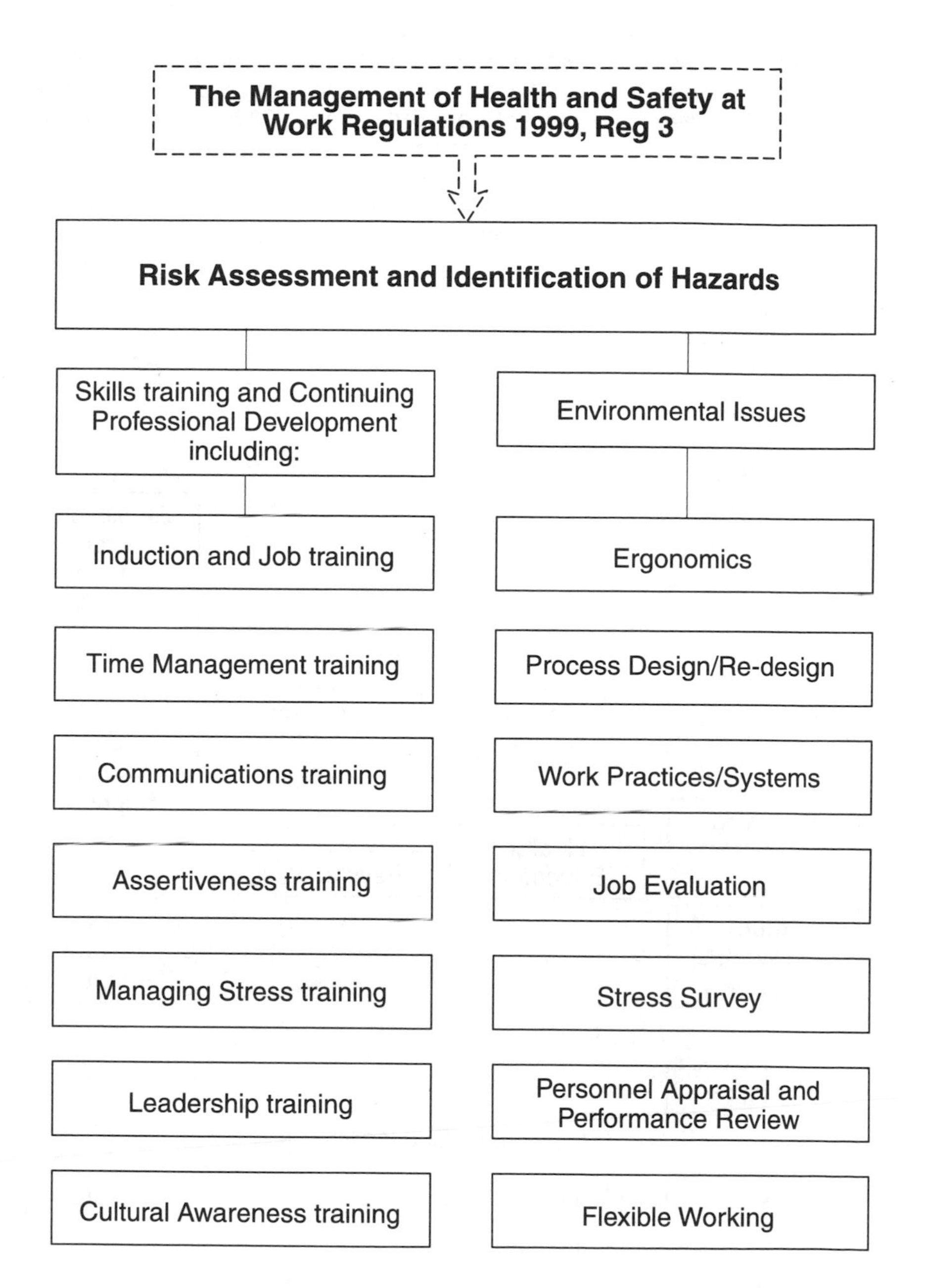

Figure 2: Proactive interventions for the prevention or reduction of work-related stress

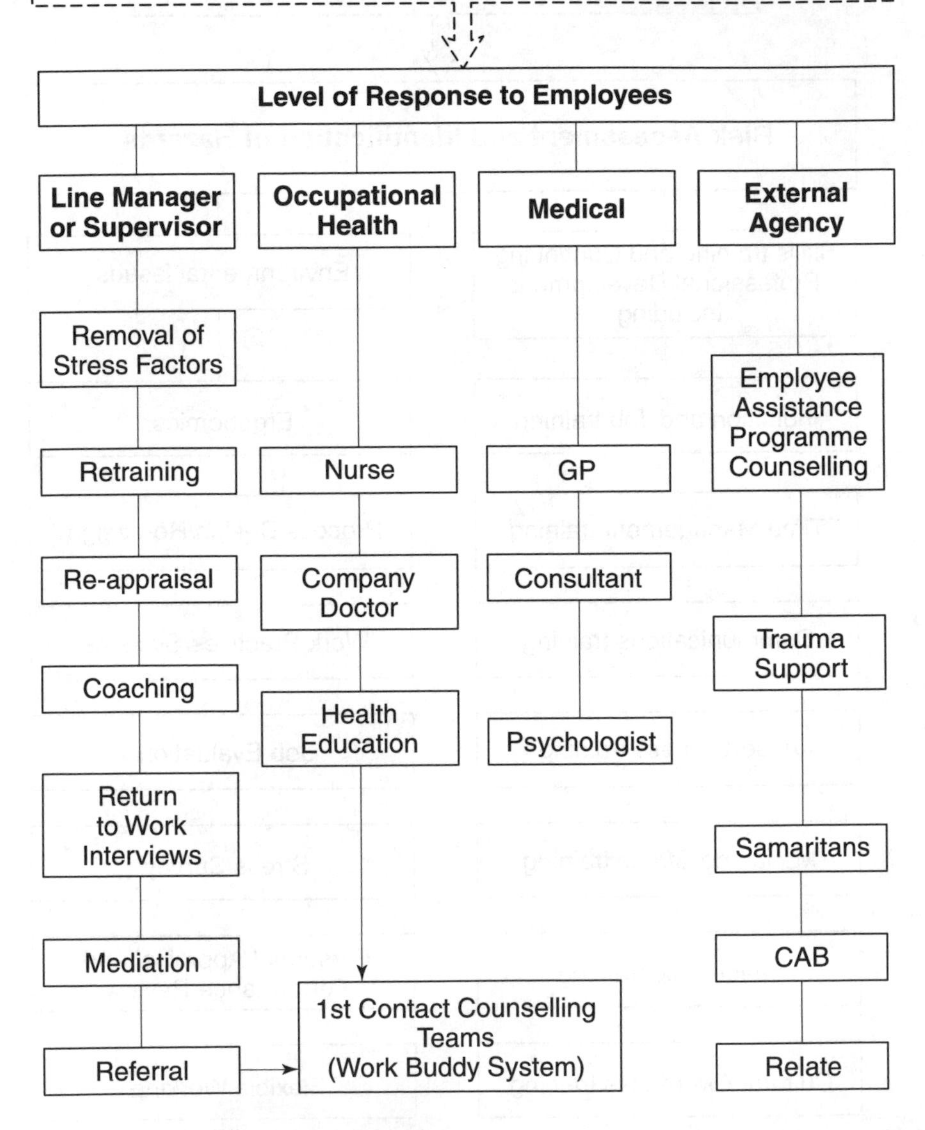

Figure 3: Reactive interventions for the minimisation or prevention of work-related stress

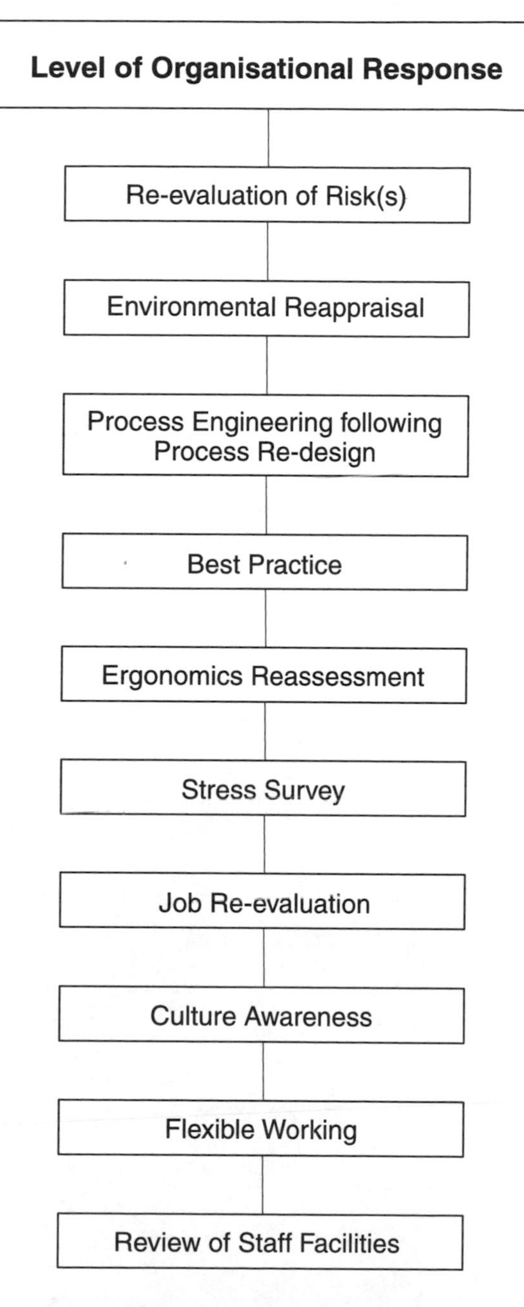

Figure 4: Reactive interventions for the minimisation or prevention of work-related stress (organisational level)

11 Personal Stress Management Strategies

Introduction

11.1

> 'I cannot teach anybody anything, I can only make them think'.
> *Socrates*

So far we have concentrated on the causes and effects of stress in relation to individuals and the organisations for which they work. In this chapter we shall concentrate on looking at personal stress management strategies, and what we, as individuals, can do to minimise the effects of stress in our day-to-day lives.

Considerable time and effort has been expended in investigating the causes and effects of the stresses of modern life, and the fact that, by definition, the vast majority of these are outside of our control. Although many of these 'stressors' may be beyond our capacity to evade or eliminate, we do have the ability to modify our reaction(s) to them. As the levels of stress to which we are exposed, continue to escalate, it is essential that we learn how to control and manage them better in order that we can lead enhanced and more productive lives.

There are a variety of actions that individuals can take to manage their stress levels and to prevent them from developing into more serious problems. In view of the fact that no single method works well for everyone, it is necessary that each individual finds an appropriate method which works best for them. In the main, the positive actions we can take, fall into three categories:

- Modification of behaviour.
- Development of a healthy lifestyle.
- Alteration of mindset.

Even something as simple as positive thinking (although it may not seem that simple at the time) will go a long way towards effective stress management.

All of us prefer to be among people who are lively, interesting and positive, rather than tired-out, negative and boring. How we are perceived by others is important for our own self-image, but when are stressed this can be easily forgotten.

There may be times when it is necessary to seek professional help, guidance or support. Where this is the case, it should not be seen as a sign of failure or weakness. On the contrary, it is a sign of strength to be able to recognise our body's warning signs and to take appropriate action. Only in that way can each of us ensure the continuation of good health, and retain control of our lives.

These points are just as relevant to our working environment as they are to our personal lives and managers should be aware that these considerations are as applicable to their teams as they are to themselves. It is important that they create an environment where stress-related issues can be acknowledged, discussed and properly addressed, at an early stage.

It is often found that those very individuals who are the most vehement in denying that they have a problem with stress are often the ones who are most in need of help, and it requires a trusting relationship to be established for these issues to be dealt with properly.

Once some success has been achieved in dealing with a particular aspect of stress, it is recommended that the method by which this was accomplished is remembered and the learning that comes from it. When future events in life overtax our coping resources, we will hopefully be able to then draw, with advantage, upon our previous experience.

Identifying sources of personal stress

11.2 In earlier chapters, we discussed the tools that can be used to identify sources of stress within the organisation, for example, risk assessment and stress mapping. With personal stress, it is just as important to adopt a similar approach and to identify the specific stressors that are the cause of the problem.

Keeping a stress diary

11.3 It follows that it may well be both beneficial and instructive to keep a diary for 2–3 weeks, to help identify the reasons for the feelings we experience that are related to excessive pressure. For many people, the act of keeping a diary may itself appear to be an additional chore. However, it can prove invaluable in helping to understand not only the primary sources of our stress, but also the frequency with which we experience the effects of them upon our daily life.

Suggested diary contents

11.4

- It is better to keep the diary entries as brief as possible, provided that all relevant details of the impact of the stress experienced are included, such as any other persons involved, events, activities, time and duration etc.

- Details should include everything that puts a strain on available resources of energy or time; that triggers anger or anxiety or which results in the physical, emotional or behavioural responses that were discussed in previous chapters.

- After a couple of weeks, the diary entries should be reviewed to identify the most important situational stressors that have caused affect, and also for determining the best way of dealing with them effectively in the future. To retain a sense of proportion, the positive events should be noted – those that are invigorating, pleasurable or which give a sense of achievement.

Figure 1: Typical 'stress diary' entries

Day/Time	Situation	Physical Response	Emotions
Monday 6 September, 10.30 am	Weekly meeting. Asked to comment on marketing proposal.	Heart started to beat faster, hands became clammy and my voice sounded high.	Really anxious. Apprehensive that my comments would be considered inadequate.
Wednesday 8 September, midday	Took client to lunch.	Felt hot and uncomfortable. Hands shaky – spilt the water.	Nervous and embarrassed. Couldn't find the right words in order to maintain the conversation.
Thursday 9 September, 09.00 am	Required to work in outer office which was freezing cold – (as mine was being re-painted).	Legs and feet cold after 10 minutes at desk entailing my getting up every half hour to walk about.	Became extremely irritable and bad-tempered. Shouted at Mary who came in to take a letter and who misheard what I had dictated.

Day/Time	Situation	Physical Response	Emotions
Friday 16 September, 4.15 pm	Suddenly told I had to give my report in tomorrow morning instead of Friday. (John's now going to Glasgow Thursday night!)	Tried not to show how I was feeling, but my shoulders tensed-up, and in the afternoon started to get a pounding headache. Found it really difficult to concentrate.	Annoyed and resentful at having to work on the report. Nervous because I didn't know if I could finish it on time.

[S Mauger (2003)]

When situations such as those described in Figure 1 above have been identified as stressful, it is important to ask the following questions:

- Were these events or activities for mine or someone else's benefit?
- Did I have any control over them or did they ostensibly control me?
- Was the handling of these situations really beyond the scope of my inherent or learned abilities?

Dependent on the answers to these questions, the next step is to try to alter the overall characteristics of the activities from being 'stress producing' to 'stress reducing'. Actually, removing or replacing people, events or activities that are causing stress is rarely practical or feasible. However, it is worth thinking about ways of reducing their negative impact and implementing positive interventions to bring about a better result and overall life-balance.

It has been suggested that the addition of more daily 'pleasurable events' has more positive effect on the immune system than reducing stressful or negative effects – implying that small daily improvements can help transform a negative, stressful existence into a more positive and productive one. Writing feelings down on paper can also be an effective way of 'unloading' frustration and taking the heat out of difficult situations.

In the following sections a wide range of proactive interventions will be examined in detail, starting with how we can try to change our usual behaviour and responses to stress.

Modification of behaviour

11.5 In terms of changed behaviour, proactive interventions that can be made, include:

- Being more assertive.
- Improving time management.
- Managing 'Type A' behaviour in others.
- Indulging in humour and laughter.
- Taking 'time out'.
- Developing better strategies for travelling and commuting.
- Utilising external help.

Being more assertive

11.6 Many people find it extremely difficult to say 'no.' They put themselves in invidious situations by accepting additional work or tasks when they have, in reality, insufficient time to complete them. The consequences are often disappointment in oneself and in others that invariably lead to both physical and psychological stress and a reinforcement of a poor self-image and reduced self-esteem.

People can exhibit four distinctively different types of behaviour: aggressive, indirectly aggressive, passive and assertive. For people who find it difficult to say 'no', assertiveness and time management training can both be extremely valuable.

Table 11.1

Being assertive is being able to:	*It is not about:*
Listen to others	Expressing anger
Verbalise what you feel and need	Aggression
Negotiate	Selfishness
Compromise	Being superior to others
Respect others	Always getting your own way
Enjoy self-confidence and control	Insisting on non-negotiability

However, being recognised as non-assertive, can allow others to 'walk all over you', because you effectively surrender control to them. By comparison, being assertive equates to standing up for your personal rights, and expressing your thoughts, feelings, wishes and beliefs directly, honestly and spontaneously in ways that are not detrimental to the rights of others.

Assertive people take responsibility for their actions and choices, and even in cases of failure, notwithstanding the obvious disappointment, their self-confidence and self-respect will remain intact. The expressing of negative feelings, at the appropriate time, also avoids the build-up of resentment, thereby helping assertive people to manage their stress more successfully.

It is important to note that it is not possible for a non-assertive person to change overnight. It takes time, practice and commitment. However, if people find it difficult to say 'no', then serious consideration should be given to attending a training course on assertiveness either trainer-led or by distance learning and there are also some excellent reference books available on the subject.

Individuals who frequently find themselves acting aggressively, ie shouting, being intimidating and generally making others frightened of their anger, might also consider taking assertiveness training. Aggressive behaviour not only alienates other people but, if persistent, can also seriously damage the person's health, by releasing too much adrenaline into the body and the consequent impact on blood pressure.

Table 11.2: Basic guidelines for assertive behaviour

- Acknowledge your own feelings, to yourself. If for example, you feel angry, it can be helpful to acknowledge that feeling, even if you decide not to express it.
- Be clear about how you feel and what you actually want.
- Be clear and direct in what you say. Misunderstandings often happen as a result of unclear messages.
- Adopt a sound inner dialogue. What are your real thoughts about the situation you are currently facing?
- If necessary, keep repeating your message. Often people are not really listening to what you are saying and may introduce 'red herrings' into the conversation. Therefore, try repeating your message in order to receive some acknowledgement of receipt.
- Use appropriate body language to back up your assertive ehaviour. Adopt an open, relaxed posture with the head erect, and face the other party square on. Establish eye contact, and keep your voice steady and firm.
- Keep calm and stick to the point. Relaxation exercises, such as deep breathing, may help.
- Respect the rights of the other person. In some situations compromise is a preferable outcome(as with negotiation).

Improving time management

11.7 How your time is managed is a key factor in determining the level of stress in your life. Many individuals complain bitterly that they are always short of time, but this can often be caused by:

- A lack of assertiveness.
- Being unable, or apprehensive, of delegating tasks to others.
- Having an excessive workload.
- Allowing time to be wasted, or not using time productively.
- Prioritising jobs and tasks incorrectly.

By comparison, good time management is about:

- Establishing priorities.
- Making a list of what *must* be done, *should* be done, and if possible, what the person would *like* to be able to do.
- Eliminating time-wasting activities.
- Getting into the habit of focusing on essentials.
- Learning to say 'no' and being able to delegate effectively.
- Scheduling daily (uninterrupted) time to organise daily activity.
- Not making excuses for *not* doing something.
- Making a list of achievable goals or targets.

One of the most difficult aspects of time management is assessing goals and whether or not they are achievable. For example, are the goals that you are trying to achieve, realistic in the time available and are they, in fact, what is really needed? This is sometimes difficult to ascertain and quantify, because individual goals can often be subsumed within the general needs of the organisation. This is why it is important to identify short-term, medium-term and long-term goals, and list tasks accordingly. It may even be that someone else is actually responsible for the problems an individual is experiencing with personal time management – in which case, they may need to work together in order to remedy the difficulty.

It is also important to set aside time for planning and relationship building. Bear in mind the 'time management matrix' (in Table 11.3 below) and reduce the time spent dealing with crises and deadlines (Quadrant 1) by employing more planning and relationship building (Quadrant 2). The time needed for this can be created by reducing unimportant activities (shown in Quadrants 3 and 4).

Do not forget that if a task is deemed important or urgent to someone else, this doesn't necessarily mean it is important to you (unless that someone is the managing director).

Table 11.3: The time management matrix

	Urgent	**Non-urgent**
Important	1. *Activities* Crises Pressing problems Deadline-driven projects	2. *Activities* Prevention PC activities Relationship building Recognising new opportunities Planning Recreation
Not Important	3. *Activities* Interruptions Some calls Some mail Some reports Some meetings Proximate Pressing matters Popular activities	4. *Activities* Trivia Busy work Some mail Some phone calls Time wasters Pleasant activities

[S R Covey *The seven habits of highly effective people* (1990)]

Asking for time

11.8 'Asking for time' is about considering yourself and valuing your time. You need to remember your right to say 'no', as there is little point in giving yourself the space to reflect on your priorities etc, if you are still unable to say 'no' at the end of it.

Before accepting any increase in workload, there may need to be discussion on any difficulties that might entail, in order to negotiate a solution that is satisfactory to both yourself and the person or organisation making the request or demand upon you.

Essentially, the best way forward is to ask for time to evaluate the request, or new instruction, in order to come to a decision that takes account of the effects of acceptance. Priorities should be evaluated together with the consequences of refusing (as highlighted in Table 11.4 below).

Table 11.4: Time guidelines

1.	Listen carefully to the details of the request or instruction.
2.	If necessary, ask for clarification to make absolutely sure of what is being asked of you.
3.	Acknowledge your understanding of the details of the request.
4.	Advise the other party that you will need time to consider and will let them have your decision as soon as possible.
5.	Specify the amount of time you will need, when you will notify the person concerned and by what means, ie verbally, by email or fax etc.

Managing 'Type A' behaviour

11.9 Many people like to blame any outburst or over-reaction they might exhibit, on their 'Type A' personalities – an issue that is best explained through a case study.

Case study

'It's the way I am and the way I think – I can't be someone else,' says Philip, the Managing Director of a UK manufacturing company. 'It's because I'm a perfectionist and proactive that I get impatient and irritable with others who are less motivated. There is also little doubt that I cannot abide fools. If other people could have the same mindset and conform to the same standards as I do, there wouldn't be any problems and I wouldn't get so frustrated and angry. To my mind, the problem is with everyone else – there is definitely nothing wrong with me.'

Philip is not helping himself, because he is battling with his own personality to achieve and maintain tight control. He cannot accept the differences in others – who, from his perspective, are either doing the job his way or are doing it wrong. He also generates unnecessary stress for himself and for those around him by excessive and continuous criticism, as a result of which, his own stress response is being continually triggered. As a consequence of this negative feedback, it will be likely that, one day, Philip will simply break down or 'burnout. This will occur as a consequence of his body having to cope with the continuous production of stress hormones that, in time, will impact and weaken his cardio-vascular system by damaging the arterial connectivity.

(cont'd)

In order for Philip to improve his life expectancy, he would be well advised to adopt some of the following fundamental techniques for better personal control.

Techniques for managing 'Type A' behaviour

11.10 Wherever possible, reduce the demands made upon you by assigning tasks to others. Delegate to people whom you trust, ensuring that they are given clear, concise instructions regarding the job required. Accept that delegation is the right approach, in this instance, and then move on with your own work. Checking at designated times can give a feeling of confidence to the other person, and will ensure that they feel supported, but not hounded.

Constructive feedback is useful, but not continual criticism. Remember that the achieved result (and time taken) will probably be absolutely acceptable notwithstanding that they are doing the job in *their* way, which may be different from your method of doing things.

- *Avoid perfectionism*

  Accept the fact that you haven't failed just because you may not have completed a job perfectly. Accept that you have done your best, and try to stop being obsessive about getting everything right, 100% of the time. Human beings are fallible.

- *Slow down – why race the clock?*

  Always allow an extra 15 minutes to ensure you have time for yourself and also to clarify your thoughts and actions.

- *Tackle one task at a time and enjoy it*

  Take your time as opposed to rushing through everything. After completing a task, sit back and reflect on what you have achieved. Has any learning come from it? Did you have problems meeting your deadline because you left some issues to the last minute? Utilise any learned experienced to draw upon the next time you meet a similar situation.

- *Learn to occupy your time whilst waiting*

  For example, the time spent whilst waiting in a queue can be used as time to think through tasks. But how many of us in reality ever stand still long enough to just think properly about the task ahead?

- *Avoid getting angry over things that you cannot influence*

  Accept that there are issues over which you have some control and those that you do not. When you have tried every possible way to achieve what you want, but without success, it may well be time to move on. It is pointless and damaging to become frustrated and angry about matters over which you have no control. Rational thinking must prevail over emotional reaction to retain good health of body and mind.

- *Accept (sometimes) the errors both of yourself and others*

  Everyone makes mistakes, but they do not necessarily have to be punished for them. Ask yourself, 'Has anyone actually suffered loss as a result of this error?' Learn to see, where possible, the amusing side when something goes wrong, and endeavour to learn from that lesson. It can often be instructive to listen to the details of someone else's mistake, and to think about what you would have done in similar circumstances.

- *Avoid immediate responses to incoming email*

  Never respond immediately to an email that has had the effect of provoking an emotional response, such as anger in you. It is better to take time to reflect rather than to reply immediately with the inevitable emotional response. It should always be remembered that once the 'send' key has been hit, no amount of wishful thinking will retrieve your message written in haste.

- *Enjoy playing without necessarily winning*

  Being competitive is fine, but not at the expense of enjoying yourself. The technique is to do your best, and by so doing, you will maximise both the enjoyment and the activity.

- *Smile and give affection*

  Many managers, particularly in the UK, walk into the office and rarely take the time to smile or say 'Good morning'. An appropriate gesture or greeting costs nothing, and can make both colleagues and subordinates, feel valued.

- *Praise*

  Being able to give praise and positive feedback, and not only criticism, will enhance any relationship.

- *Practice being a good listener*

  Learn not to finish sentences for others, as apart from being discourteous you cannot necessarily know what they are going to say. One who always interrupts and never waits for a response will simply be regarded as a poor listener. Consequently, others will be loath to share their thoughts with you, as there would be little point. Learn to pose basic, open questions – 'what, why, when, who, how, where', in order to elicit a meaningful and interesting reply. Listen to what is being said and follow the subject through. Keep an open mind and you will gain far more from the conversation!

- *Learn to relax*

  It is essential to schedule relaxation time for yourself. Set yourself realistic relaxation goals and keep to them.

- *Learn to turn stressful life events into challenges for personal growth*

  Stand back and reflect. Commit to paper the learning gained from the experience so that the knowledge obtained can be turned to future advantage.

Indulging in humour and laughter

11.11 Humour gives us a different perspective on our problems. If we can make light of a situation, then it becomes less threatening. Humour is a wonderful stress-reducer and antidote to bad temper and tension. It is known that laughing relaxes tense muscles, speeds more oxygen into the system and lowers blood pressure.

It is also believed that laughter reduces the levels of certain stress hormones that tend to weaken the immune system. Laughter potentiates the ability of defensive cells to destroy tumours and viruses, as does gamma-interferon (a disease-fighting protein), T-cells (which are a major part of the immune response) and B-cells (which make disease-destroying antibodies).

In 1979, the American, Norman Cousins, then editor of the *Saturday Review*, published a book, *Anatomy on an illness as perceived by the patient: Reflections on healing and regeneration*. The book describes his experience of the effects of humour on his severe illness, ankylosing spondylitis, that he contracted in 1964. Deciding that the hospital regime of strong medication, unimaginative food and the institutional system was too depressing for any real benefits to be gained, he developed his own recovery programme comprising a mix of vitamin C and regular doses of laughter stimulated by re-runs of Marx Brothers films and 'Candid Camera'. The treatment appeared to relieve his pain considerably, and indeed, when his levels of inflammation were tested, they were found to have dramatically reduced.

According to members of the Association for Applied and Therapeutic Humor, the psychological benefits of humour are quite amazing. People often store negative emotions such as anger, sadness and fear, rather than expressing them. Laughter provides a way for these emotions to be harmlessly released, and laughing with someone can be an equalising experience that can defuse many a tense or awkward situation. (See the Association's website at: www.aath.org; see also: www.howstuffworks.com)

Overall, the advice is to try to surround yourself with people who are able to make you laugh and to feel good about yourself.

Taking 'time out'

11.12 Physical activities such as walking, stretching or even working in the garden, can be effective 'stress-busters'. This is why it is so important to take regular breaks away from situations or tasks that are sources of stress and frustration. Although this will not solve the root cause of stress, it gives an opportunity to think about the situation more objectively and may even help to resolve it by looking at it from a different perspective.

Developing better strategies for travelling and commuting

11.13 Travel and commuting can be major stressors in modern living. The following techniques may help to ameliorate their effects:

- Always leave an extra 15 minutes for contingencies. If transport schedules are unreliable, extra time in hand will be needed. However, if everything goes smoothly and arrival is earlier, there will be more time for thinking or for relaxation before the appointment.
- Do not make back-to-back appointments, as meetings will frequently run over time. Also remember the need for 'you' time between meetings.
- Check the route before leaving, especially if travelling to an unfamiliar area, and if possible talk to others who make the journey on a regular basis.
- Make sure the car has sufficient fuel the day before the journey, and if necessary check tyres, oil etc, rather than leaving this until just before leaving.
- Never drive immediately after an emotional upset such as receiving notice of a bereavement or being involved in an argument, as the mindset and feelings might increase the risk of an accident.
- Have small change available in the car for parking meters etc, and carry a copy of the insurance certificate in the glove compartment, together with details of an emergency motoring organisation, in case you are involved in an accident or stopped by the police.
- Before starting out, adopt a comfortable seating position and adjust the seat and driving mirrors
- On long journeys, remember to take short breaks to help avoid becoming over-tired. Keep arms slightly bent and in a '10 to 2' position on the steering wheel for optimum control and to avoid fatigue.
- Whilst stationary or sitting in heavy traffic, try to perform simple relaxation techniques to help reduce tension in the body, especially vulnerable areas such as the neck, shoulders and arms.
- Have tapes or CDs in the car which you enjoy listening to so that time can be used effectively. View the car as a personal time and space capsule away from everyone else – an environment which provides your own selection of music and radio channels.

Remember:

- Travelling, by its very nature, takes place in a spatial context that is outside your control.
- It can be advantageous, following a near accident, to re-examine the incident in your mind to ascertain what action could have been taken to avoid or pre-empt its occurrence.
- Be tolerant of others. Shouting at other drivers following what has clearly been their mistake will not change anything, but your anger will affect your judgement for some time afterwards. Remember that the other driver does not know you – bad driving by other road users is not a personal attack. The personalities of some people change when they get behind a wheel, and a calm 'Type B' personality can become a supercharged 'Type A' rally driver as soon as they sit in their car.
- Continual clock watching will not help you get to your destination any quicker, but will increase stress levels and you will arrive more stressed than when you set out, and certainly not in a fit state to conduct yourself effectively at a business meeting.
- Accept that drivers (including you) make honest mistakes and have occasional lapses in concentration.
- Be courteous and thank others for their courtesy. (How many times do we become aggrieved because we let another driver out into a stream of traffic and did not even get a 'thank you' wave from them?)

Utilising external help

11.14 There are occasions when external help can be beneficial for many individuals exhibiting signs of stress. The following are just some of the many sources of external help available.

Counselling

11.15 As the proverb says, 'a problem shared is a problem halved'. However, there are occasions when friends or family, however supportive, may not be the people who an individual feels he or she can turn to. People are often embarrassed about discussing their worries particularly if these are intimate, or may be concerned that they will be thought of as weak or unable to cope. They may also be worried about confidentiality, not wanting the entire world to know about their problems.

In the workplace, individuals may also have a need for someone with whom to discuss problems, but they can be wary about talking to just anyone, in case they are perceived to be weak or unable to cope. They therefore may turn to professional counsellors for confidential help.

Cognitive behavioural therapy

11.16 Cognitive-behavioural therapy ('CBT') combines cognitive techniques (those related to what we think) with behavioural techniques (those related to what we do). CBT is a scientific approach that focuses on what is happening now, and its main purpose is to help solve particular problems by identifying specific goals, together with the means by which these goals can be achieved.

As a result, CBT can be extremely effective in the reduction of personal stress, by helping individuals to identify the source(s), to change their automatic response(s), to revisit their priorities and to identify and adopt the most effective method(s) that will enable them to manage the problem.

Employee assistance programmes

11.17 An employee assistance programme (EAP) offers employees access to a confidential, professional, counselling service that can give support to both employees and often their families. It can also provide information and advice on a wide variety of practical matters (see CHAPTER 10).

Strengthening or establishing a support network

11.18 It is recognised that a majority of those individuals who appear to remain happy, healthy and are able to withstand multiple life stresses, have the advantage of possessing a good network of social support, usually comprised of family and close friends. In addition, professional support from outside sources such as Samaritans, can also be extremely helpful, as there are times when individuals might rather speak to a stranger than a friend. Confiding in a stranger who knows absolutely nothing about you and has no expectations of you, enables feelings to be expressed within a safe framework. When the person with the problem wants to move on and not discuss it any further, they may not wish to be reminded of the time when they were experiencing difficulties.

Whichever means are chosen, it is important to be able to express feelings with someone in whom there is trust. However, this does not necessarily have to involve an actual person. Venting feelings by writing a diary, or perhaps a letter that may never be posted, may have the desired effect of releasing and/ or expanding thoughts and feelings.

Development of a healthy lifestyle

11.19 Proactive lifestyles that can be adopted to help maintain the health of mind and body include:

- A healthy diet.
- Regular exercise.
- Relaxation.
- Uninterrupted and sufficient sleep.
- Increasing leisure time and taking up hobbies.

A healthy diet

11.20 Our bodies are remarkably efficient at extracting the nutrients we need, but at times of stress these needs increase and a well balanced diet is, therefore, essential in preserving health.

- *Alcohol*

  When taken in moderation, alcohol has been shown to benefit the cardiovascular system. Many people, however, take alcohol to combat stress, but in fact, by doing so, make matters worse. Alcohol is a depressant, and excess alcohol increases fat deposits in the body and decreases the immune function. It also reduces the ability of the liver, over a period of time, to remove toxins from the body. When experiencing stress, the body produces several toxins which, in the absence of efficient filtering, remain in the body as potential causes of damage (see CHAPTER 7).

  Consuming large amounts of alcohol neither solves problems nor improves health. In addition, it is not a good idea to concentrate alcohol intake into Friday and Saturday nights! It is important to remember the recommended safe levels of alcohol *per week* are 28 units for males and 21 units for females (a unit is equal to just one normal size glass of wine).

- *Caffeine*

  Caffeine intake should be limited, as this can 'kick start' the stress reaction. Caffeine is found in coffee, tea, chocolate, cola drinks and some headache remedies. When taken in moderation, caffeine can increase alertness. However, consuming an excess of caffeine, whether in coffee or cola, can be addictive and lead to irritability, sleeplessness and impatience. Caffeine also acts as a diuretic, an excess of which can lead to dehydration. It is recommended to reduce coffee and caffeine consumption slowly over a period of time, as stopping abruptly can result in withdrawal symptoms.

- *Carbohydrates*

  Complex carbohydrates such as rice, pasta, potatoes and bread result in a slow release of energy, which is important in maintaining a constant blood sugar level, particularly so for diabetics – as opposed to the 'quick fix' provided by sugar. Carbohydrates also trigger the release of the powerful neurotransmitter, serotonin, that has an important role in the maintenance of mood control.

- *Fat*

  Avoid the consumption of foods rich in animal (saturated) fats that cause obesity and increase cholesterol levels. Both these factors adversely impact the cardiovascular system and are a contributing factor in coronary heart disease (CHD).

- *Fibre*

  Fruits, vegetables and grains are excellent sources of fibre. These bulking agents are important for a healthy digestive system and it is recommended that a normal diet should include at least 25 grams of fibre per day.

- *Sugar*

  Sugar has no essential nutrients and it provides a short-term boost of energy to the body, resulting in excess fat deposits. High sugar consumption over time can lead to overweight, glucose intolerance and, eventually, type 2 diabetes.

- *Salt*

  The consumption of 'convenience' foods should be kept to a minimum, as most contain either large amounts of sugar, salt, fat and/ or preservatives. Salt increases blood pressure, affects the adrenal gland and causes dehydration, thirst and possible emotional instability. Sufferers from stress should use a low salt substitute that has potassium rather than sodium, and avoid junk foods high in salt such as ham, pickles, sausages, burgers etc.

- *Vegetables*

  Eat a balanced diet. In periods of high pressure, eat food that is high in Vitamin B (wholemeal bread, whole grains, pasta and jacket potatoes) and vitamin C (fresh fruit and vegetables). Green, yellow and orange vegetables are all rich in minerals, vitamins and phytochemicals that boost immune response and protect against disease.

- *Water*

  Water is essential to maintain life and for our bodies to operate efficiently. This includes temperature regulation, nerve impulse conduction, circulation, metabolism, the immune system, eliminative processes, sensory awareness and perceptive thinking.

  Many people drink insufficient water. One glassful a day is *not* enough as the many chemical processes inside the body require more than this for optimal completion of reactions.

  It only takes a 1% fluid loss for the body to become dehydrated and an insufficiency of water can seriously disrupt the body's biochemistry. This generally happens without any conscious sensation of being thirsty.

Stress, alcohol and caffeine can all influence the amount of water available to the body's systems and the speed with which the body loses it. Any of these factors, alone or in combination, may cause small but critical changes in the brain, that can impair neuromuscular coordination, decrease concentration and slow down the thought processes.

The average amount of water loss per day is equivalent to two cups through breathing, two cups through invisible perspiration, and six cups through urination and bowel movements. This equates to a total of ten lost cups per day that need replacing – without taking into account perspiration from exercise or hard work, excessively dry air, alcohol or caffeine consumption.

Furthermore, travelling by air can entail a loss of as much as one litre of water during a three to four hour flight! It is hardly surprising, therefore, that the daily recommended intake of water is 1.5 to 2 litres.

Regular exercise

11.21 Exercise not only improves general fitness and increases overall strength, stamina and suppleness, but also has many other additional benefits. Self-esteem can be raised and sleep patterns can improve – meaning that exercise is usually good for both the body and the mind.

Exercise improves cardiovascular function by strengthening the heart, causing greater elasticity of the blood vessels, increasing oxygen throughout the body, and lowering the blood levels of harmful fats such as cholesterol and triglycerides.

Exercise provides a physical outlet for negative emotions such as frustration, anger and irritability, thereby promoting a more positive mood and outlook. Exercise improves mood by producing positive biochemical changes in the body and brain, as well as reducing the amount of 'fight or flight' hormones the body releases in response to stress. The body also releases greater amounts of endorphins during exercise – the powerful, pain-relieving, mood-elevating chemicals in the brain, that are often lacking in people who are depressed.

Exercise is also an excellent distraction from stressful events and circumstances, and it is thought that stress itself poses significantly less danger to the overall health of people who are physically fit. This is because their heart and circulation are able to work harder for longer periods, and being physically stronger, they are less susceptible to musculoskeletal injury.

Exercise will therefore keep the body functioning properly; helps feelings to be relaxed and refreshed; promotes deep, restful sleep; and is a good stress management technique, because it:

- Reduces muscle tension, and uses up the adrenaline and energy released by the 'fight or flight' response.
- Makes the body stronger and better able to cope with the debilitating effects of stress.
- Increases energy and stamina.
- Maintains self-image, appearance, and tends to control weight.
- Helps to clear the mind of worrying thoughts.

Feeling fit increases the overall feeling of well-being and a commitment to exercise will increase feelings of control and self-respect.

If a person is considering regular exercise, the following points are important to bear in mind:

- Physical exercise is an excellent way of getting the necessary relief and increasing the coping resources – but more time does need to be made for it.
- The choice of exercise is personal. It is best to find something that suits the person and fits in easily with their daily life.
- Begin with an exercise that is enjoyable. Find a regime that is interesting, challenging and satisfying, and one that preferably also brings contact with other people.
- 20 games of squash or 15 aerobic workouts are unnecessary and can be dangerous!

It takes as little as three 20-minute sessions per week to increase physical fitness and it will also stimulate mental acuity and help to combat the adverse effects of stress. It is, however, recommended that if a person is not used to exercising and is over 35 years old, they should visit their doctor prior to embarking on any exercise programme, in order to check that it is suitable for them.

The value of exercise cannot be over-emphasised. Problems appear less important when walking, swimming, running, cycling, or being involved in any other physical pursuit. This is because the mind is better able to maintain a proper perspective over events and situations when tension is released. Any activity that concentrates the attention on an interesting and enjoyable subject that is divorced from life's inherent problems, will be beneficial in renewing inner strengths, natural resistance and coping resources.

Relaxation

11.22 Regular periods of relaxation in between or away from work schedules are extremely important as they help to:

- Switch off the stress response.
- Improve sleep patterns.
- Reduce fatigue.
- Ameliorate pain.
- Increase self-esteem.
- Assist the body to heal and repair.

Relaxation gives a chance to 'recharge your batteries'. In medical terms, during periods of relaxation the sympathetic nervous system activity is at its lowest, allowing the parasympathetic nervous system to increase its influence over body functions. Accordingly, levels of noradrenaline, adrenaline and cortisol are low, and physical functions such as heart rate and breathing also decrease.

When stressed, the muscles in our bodies tense and this muscular tension can cause headache, neck and shoulder discomfort, backache, etc. These aches and pains can in turn increase tension leading to a vicious circle of stress/ tension and worry.

Tension and relaxation being two sides of the same coin, mean that we cannot experience both at the same time. We therefore need to learn to relax in order to switch off the effects of tension.

Relaxation techniques

11.23 Some people relax by doing something they have already discovered to be enjoyable, for example:

- Listening to music.
- Reading.
- Having a bath – perhaps with the addition of essential oils.
- Watching a favourite film.

There are also a range of more specific relaxation techniques that individuals can learn and use, some of the most popular of which include the following.

- *Progressive and deep muscular relaxation*

  This aims to reduce anxiety by emphasising physical relaxation. It consists of first tensing and then releasing all 16 major skeletal muscle groups in sequence. At each stage, the mind concentrates initially on the feeling of tension and then on relaxation. It should be noted that these techniques are not recommended for sufferers of hypertension (high blood pressure) since tensing of the muscles can elevate blood pressure. Instructional tapes and CDs are available, and although the technique requires practice to become proficient, it enables the body to relax in stressful situations.

- *Meditation*

  Meditation is based on the belief that we all have the capacity to interact with our inner self, and that meditation can be the key to gaining access to this centre of stillness. In the context of stress management, meditation therefore concentrates on relaxing the mind. There are a variety of methods, but the simplest one involves focusing for 20 minutes or more on a single word or sound, that is repeated over and over again in the mind, as a mantra.

  Transcendental meditation ('TM') has been widely analysed in terms of its relief of stress, and has been reported to help with the reduction of tension. Approved TM instructors recommend two daily 20-minute sessions in a quiet, comfortable place, silently repeating the 'mantra' given by the instructor. The aim is to enter a (temporary) plane of consciousness that is 'above' our normal state of mind, where we will tend to feel peaceful, content and experience a stillness of thought.

- *Autogenic training*

  Autogenics was developed as a deep relaxation technique more than 50 years ago, based on the principles of eastern meditation. It is a technique for obtaining voluntary control over the involuntary nervous system that can, in theory, modify an individual's reaction to stress. The relaxed state achieved by autogenic training is described as a state of 'passive concentration'. Autogenics is taught as a series of easy mental exercises designed to switch off the stress reaction, and once the technique has been learnt, it then becomes part of the individual's lifestyle.

- *Self-hypnosis*

  Hypnosis involves entering into an altered state of consciousness in which all concentration is focused on a single objective or image, with all other stimuli blocked out. Doctors who use hypnosis as a therapeutic tool can teach their patients self-hypnosis, so that at a time of stress they can immediately produce in themselves a relaxed and altered state, free from anxiety.

  To succeed with self-hypnosis, the individual needs to possess a mind that is open to suggestion, a quiet time and place and an effective set of hypnotic suggestions. The aim is to feel free and receptive to suggestions from within. Many people believe that they cannot be hypnotised, but anyone who can lose themselves in a engrossing book or enthralling film, or become so absorbed in a task that they are oblivious to their surroundings, is actually practicing a form of self-hypnosis. Once learned, self-hypnosis can be used to relieve tension, anxiety and feelings of stress.

- *Imagery*

  We may be able to imagine sitting on a beach, listening to the crashing waves and watching the sunset. We can sit there and forget all our problems, and in effect take ourselves off to another world – a perfect way of relieving stress from our bodies. It is, in effect, a type of auto-suggestion that allows us to indulge our senses and let the pleasure bring relief from our problems. This is a scene that we can bring to the forefront of our minds at any time – even when sitting behind a desk.

- *'You' time*

  It is important for all of us to ensure that we make at least a modicum of time for ourselves each day. With 16 waking hours in a day, it should be possible to reserve 20 minutes solely for ourselves! When you look in your diary, it is unlikely you will find your own name appearing in it because it will be full of everyone else's – so allocate time to yourself and make sure you keep your appointment!

- *Having a warm bath*

  Water seems to have special powers in minimising stress and rejuvenating our bodies. It has a beneficial effect on relaxing the skin and muscles, and calms the lungs, heart, stomach and endocrine system by stimulating nerve reflexes. Heat generally quiets and soothes the body, slowing down the activity of internal organs. Cold stimulates and invigorates, increasing internal activity. If you are experiencing tense muscles and anxiety from stress, a hot shower or bath will very often help. Alternatively, if feeling tired and stressed out, try a warm shower or bath followed by a short, invigorating cold shower to help stimulate the body and mind. Experiment with different water temperatures and times, to determine that which is best for you and your body.

The following are simple instructions for a technique that is especially beneficial for dealing with long-term stress and stressful situations; and can also help with panic attacks, hyperventilation, breathlessness, dizziness, headaches and tension.

As the first step in learning the technique, you will need to set aside at least 10 minutes twice a day to practice 'paced breathing'.

Basic paced breathing

11.24

- Sit or lie down in a comfortable position away from the distractions of everyday life.

- Support the head with soft cushions so that the weight of the head is taken off the neck.
- Start to breathe regularly and slowly as if you were going to sleep, moving your stomach in and out (only your stomach should be moving, not your shoulders).
- It may be helpful to place a hand on your stomach to feel it moving as you breathe. Sometimes it can also be useful to practise in front of a mirror first.
- Place both hands on your diaphragm with your fingertips just touching. As you breathe you should be able to see your fingertips parting slightly as your diaphragm expands.
- In order to pace your breathing, it is advisable to practice this by breathing in to the count of 3 and then breathe out to the count of 3 (this should take 6 seconds).
- Continue this paced breathing for 2 minutes. Some people may find that initially this pattern of breathing may make them dizzy. If so, continue with the previous steps for a bit longer before introducing paced breathing.
- Gradually, you will be able to keep this paced breathing up for longer than 2 minutes. When you can keep this breathing rhythm for longer than 5 minutes you can begin to introduce some progressive muscle relaxation techniques.

Progressive muscle relaxation techniques

11.25 Some people like to tape this sequence and play it back to help them follow the instructions.

- Curl the toes hard and press feet down. Tense up on the 'in' breath for 3 seconds and then relax the muscles over 3 seconds on the 'out' breath.
- Press heels down and bend feet up – tense on the 'in' breath for 3 seconds and then relax over 3 seconds on the 'out' breath.
- Now tense up the calf muscles over 3 seconds on the 'in' breath and relax them over 3 seconds on the 'out' breath.
- Tense the thigh muscles over 3 seconds on the 'in' breath and relax them over 3 seconds on the 'out' breath.
- Now clench the buttocks tight on an in' breath for 3 seconds. Unclench them on the 'out' breath over 3 seconds.
- Now tense the stomach as if to receive a punch during an 'in' breath for 3 seconds. Relax during an 'out' breath over 3 seconds.

- Bend the elbows and tense up the muscles in the arms – again during an 'in' breath over 3 seconds. Relax on an 'out' breath for 3 seconds.
- Now hunch the shoulders and press the head back into the cushions. Tense up on the 'in' breath for 3 seconds and relax on the 'out' breath for 3 seconds.
- Now clench the jaws, frown and screw the eyes up really tight – tense on the 'in' breath and relax on the 'out' breath for 3 seconds each.
- Now tense all the muscles in the body together – tense on the 'in' breath for 3 seconds and relax on the 'out' breath for 3 seconds.
- Keeping the eyes shut, breathe deeply and be aware of the feeling of physical wellbeing and heaviness spreading through the body. Let yourself be aware of the warm sensation in your limbs and feelings of heaviness, peace and relaxation.

Everyday quick relaxation techniques

11.26 Even if you do not have time for the techniques described above, there are a number of techniques can be used to help the body to stay relaxed.

- When you feel the urge to *stretch*, doing so will help to release tension. Trust your urge to stretch and give yourself some quick relaxation.
- Try not to suppress the desire to *yawn*. A good yawn will stretch and relax your face, neck and shoulder muscles. It is also nature's way of telling you your body is tired, and will help to give you more oxygen with which to re-energise your body.
- When your legs and feet feel tired after a long day, you naturally want to rub your feet. Follow this urge and massage your feet. Rotate your feet and ankles slowly, as this will help to relax your feet naturally if a massage is not available. Alternatively, place your feet under warm running water, or in a bowl of warm water with smooth pebbles to massage the reflex points on your soles.
- If you feel anger rising in your chest or tears beginning, *breathe* in to the count of four and then breathe out to the count of four or even more. Be sure to breathe from the diaphragm (as previously mentioned). Many people take a bit longer to breathe out than to breathe in, and this is even more relaxing. Repeat this six times or more and you will be more in control of your emotions, more relaxed and better able to deal with the situation.
- Anywhere, and at any time of the day, stop and take one *deep breath*. This will help to revitalise your body and strengthen a good habit of breathing properly.

- An excellent de-tenser and refresher is a 15–30 minute brisk walk in the open air. As your breathing deepens so more oxygen can be supplied, increasing your body's ability to work properly for you.

- Any *change* of activity is helpful when you are feeling stressed or emotional. Get up and get a glass of water, cup of tea, coffee or a soft drink (remembering to keep your caffeine levels down), or go and talk to someone.

Uninterrupted and sufficient sleep

11.27 We have already looked at the role of sleep in helping the body to repair itself both physically and mentally, so not surprisingly it is widely acknowledged that sleep deprivation can be a major cause of stress.

The following will help you obtain benefit from 'good sleep' that you need to function at maximum levels:

- Try not to just lie in bed fretting when you cannot sleep. Get out of bed and perhaps make a warm drink, or do something you enjoy and that relaxes you.

- If you have a tendency to wake up with your 'to do' list in your mind, try getting up and writing about what may be bothering you. If your 'to do' list continues to go round and round in your head, try writing lists *before* you leave work or going to bed.

- If you wake in the night, don't keep looking at the clock. Watching the time go by will only increase your anxiety and postpone sleep for even longer.

- Ensure caffeine consumption is kept to a minimum. More than five cups of coffee a day, or ten cups of tea, can increase the pulse rate and disturb sleep patterns.

- Prior to going to bed, try to take your mind off the problems of the day. Read a book, listen to some soothing music, watch an amusing film, etc.

- Poor eating habits can cause poor sleep patterns. Eating heavy meals late at night or going to bed hungry are not be to encouraged. There should be at least a two-hour space between finishing a meal and going to sleep.

- Small amounts of alcohol can be relaxing. Alcohol can however over-stimulate some individuals to such a degree that they find it difficult to sleep unless they consume large amounts. This is likely to lead to waking in the small hours as the alcohol leaves the system, so that withdrawal plus dehydration will wake the sleeper.

- A warm bath prior to going to bed can be very relaxing – perhaps with the addition of essential oils such as lavender or camomile. Valerian root oil is particularly effective as it is very 'grounding' for an overactive mind.

Case study

11.28

Insomnia

Zak sought help from Lisa (an aromatherapist) for his insomnia, that had been a problem for him over the previous nine months. He had attended various sleep clinics and had been prescribed strong sedatives, which he did not like to take as they tended to make him feel very drowsy the next day. As a first step, Lisa took a detailed case history, and it became apparent that Zak was not sleeping properly owing to work-related stress. This caused him excessive worry, leading to an inability to switch off at night when trying to sleep.

Over time the stress had led to a weakening of Zak's immune system, making him very prone to coughs and colds. This in turn meant that he had to take time off work – leading to even more anxiety and sleeplessness.

Zak was treated by Lisa with a blend of valerian, camomile, lavender and vetiver oils, which were applied during a relaxing body massage. He was also given the oils to take home and use every night for a week. The following week, Zak reported that he had slept very well for three nights but slept intermittently for the remainder. This was a definite improvement over his experience during the previous nine months. Zak subsequently followed the same treatment plan for a further three months, with great benefit and regained a normal sleeping pattern that allowed him to feel completely refreshed upon waking.

Increasing leisure time and taking up hobbies

11.29 A very important way to relax is to take time out to enjoy hobbies or increase leisure time. It does not matter what is done – it is the making time away from work or family stresses that is important.

Some hobbies can give a wonderful sense of achievement that is not available in other areas of life, for example, learning to play an instrument, climbing, flying an aeroplane. These activities allow the mind to focus on achieving an enjoyable goal and ceasing to dwell on everyday stress.

It is common nowadays, however, to hear someone say, 'I really enjoy singing/ dancing/ rowing etc, but I do not seem to have the time to do it anymore.' To have a well-balanced work and home life though, it is important to make time for activities that you enjoy. Some people feel incredibly guilty in making time for their hobby or themselves. This is particularly true in working parents, who feel that every spare moment should be spent with their children. So why not get the whole family involved in a hobby or leisure activity?

By focusing on something completely different, it is entirely possible to feel more energised. Many people have even turned their hobbies into successful careers – proving the saying that if you enjoy something you are probably very good at it!

Alteration of mindset

11.30 The way in which we *perceive* situations is as important as to how we *respond* to them. How we think is therefore the third major area in which we can make proactive interventions, including:

- Modifying our perspectives.
- Positive thinking.
- Self-talk.

Modifying our perspective

11.31 It is very helpful to be able to reflect and modify our thinking so that we can improve the way we perceive events and relate to others. There are many ways of interpreting a situation – one of the best known of which is 'is the glass half empty or half full?'. Some people will always see the negative aspect and others the positive. It can therefore be useful to ask, 'is there another way of seeing this situation?'. As always it is important to learn from each event – what was helpful and unhelpful about the situation and how we viewed it.

Positive thinking

11.32 The power of positive thinking can help us resist feelings of hopelessness, desperation and failure. That is why it is important to focus on our strengths and to look for, and seize, opportunities that could result from a particular situation or set of circumstances.

It is natural to think about what the future may hold, but it is important to retain a proper perspective. It is pointless to worry excessively about future events – 'what might happen'. The future has not taken place yet and there will inevitably be large parts of it over which we have no control. In addition, worrying about a possible negative outcome will simply increase anxiety and tension, whereas focusing on a positive outcome will reduce this tension and so help us to achieve our goals. A practical way of doing this is as follows:

- Write down the worst scenario possible.
- Consider (truthfully!) the likelihood of it occurring.
- Imagine the best scenario possible.
- Put together a plan that maximises the likelihood of the best scenario and minimises the likelihood of the worst.
- Try to remember similar situations in the past which at first seemed just as threatening, but ultimately turned out well.

The learning curve is to try to recognise and accept when a situation is beyond our control, focus on what we can do positively and keep on practising.

Self-talk

11.33 Much of the distress that individuals experience is caused by negative thoughts, their expectations of themselves and other people, and what they think others are expecting of them.

Many individuals cause themselves undue stress by inappropriate 'self talk'. It is often very constructive, therefore, for people to think about what they tell themselves, and the internal language they use. Even simple phrases such as, 'I shouldn't be saying this, but . . . ' or 'I know I haven't got the right to say this, but . . . ' reveal a lot about how people perceive themselves in relation to those around them, their self-esteem, and aspects of their lives that may therefore be causing them stress.

Alternative therapies

11.34 In recent years, a wide range of 'alternative' therapies have been offered for the treatment of stress. Some are more effective than others depending on the method and mode of treatment and, of course, the individual. Examples of these therapies include:

- Acupuncture.
- Alexander technique.
- Aromatherapy.
- Chiropractic.
- Homeopathy.

- Massage.
- Osteopathy.
- Physiotherapy.
- Pilates.
- Reflexology.
- Shiatsu.
- Yoga and Tai Chi.

Acupuncture

11.35 Acupuncture literally means 'needle piercing' – the practice of inserting very fine needles into the skin to stimulate specific anatomical points in the body (called acupoints) for therapeutic purposes. Along with mild surface puncturing of the skin, practitioners use heat, pressure, friction, suction or impulses of electromagnetic energy to stimulate these points and so balance the movement of energy in the body to restore health.

To really understand how acupuncture works, it is necessary to become familiar with the basics of Chinese philosophy, which are fundamental to traditional Chinese acupuncture and to its specific role in helping to maintain good health and a person's well-being.

Alexander technique

11.36 This aims to treat and prevent a range of disorders, and so improve health and well-being through a system of postural change involving a new approach to coordination and movement. There are an ever-expanding group of teachers of the technique worldwide, and many doctors and specialists refer patients for lessons.

Aromatherapy

11.37 Aromatherapy infers treatment with essential oils, which are massaged into the skin, inhaled, added to bath water, used in compresses for affected areas, or vaporised into a room's atmosphere. Some essential oils have anti-viral, antibacterial and fungicidal properties, and can be used to disinfect the air; in facial washes to combat skin problems such as acne; as insect repellents; or in the treatment of burns. Examples of relaxing oils include cedarwood, geranium, jasmine, lavender, rose and ylang ylang.

Chiropractic

11.38 A chiropractor manipulates the joints of the body, and more specifically the spine, in order to relieve pain. They work on the principle that pain is often caused by a nerve dysfunction or compression and thus the spine is manipulated as it is the main component of the central nervous system. Chiropractic is very useful for back pain, neck pain and headaches, which are common ailments for stressed office workers who often spend hours at their desks.

Homeopathy

11.39 Homeopathy is the art of treating like with like. For example, if a bacteria, virus or substance such as cat hair causes an allergic reaction in a person, a homeopath will treat that person with minute quantities of that allergen. This builds up the patient's resistance and immunity to the allergen. Many homeopathic remedies have to be used and stored well away from strong odours as this can reduce their effectiveness.

Massage

11.40 Massage may be the oldest and simplest form of medical care. Egyptian tomb paintings show people being massaged, and in Eastern cultures massage has been practiced since ancient times. Massage was also one of the principal methods of relieving pain used by Greek and Roman physicians. All of us practice massage when part of our body hurts or is sore, because we will instinctively rub the affected area.

Massage improves the functioning of the circulatory, lymphatic, muscular, skeletal and nervous systems, and may improve the rate at which the body recovers from injury and illness. Physically, massage involves holding, causing movement of soft tissue, and/ or applying pressure to the body. Eastern massage also incorporates a more philosophical base, and many therapists now practice a combination of the two – treating the 'whole' person through holistic massage.

As a therapy, massage has proved very beneficial to well-being, by enabling relaxation and alleviating the effects of stress. A variety of conditions benefit from massage including asthma, backache, joint and muscle problems and headaches. Massage also improves the function of the lymphatic system, thereby allowing the body to dispose of toxins which accumulate during stressful periods.

Different types of massage commonly practiced include:

- *Relaxation massage:* A smooth, flowing style that promotes general relaxation, improves circulation and range of movement, and relieves muscular tension.
- *Remedial massage:* A treatment that helps restore function to injured 'soft tissues' (muscles, tendons and ligaments). This therapy may involve the use of various types of massage such as neuromuscular techniques, soft tissue manipulation and muscle energy techniques, as well as a range of other physical treatments to assist recovery.
- *Sports massage:* Combines deeper, firmer massage techniques to enhance sports performance and recuperation. It can be used before exercise to warm and prepare the muscles for activity, or after exercise to relax tired muscles and disperse lactic acid deposits – and so help prevent aches and pains.
- *Aromatherapy massage:* Combines the therapeutic properties of essential oils with specific massage techniques to promote health and well-being.

  Essential oils are diluted in a carrier oil, and are usually applied in a relaxing massage technique. Essential oils can also be used in remedial and sports massage techniques, as the anti-spasmodic and healing properties contained in some oils will greatly aid tissue regeneration and muscle relaxation.
- *Indian head massage:* This is a deeply relaxing form of massage that works on both physical and mental levels, and should only be carried out by a qualified therapist. It is a specific therapy for the head, neck, shoulders, upper arms, upper back and face, and therefore covers the vulnerable areas where we are most likely to suffer from stress and tension. The origin of this massage has its roots in Indian Ayervadic medicine, and works on the marma points (the Indian system of energy centres of the body, equivalent to the oriental acupuncture points).
- *Acupressure massage:* In traditional Chinese medicine, energy flows through the body in invisible channels called meridians. There are 12 major meridians in the body, each associated with an organ system with over 350 acupressure points and a further 250 non-meridian points. An acupressure point is an 'access point' just under the skin. These are generally located at a weak spot on the body (ie between muscles and bones, or near the tip of a nerve), and are often in a slight depression or indentation on the body. During times of disease, stress or injury, energy can become blocked and cause an imbalance within the body. Stimulating these points with the finger or elbow can remove the stagnant Qi (pronounced 'chee') that collects in these points. This type of treatment can be incorporated either into a relaxing full body massage with essential oils, or, particularly more recently, introduced into a working environment with a seated, fully clothed acupressure massage.
- *Osteopathy:* An osteopath manipulates the joints of the body. The

treatment works on the basis that the body's structure and function are inter-dependent. If the structure is injured then it will affect the function. The main benefits of this treatment are postural and structural correction which will aid body function. Osteopathy is very useful for backache, neck and shoulder problems, which are indicative of today's long working hours spent sitting at a desk and using a computer.

- *Physiotherapy:* This involves the use of physical exercise, massage and pressure (both manual and automatic) to relieve pain from injury, illness and muscular tension. It also uses the application of heat in the form of UV lamps and ultrasonic sound, pulse machines, TENS (Transcutaneous Electrical Nerve Stimulation) machines, etc.

- *Pilates:* Developed by Joseph H Pilates in the 1920's, this technique is a fusion of western and eastern philosophies, teaching people about breathing with movement, body mechanics, balance, coordination, positioning of the body, spatial awareness, strength and flexibility. The pilates system works the body as a whole, coordinating upper and lower musculature with the body's centre.

- *Reflexology:* This was initially used by an American ear, nose and throat surgeon, William R Fitzgerald, who applied pressure on certain points which produced an anaesthetic effect so that minor surgery could be performed without pain. The science deals with the principle that there are reflexes in the feet which correspond to all the glands, organs and parts of the body. The uniqueness of reflexology is the proper method of using the thumb and fingers on the reflex points. The reflexes of the feet can also be found to correspond to those in the hands.

It is emphasised that prior to undertaking any alternative therapy, it is important to make sure that the therapist holds a certificate of competency from a recognised professional institution.

Practitioners should correctly take a full case history before starting any therapy including a list of any medication being taken. Individuals should also seek advice from their doctor prior to treatment if they are pregnant or have any of the following:

- An abnormal body temperature.
- A heavy cold.
- Recent scar tissue.
- Undiagnosed 'lumps' or 'bumps'.
- Any kind of infection.
- A history of epilepsy.
- Varicose veins.
- An open wound.

- A recent fracture.
- Or, in the case of aromatherapy massage, are undergoing chemotherapy or radiotherapy

- *Shiatsu:* This originated in China at least 2,000 years ago, and uses a combination of pressure and assisted-stretching techniques – some of which are common to other therapies such as massage, physiotherapy, acupressure, osteopathy, lymphatic drainage and others. The treatment stimulates both the circulation and flow of lymphatic fluid, releases toxins and deep seated tensions from the muscles, stimulates the hormonal and immune systems, and acts on the autonomic nervous system, allowing the recipient to relax deeply.

- *Yoga and Tai Chi:* Yoga and Tai Chi combine many of the benefits of breathing, muscle relaxation and meditation, while also toning and stretching the muscles. They can also elevate mood, and improve concentration and ability to focus. If you join a yoga class you may find that the breathing techniques you will learn are very effective for releasing negative emotions such as anger and fear as they arise.

Stretching and useful exercises to do at work

11.41 The following are useful exercises to perform at work – they are easy to follow and can easily be done at the desk. They help to stretch the muscles, which can shorten due to overuse and repetitive strain, boost circulation, and help ease tension and stress by building-up overused muscle groups.

Hand warm-up routine

11.42 This is an excellent routine for keyboard users or those who do a lot of writing.

- Shake your hands and wrists vigorously to increase mobility and circulation. Press the fingers back from the palms to the fullest limits, with your fingers held together.

- Gently, press each finger back separately.

- Clench and relax your fists.

- Rotate your hands from the wrists with your fists clenched, and your elbows at the sides of your body.

- Always use a wrist protector band to alleviate pressure from keyboard edges. Some people like to use a long heated wheat bag to provide heat to sore and tired forearm muscles whilst typing.

Stretches to be done during a break away from your desk

Neck and head

11.43

- Gently let your head come down onto your chest and bring it back to the centre. Slowly rotate to the left so your chin rests on your left shoulder. Bring back to the centre and then rotate to the right and repeat the same procedure.

Shoulders

11.44

- Clasp your hands behind your back and raise your arms towards your shoulders. You should feel a stretch at the front of your chest and shoulder area. This is a good exercise for opening up the chest and diffusing tension.

- Raise your right arm so that it is against your right ear. Bend your arm so that your right hand is placed in the centre of your back between your shoulder blades. With your left hand, gently pull the arm further down so that your hand travels downwards. The stretch should be felt in your right arm and shoulders. Repeat on your left side.
- Clasp your hands in front of you and imagine that you are hugging a tree! Push outwards from the shoulders and you will feel a stretch across the back of your shoulders

- Bring both arms out in front of you and then bring the left arm round to the left side and the right arm round to the right. Push behind with both arms and squeeze the shoulder blades together. The stretch should be felt on the front of your chest.

Lower back

11.45

- Lie on the floor and bring your knees into your chest. Hold for a few seconds and then relax.

- Bring the right knee over to the left shoulder and repeat on the other side. This stretches the sides, back and stomach muscles, and is very good for lumber pain and sciatica.

Legs

11.46 The best exercise for the legs is to take regular breaks away from your desk.

Walk around the office, take a trip to a colleague's office instead of phoning, and take the stairs instead of the lift. The calf muscle is vital in pumping and returning the blood back up from the feet to the trunk of the body, so any exercise which flexes the calf muscle is beneficial.

Foot rotation

11.47 Rotate and flex both feet. This will help flex the calf muscle and help with return of blood back to the heart via the veins.

Don't forget the eyes!

11.48 Rub your hands together to warm them, and then cup the eyes without pressing your hands to your eyes. Close your eyes and breathe deeply and slowly, visualising that you are looking into darkness. This helps to relax the internal and external muscles of the eye. Also remember to take regular breaks away from your computer screen.

Key learning points

11.49 In order to get the most from life, it is important for all of us to try to maintain a balance between stimulation and relaxation; exercise and rest; responsibility and freedom; work and play; laughter and tears. This is not easy, but as we have seen, there are an extraordinarily wide range of proactive interventions that we can make in order to improve the balance in our lives.

- A 'stress diary' can be invaluable in helping to understand not only the major sources of stress, but also the frequency with which they are being experienced.
- It is important to try to alter the balance in our activities from being 'stress producing' to 'stress reducing'.
- For people who find it difficult to say 'no', assertiveness and time management training can both be extremely valuable.

(cont'd)

- How we manage our time is a key factor in determining how stressful our lives can be.
- Humour gives us a different perspective on our problems. If we can make light of a situation, then it can become less threatening.
- It is important to take regular breaks away from situations or tasks that are the source of stress and frustration.
- The proactive interventions we can make to help develop a more healthy lifestyle include changes related to diet, exercise, relaxation, sleep and leisure.
- A well balanced diet is crucial in preserving health and helping to reduce stress, but there are certain foods and drinks that act as powerful stimulants to the body and can contribute to stress.
- Consuming large amounts of alcohol will neither solve our problems nor improve our health.
- It is important to limit our caffeine intake, as this 'kick starts' the stress reaction.
- We should aim to keep the consumption of 'convenience' foods to the minimum, as many contain large amounts of sugar, salt, fat and preservatives.
- It is essential that we drink at least 2 litres of water per day.
- Exercise can be good for both the body and the mind, and it is thought that stress poses significantly less danger to the overall health of people who are physically fit.
- It is extremely important that all of us include a daily period of relaxation in our lives.
- Tension and relaxation are two sides of the same coin – you cannot experience both at the same time. We therefore need to learn to turn *on* the bodily effects of relaxation so that we can turn *off* the symptoms of tension.
- Proactive interventions in terms of how we think about things include changing our perspective, positive thinking and self-talk.
- Exercise provides an outlet for negative emotions (such as frustration and anger) to be dispersed. It is important to choose an activity that you enjoy so that you increase your chances of maintaining it.

- In recent years, a wide range of 'alternative' therapies have been offered for the treatment of stress. Depending on method, mode and the individual, some will be more effective than others.

12 The Future of Stress

Introduction

12.1

> 'More and more British workers are recognising that balancing the quality of life and family is as important as a fulfilling career. People clearly want greater control and choice over their working hours but lack the confidence and the knowledge to do anything about it.' *Patricia Hewitt, Secretary of State for Trade and Industry (2003)*

In the previous chapters, we have examined in detail the causes and effects of stress within the work environment, and have tried to explain and clarify the complexity of the issues surrounding the problem in today's society. In particular, we have looked at the highly individual nature of the varying responses to stress, and the absolute need for proactive management in order to reduce the impact of it on both our health and our behaviour.

There are still those who argue that there is no such thing as 'stress', and that people are simply responding to media hype surrounding the subject by convincing themselves that they are suffering from this ('hypothetical') condition – when in fact, they are not. There are others in the world of work, who even though they accept that 'stress' exists, believe that 'a certain amount (of it) is good for you' – and will then use this as an excuse to impose a highly pressurised regime in the hope of increasing productivity and obtaining a competitive advantage. In fact, the converse is true, because a work environment that induces stress is more likely to confer a competitive disadvantage on the organisation that operates it.

The true position is given in the summary report 'The state of occupational health in the European Union: Pilot Study' (September 2000), conducted by the European Agency for Safety & Health at Work. The report concluded that 'no other issue, considered in the pilot study, had as many responses for the need for further actions, than stress'.

The simple fact is that there are numerous adverse trends in modern society that are, unfortunately, increasing. The prevalence of stress-related issues is one of these, and unless individuals and organisations have both the motivation and the knowledge to improve their understanding of how to manage stress better, the problems resulting from these issues can only worsen. In considering the future trends in work-related stress, we shall therefore look at the implications of these issues.

Legislative pressures

12.2 In January 2003, the Health & Safety Executive ('HSE') announced its intention to include work-related stress into its routine health and safety inspections by the end of 2003. It also stated that it hoped ultimately to have the power to impose fines on organisations that fail to introduce minimum standards, and also to investigate specific complaints about stress in the workplace raised by trade unions or individuals. The Health & Safety Commission Chairman, Bill Callaghan, commented: 'Stress seems to be endemic in modern society and the rate of increase in recent years has been considerable. The key to reversing the upward trend is to avoid stress in the first place'.

By 2006, the UK will be required to introduce anti-age discrimination laws alongside those on disability, race, gender, sexual orientation and religion. Businesses that fail to comply will face prosecution and many employers are worrying about how this law will affect them. It could turn the conventional view of ageism upside down, with benefits such as additional holidays and long service awards for older members of staff being seen as discriminatory against younger members. Employers will therefore need to try to stay 'age neutral'.

Given today's increasing emphasis on performance, bullying is becoming a growing problem. In many parts of the world, legislative bills are being drafted against bullying and intimidation. For example, there is the intention to revive the Dignity at Work Bill in Parliament (which failed to get a second reading in 2002); the US Campaign Against Bullying is planning to lobby for state legislation in California, Colorado and other US states; and the research into violence in the workplace commissioned by the European Union and completed in 2001, is a first step towards proposed legislation for member states in this area.

Other legislation which has been implemented/ or is coming forward, includes:

- New flexible working regulations that came into force on 6 April 2003;
- *Working Time (Amendment) Regulations 2002 (SI 2002/3128)* that came into force in April 2003;
- New disability discrimination legislation which comes into force in October 2004;
- Sexual and sex-related harassment legislation which comes into force in October 2005;
- Age discrimination legislation which comes into force in December 2006.

Globalisation and global competition

12.3 Globalisation and global competition have transformed the industrial landscape over the last 20 years. The relentless drive for increased performance and productivity has changed the way that organisations operate. Whereas some years ago the responsibility was on management to deliver these improvements, there has been a paradigm shift whereby the shop floor is now typically at the sharp end of increasing productivity and lowering costs, through the embracement of new work practices and the extended use of information technology.

We have emphasised that the extent to which individuals have some control over their own work patterns can be crucial in minimising stress. An environment that 'empowers' employees, for example through self-managed or self-directed work teams, would appear to embody some of the main principles of positive stress management. Yet empowerment may, in some cases, be a double-edged sword – giving workers greater control over their day-to-day work, but possibly changing their job description into something quantifiably different from that contained in their original contract of employment.

In an organisational culture where shop floor workers are frequently those who are tasked with identifying process improvements, and management's role is increasingly one of facilitating the implementation of these changes, no-one is exempt from the pressure of coming up with new ideas, cost-justifying them and making them work. The idea that an employee can simply come into work, do their job and go home again has all but disappeared. So while on the one hand, 'empowered' employees may have more control over how they work, they also have to contend with an increasingly pressurised work environment in which they are expected to increase, and to keep increasing, levels of personal performance.

Employers may seek to justify these pressures by implying that they are making work more interesting and fulfilling, and enabling employees to contribute to the ultimate success of the business. While this may be true, and the constant drive for competitive advantage requires employers to utilise fully the talents and creativity of all their employees, they must nevertheless comply with their statutory duty of care to minimise any excessive pressure and the consequent adverse effects of this.

The pressures of competition will continue to intensify for the foreseeable future – which is why it is vital that employers recognise the potential for damage to health and safety, where statutory standards of health and safety are not rigorously met.

The accelerating pace of technology

12.4 According to 'Moore's Law', the pace of technological change is such that the power of microchip technology is currently doubling every

18 months. Many high technology organisations pride themselves on their ability to develop new products at rates that exceed even this meteoric growth, and most experts are agreed that this is a trend that will be likely to continue over the next two decades.

This extraordinary rate of technological development, combined with the extent to which the successful deployment and exploitation of new technology is a key driver of business performance, means that employees are having to learn new skills and discard 'redundant' ones, faster than ever before. This is also changing the way that organisations operate and manage their workforce.

As little as five years ago, team briefings, staff/ management meetings, annual appraisals and a monthly newsletter, would have been the norm for the dissemination of information between management and staff in many large organisations. Now, however, most people working within an organisation will maintain contact electronically, through various corporate intranets (ie restricted access networks), rather than by face-to-face meetings.

We have touched on the 'depersonalisation' of the work environment that is having such a dramatic effect on the relationships between employees and employers. Against a backdrop of ongoing redundancies, pressure to cut costs, and an end to the concept of 'a job for life', individuals can only successfully survive in the work environment if they accept that change and uncertainty are the norm, rather than the exception.

In this respect, while even the most determined Luddite would find it hard to justify a move away from technological advancement, organisations that depend on the adoption of new technology have to accept that there may be some effects of the implementation and use of new systems that are not all beneficial to the individual. Technology can be a great enabler of innovation and increased productivity, but there are downsides, not least when the result of one day out of the office may be two days spent trying to get to grips with an overflowing 'inbox', while at the same time running hard just to get back into the (work) position that the employee was in a few days previously.

The ageing profile of the working population

12.5 While many employers would appear to prefer a world where they can take advantage of an endless pool of fresh young talent, better able (they often mistakenly believe) than their older colleagues to cope with the demands of technological advance – the fact is that by adopting this approach they are facing a 'demographic time bomb'.

By 2010, increased life expectancies, which are currently growing by two years per decade, mean that nearly 40% of the UK population will be over

45. By discriminating against this ageing workforce, employers risk ignoring the skills and talents of a significant proportion of the UK population. If they continue with this stance, they will very soon be in breach of government legislation which (as mentioned above at 12.2), following an EU directive, is due to be implemented in 2006, and will introduce anti-ageist discrimination laws alongside those already in place regarding disability, race, gender, sexual orientation and religion.

According to research by the Department for Work and Pensions (2001) (reported in *People Management*, 5 December 2002), based on a random survey of 800 British enterprises in a range of industry sectors, and a representative sample of 500 people aged 50–69, one in four 'older' people believed that they had suffered discrimination when applying for a job. Nearly half the organisations surveyed employed no staff aged 60 or over; line managers, while insisting they were 'age-friendly', were often ignorant of relevant guidelines; and ageism was seen as 'more acceptable' than other types of discrimination.

So why does this discrimination continue to take place – especially when (according to a survey by the charity Age Concern, also reported in *People Management*, 5 December 2002) 97% of Britons believe that age should be ranked as the least important criterion when recruiting a new member of staff, with ability (57%) and a good track record (40%) counting for far more? Skills, knowledge and experience cannot be created overnight, yet many human resources managers seem to see age (or preferably, lack of it) as a far more important factor in selecting employees to fill their vacancies.

This preference for younger employees becomes even more difficult to understand when one considers the length of time that employees of different ages expect to stay in their current jobs. Research by the International Stress Management Association (2002) found that 54% of 18–24 year olds expected to stay in their current job for no more than the next two years (and 32% for no more than the next 12 months). This compared to 76% of 35–44 year olds and 74% of 45–54 year olds who expected to stay in their current job for at least the next five years.

One way of interpreting these results is that employers appear to have a preference for employing individuals who are less experienced, require more training, and are less loyal than others in the labour market, simply on the grounds of age. If they continue with this strategy, given that the ageing population profile will mean that 'younger' talent will become an increasingly scarce commodity, they will also, presumably, end up paying more and more for the privilege. Another interpretation is that as the employee becomes older, they know that realistically their chances of changing their job is significantly less, mainly due to ageism.

Sooner or later, employers will have to realise that by failing to employ the best person for the job, for whatever (perceived) reason, the result will be a negative impact on their profit margins, and ultimately the viability of their business, with yet more implications for stress as a result.

Call centres – one example of the shape of things to come?

12.6 Over the past few years, brown field sites in many former heavily industrialised areas have been redeveloped to be used as call centres, employing many hundreds of men and women – many of whom were originally made redundant by the closure of companies in the engineering and manufacturing sectors. In fact, according to Datamonitor in their report 'Call centers in EMEA to 2007' (Datamonitor (2002)), call centres have now become one of the fastest growing industrial sectors in the UK.

However, call centres are considered by some to be one of the most stressful environments yet created by modern business practice. The pressure to achieve very tough targets (that can lead to high levels of stress) is a typical work-related issue, while staff not being able to leave their desks without permission (and when doing so, having their breaks timed) is reminiscent of the school room.

In addition to the constant pressure on employees to meet tough call handling targets, they are often subject to work practices involving many other stressors. A large proportion of call centre business is conducted via the telephone while using screen display equipment – a way of working that is often implicated in musculoskeletal problems. Being seated for most of the working day also involves risk to the back and upper limbs if workstations are not designed to suit the individual worker – something that can be difficult to ensure when people on different shifts share the same workstations at different times of the day (a policy often termed 'hot desking').

This need to share desks, together with the accompanying lack of personal space, can leave call centre workers with a diminished sense of identity. Work involving constant use of a keyboard, over many hours can also result in repetitive strain injury (RSI). Although it is important for individuals to take time away from their work positions at regular intervals in order to avoid problems relating to posture, this is unfortunately an effective stress management intervention that many call centre managers appear to discourage.

A report published by the Work Foundation (2001) (formerly The Industrial Society) also suggested that call centres could be bad for the mental health of employees. The report, 'New Work, New Stress', said that the current trend for creating jobs intended to improve productivity and efficiency, gave employees little job control, which in turn was not conducive to their mental health.

To help combat these types of issues, forward-thinking call centres have already started to invest in prevention and remedial support such as chair massage and reflexology. Some have even come to realise that what is most important for call centres is not simply the number of calls that are being handled, but that the right person is handling the right call in the right manner. Rather than operating 'sweatshops', they are now starting to invest in creating better work environments, with clearly defined career paths for their employees.

One of the most remarkable results of this, particularly in view of the comments in the previous section at 12.5, is the realisation of the benefits of employing 'older' workers in call centre environments. Older employees have been shown to demonstrate better customer service skills than their younger colleagues, and some women have been found to have higher levels of sensitivity and conscientiousness than their male counterparts.

All of this simply amplifies the fact that the real potential for increased productivity in call centres lies with the quality of the team of people who actually talk to customers. In order to motivate employees, it is vital to give people a reason to do better in order to gain job satisfaction. If instead of bringing only 80% of themselves to their job, employees decide that it is fun, interesting and fulfilling, and so bring 100% of their potential, this equates to a 20% increase in productivity, all at no additional cost to the employer.

Managing work-related stress – 2003 and beyond

12.7 So how can we pull all these various strands together? On the one hand, we have figures showing a dramatic increase in the rates of stress-related sickness absence; an ever-more competitive business environment; an ageing working population; and businesses whose policies can result in the wrong people being employed, often for the wrong reasons.

On the other hand, we have an increasing realisation by government of the need to tackle stress and discrimination (including that based on age), in addition to a broader implementation of the Working Time Directive – resulting in a series of initiatives designed to ensure that employers take their responsibilities in these areas seriously. In 2002, for example, BP Grangemouth was one of 20 companies in Europe to receive an award from the European Agency for Safety & Health at Work in recognition of outstanding and innovative contributions to the prevention of psychosocial risks, especially work-related stress. BP won the award through its use of risk management to prevent potential stress from a plant-commissioning project.

Yet among the positive responses to the HSE's announcement in January 2003, there were still voices that expressed concerns over what the results of new policies such as those associated with stress-related risk assessments would actually be.

Some people, for example, felt that the HSE risked creating a 'checklist mentality', leading employers to tackle stress from a regulatory point of view, rather than addressing the issue from a broader perspective. Others also doubted how effective they would be in reducing work-related stress, due to the lack of definitive evidence related to identifying the causes of workplace stress, and the impact of these on individual employees.

To put these comments into the context of this book, in CHAPTER 10 we looked at a very broad range of stress management interventions and best practices that individuals and organisations can use in order to alleviate problems. It would be unreasonable to expect anybody to attempt to implement all of these. However, what is most important is that individuals adopt whichever strategies 'suit them best', and endeavour to use them accordingly.

From a stress management point of view, it is important that whatever steps are taken, the individual or organisation makes a proper record of these and revisits them at a later date to assess the extent to which they have been effective. Too many organisations, for example, undertake either assertiveness or time management training in the belief that having done so, stress will simply disappear – notwithstanding that they have failed to grasp the fact that the underlying causes of stress have been left unchanged.

While there are many ideas and strategies for better working environments, stress management professionals need to work with their clients to identify which processes are effective and which are not, and to establish best practices that can be of benefit to all. Many of these, such as the need for clear communication, clearly defined roles and responsibilities etc, have already been identified, but what is often missing are the measurement systems and assessment tools that will enable us to move stress management forward onto the next level.

The ongoing research into work-related stress gives us confidence that attitudes are changing, and that both employers and employees will see clearly that anything in the workplace that may lead to stress and stress-related illness is unacceptable in the commercial and industrial environment of the 21st century: unacceptable in relation to the provisions of employment legislation; and unacceptable in relation to the substantial costs of compensation and legal fees in the event that management are found guilty of negligence, or are in breach of their duty of care. These are also matters that will inevitably reduce shareholder confidence and possibly affect external capital investment.

Unlike working conditions during and after the industrial revolution, a healthy environment at work today is mandatory for all organisations that employ any person within the United Kingdom and are subject to UK law.

Alongside these legal requirements, however, *all* organisations need to remember that essentially what we are talking about is *people*, and that their value should never be underestimated. They are an organisation's most valued resource, and it is important that managers motivate, nurture and appreciate their staff, encouraging them to give of their best – which will ultimately be reflected in the success of the organisation and its position in the market place.

The interpersonal skills necessary for good management may not come naturally to some managers. It is therefore vitally important, when recruiting a new manager or developing an existing one, to ensure that they have well-honed people management skills. No one training course is likely to meet the needs of any one manager. It is therefore well worth investing in coaching or mentoring to enable managers to acquire a high level of interpersonal skills for their own personal development and the success of the organisation.

If managers do this, they will receive good will, loyalty and commitment from their teams – and their organisations will clearly benefit as a result.

Bibliography

Chapter 1 – The Nature of Stress

European Commission, 'Guidance on work-related stress: Spice of life or kiss of death?' Employment & Social Affairs (ISBN 92828 9806 7)

Friedman, M, Rosenman, R H, *Type A behaviour and your heart* (1974) Wildwood House, London

French, J P R, Caplan, R D, R van Harrison, *The mechanisms of job stress and strain* (1982) Wiley, New York

Karasek, R A, Job demands, job decision latitude and mental strain: Implications for job redesign (1979) *Administrative Science Quarterly*, pp 285–308

Lazarus, R S, *Psychological stress and the coping process* (1966) Jossey Publications, London

Lazarus, R S, Folkman, S, *Stress, appraisal and coping* (1984) Springer, New York

National Institute of Occupational Safety and Health, 'Stress at work' (1999) *DHSS (NIOSH)* No 99–101

Rotter, J, 'Generalised expectancies for internal versus external control of reinforcement' (1966) *Psychological Monographs*, Vol 80

Holmes, T, Rahe, R, 'The Social Readjustment Rating Scale (SRRS)' (1967) *Journal of Psychosomatic Research* Vol 2

Chapter 3 – The Health and Safety Framework

Health & Safety Commission, 'Securing Health Together' (2001) Occupational Health Strategy for England, Scotland and Wales

Health & Safety Executive, 'The five steps to risk assessment' (1999) (INDG163 (rev1))

Health & Safety Executive, 'A critical review of psychosocial hazard measures' (2001) Contract Research Report 356/2001

Health & Safety Executive, 'Tackling work-related stress: A manager's guide to improving and maintaining employee health and wellbeing' (2001) (ISBN 0 7176 2050 6)

Van Emelen, J, 'Bilan et perspectives de médecine de travail' *Cahiers de médecine de travail* (1996) Vol XXXII, pp 9–15

Chapter 4 – Identifying Current Workplace Stressors

Cox, 'Stress research and stress management: Putting theory to work' (1993) HSE Books

Chartered Institute of Personnel and Development (CIPD), 'Married to the job?' (2001) (Survey Report)

Health & Safety Executive, 'Tackling work-related stress' (2001) (series of booklets) HSE Books

Health & Safety Executive, 'Work-related factors and ill health: The Whitehall II study', Contract Research Report HSE266/2000 (website: www.hse.gov.uk/research)

International Labour Organisation, *Preventing Stress at Work* (1992) (website: www.ilo.org)

International Stress Management Association UK (ISMA^{UK}), Research published for National Stress Awareness Day (2000)

Johnson & Hall (1988), Karasek & Theorell (1990), The demand and control model

University of Surrey, 'Report on irregular hours' (2001) *The Lancet*

Wheatley, R, 'Taking the strain: A survey of managers and workplace stress' (2000) Institute of Management, PPP Healthcare (ISBN 0859 463 133)

Chapter 5 – The Effects of Stress on an Organisation

ACAS (website: www.acas.org.uk)

British Crime Survey, published by Home Office

Confederation of British Industry (CBI), 'Employee absence: A survey of management policy and practice' (2002)

Chartered Institute of Personnel Development, 'Employee Absence Survey' (2002)

De Becker, G, *The gift of fear* (1997) Bloomsbury (ISBN 0747538352)

Handy, C, *Understanding organisations* (1999) Penguin Books (ISBN 01 95087321)

Health & Safety Executive, *Safety Statistics Bulletin* (2000/01)

Health & Safety Executive, Press Release E108:01 (June 2001)

Health & Safety Executive, Press Release C056:02 (December 2002)

International Stress Management Association UK (ISMA^{UK}), Research published for National Stress Awareness Day (2002)

Samaritans (website: www.samaritans.org.uk; email: jo@samaritans.org)

Trade Unions Congress (TUC), *Violent times* (2002)

Chapter 6 – Bullying at Work

Adams, A, *Bullying at work: How to confront and overcome it* (1992) Virago, London (ISBN 185381542X)

Andrea Adams Trust, Hova Villas, 1 Hova Villas, Hove, East Sussex BN3 4DH (tel: 01273 417 850; fax: 01273 704 901; email: mail@andreaadamstrust.org; website: www.andreaadamstrust.org)

AMICUS, 'Bullying at work: How to tackle it' (www. amicus-m.org)

Einarsen, S, and Skogstad, A, 'Bullying at work: Epidemiological findings in public and private organizations' (1996) *European Journal of Work and Organizational Psychology* Vol 5 No 2, pp 185–202

Hoel, H, and Cooper, C L, 'Destructive conflict and bullying at work' (November 2000) (Unpublished report), UMIST (explored by Rayner, Hoel & Cooper in *Workplace bullying* (2002))

Liefooghe, A P D, and Mackenzie Davey, 'Explaining bullying at work: Why should we listen to employee accounts?' (2001) Paper to British Academy of Management, Cardiff

Lewis, D, 'Workplace bullying: Interim findings of a study in further and higher education in Wales' (1999) *International Journal of Manpower* Vol 20 Nos 1 and 2, pp 106–118

Leymann, H, 'The content and development of mobbing at work' (1996) *European Journal of Work and Organizational Psychology* Vol 5 No 2, pp 165–184

McCarthy, P, Sheehan, M, and Kearns, D, 'Managerial styles and their effects on employees' health and wellbeing in organisations undergoing restructuring' (1996) Report for Worksafe Australia, Brisbane, Griffith University

Rayner, C, 'Workplace bullying: Do something!' (1998) *The Journal of Occupational Health and Safety – Australia and New Zealand* Vol 14 No 6, pp 581–585

Rayner, C, 'Incidence of workplace bullying' (1997) *Journal of Community and Applied Social Psychology* Vol 7 No 3, pp 199–208

Rayner, C, Hoel, H, and Cooper, C L, *Workplace bullying: What we know, who is to blame and what can we do?* (2002) Taylor Francis, London,

Tehrani, N, *Building a culture of respect* (2001) Taylor Francis, London

UNISON, 'UNISON police section members experience of bullying at work' (2000) UNISON, London

UNISON, 'UNISON members' experience of bullying at work' (1997) UNISON, London

Chapter 7 – The Effects of Stress on the Individual

Charlton, J, Kelly, S, Dunnell, K, Evans B, and Jenkins, R, *The prevention of suicide* (1994) HMSO, London

Department of Health, 'Saving lives: Our healthier nation' (2002) National Suicide Prevention Strategy for England No 27761

Department of Health, 'National service frameworks: Modern standards and service models: Mental health (1999)

Freudenberger, H J, 'The staff burn-out syndrome in alternative institutions' (1975) *Psychotherapy: Theory, Research and Practice* 12(1), pp 35–45

Gleick J, *Faster: The acceleration of just about everything* (1999) Abacus

Kelly, S, and Bunting, J, 'Trends in suicide in England and Wales: 1982–96' (1998) *Population trends* No 92

Maslach, C, 'Burned-out' (1976) *Human Behaviour* 5(9), pp 16–22, Manson Western Corp

McDonald, V, 'Suicides in young men on the increase' (1992) *Sunday Telegraph*

Parrott, A, University of East London, *American Psychological Association* Oct 1999, Vol 54 No 10, pp 817–820

Platt, S, 'Parasuicide and unemployment' (1986) *British Journal of Psychiatry* 140, pp 401–405

Sillaber, I, Rammes, G, Zimmermann, S, Mahal, B, Zieglagaensberger, W, Whrst, W, Holsboer, F, and Spanagel, R, 'Enhanced and delayed stress-induced alcohol drinking in mice lacking functional CRH1 receptors' (May 2002) *Science*

Selye, H, *Stress in health and disease* (1976) LexisNexis Butterworths Tolley

Schneidman Edwin, S, 'Suicide' (1973) *Encyclopaedia Britannica*

Support groups:

Rural Stress Information Network (website: www.ruralnet.org.uk)

Samaritans (website: www.samaritans.org.uk)

Survivors of Suicide (website: www.survivorsofsuicide.com)

Chapter 8 – Trauma and Coping with the Aftermath of a Critical Incident

American Psychiatric Association (APA), 'Diagnostic and statistical manual of mental disorders' (DSM-III) (3rd edition), APA, Washington DC

American Psychiatric Association (APA), 'Diagnostic and statistical manual of mental disorders' (DSM-IV) (4th edition), APA, Washington DC

Armstrong, B, O'Callahan, W, and Marmar, C R, 'Debriefing Red Cross Disaster Personnel: The multiple stressor debriefing model' (1991) *Journal of Traumatic Stress* 4(4), pp 581–593

Brom, D, Dleber, R J, and Wiztum, E, 'The 'prevelance of post-traumatic psychopathology in the general and the clinical population' (1991) *Journal of Psychiatry and Related Sciences* Vol 28 No 4, pp 53–63 (Israel)

Dyregrov, A, 'Caring for helpers in disaster situations: Psychological debriefing' (1989) *Disaster Management* 2, pp 25–30

Freeman, C, Flitcroft, A, and Weeple, P, 'Psychological first-aid: A replacement for psychological debriefing' (2001) The Rivers Centre, Edinburgh

Health & Safety Executive, 'Managing crowds safely' (HSG 154) (2000) HSE Publications (ISBN 071761834X)

Kennardy, J, 'The current state of psychological debriefing' (2000) *British Medical Journal* 321, pp 1032–1033

Mitchell, J T, 'Guidelines for psychological debriefing: Emergency management course manual' (1983) MD: Federal Emergency Management Agency, Emergency Management Institute, Emmitsburg

Mitchell, J T, and Everly, G S, 'Critical incident stress debriefing: An operations manual for CISD, defusing and other group crisis intervention series' (3rd edition) MD: Chevron Publishing Corporation, Ellicott City

Orner, R J , Avery A, and King, S, 'An evidence based rationale for early intervention after critical incidents' (1999) Paper presented at European Regional Conferences on Early Interventions and Psychological Debriefing, Aberdeen, Scotland

Raphael, B, *When disaster strikes: A handbook for caring professions* (1986) Hutchinson, London

Rose, S, 'Evidence-based practice will affect the way we work' (2000) *Counselling* March 2000, pp 105–107

The Royal College of Psychiatrists, 'Trauma Screening Questionnaire' *British Journal of Psychiatry* 2002 Vol 181, pp 158–162

Solomon, S D, Gerrity, E T, and Muff, A M, Efficacy of treatments for post traumatic stress disorder (1992) *Journal of the American Medical Association*, pp 265, 633–637

World Health Organisation (WHO), 'International classification of mental and behavioural disorders: Clinical descriptions and diagnostic guidelines' (ICD-10) (1992) WHO, Geneva

Useful trauma support contacts:

UK Trauma Group
Website: www.uktrauma.org.uk

British Association for Counselling and Psychotherapy
Website: www.bac.co.uk

The British Psychological Society
Website: www.bps.org.uk

The Royal College of Psychiatrists
Website: www.rcpsych.ac.uk

Priory Healthcare Services
The Priory Ticehurst House
Ticehurst
Wadhurst
East Sussex
TN5 7HU
Tel: 01580 200 391
Fax: 01580 201 006
Website: www.prioryhealthcare.com

Traumatic Stress Clinic (UK-wide, referrals by GP only)
73 Charlotte Street
London
W1P 1LB
Tel: 020 7530 3666
Website: www.traumaclinic.org.uk

Traumatic Stress Clinic (Institute of Psychiatry)
Maudsley Hospital
Denmark Hill
London
SE5 8AF
Tel: 020 7919 2969
Fax: 020 7919 3573
Website: www.iopkcl.ac.uk

Ty Gwyn (Nationwide clinics – for ex-service personnel, and drop in centres for non ex-service too)
Post-Traumatic Stress Disorder Centre
21 Bryn-y-Bia Road
Llandudno
Conway
LL30 3AS
Tel: 01492 544 081
Fax: 01492 544 372
Website: www. tygwyn.com

Roderick Orner
Department of Psychological Services
Baverstock House
St Anne's Road
Lincoln
LN2 5RL
Tel: 01522 560 617
Fax: 01522 546 337

The above offers early intervention services but does not include psychological debriefing.

Chapter 9 – Stress and Health

Andersen, B L, Farrar, W B, Golden-Kreutz, D, et al, 'Stress and immune responses after surgical treatment for regional breast cancer,' (1998) *Journal of the National Cancer Institute*

British Medical Journal (2002) 325: 857

British Heart Foundation, 'Coronary Heart Disease Statistics' (2002)

Confederation of British Industry and Department of Health, 'Promoting mental health at work' (1997)

Department of Health, 'Our healthier nation: A contract for health', The Stationery Office

Department of Health, 'The prevalence of back pain in Great Britain' (1998)

Department of Health, 'The journey to recover: The Government's vision for mental health care' (2001)

Everson, S A, Lynch, J W, Kaplan, G A, Lakka, T A, Sivenius, J, and Salonen, J T, 'Stress-induced blood pressure reactivity and incident stroke in middle-aged men' (2001) *Stroke* (June 2001) 32(6), pp 1263–1270

Faulkner, A, 'Knowing our own minds' (1997) Mental Health Foundation

Health & Safety Executive, 'The cost to Britain of workplace accidents and work-related ill-health 1995/1996' (1999) HSE Publications

Health & Safety Executive, 'An assessment of employee assistance and workplace counselling programmes in British organisations' (1998) HSE Publications

Health & Safety Executive, 'Musculoskeletal disorders in supermarket cashiers' (1998) HSE Publications

Health & Safety Executive, 'Key sheet on back injuries to employees 1987–1995/6' (1997) HSE Publications

Health Education Authority, 'Workplace health' (1997)

Helliwell P S, Smeathers, J R, 'Driving posture, vibration and psycho-social factors for back pain in long distance drivers' (1998) Leeds University

Kendrick, A, *British Medical Journal* (1991) 302, pp 508–510

Mann, A, et al, 'British psychiatric morbidity' (1998) *British Journal of Psychiatry* 173, pp 4–7

Office for National Statistics, 'Labour Force Survey' (1999) HMSO

Patel, A, and Knapp, M, 'The cost of mental health in England' (1998) Centre for Economics and Mental Health *Mental Health Research Review* No 5

Reuters Health Information Inc (www.reutershealth.com)

Samaritans (See references under CHAPTER 10)

World Health Organisation (WHO), 'International classification of mental and behavioural disorders: Clinical descriptions and diagnostic guidelines' (ICD-10) (1992) WHO, Geneva

Examples of the many studies from around the world linking back pain to psychological stress at work:

Ahlberg-Hulten, G K, Theorell, T, and Sigala, F, 'Social support, job strain and musculoskeletal pain among female health care personnel' (1995) *Scand J of Work, Environ & Health* (Dec 1995) 21(6), pp 435–439 (Sweden)

Papageorgiou, A C, Macfarlane G J, Thomas, E, Croft P R, Jayson M I, and Silman A J, 'Psychosocial factors in the workplace: Do they predict new episodes of low back pain?' Evidence from the South Manchester Back Pain Study (1997) *Spine* (15 May 1997) 22(10), pp 1137–1142 (Britain)

Van Poppel, M N M, Koes, B W, Deville, W, Smid, T, and Bouter, L M, 'Risk factors for back pain incidence in industry: A prospective study' (1998) *Pain* (July 1998) 77(1), pp 81–86 (Netherlands)

Williams, R A, Pruitt, S D, Doctor, J N, Epping-Jordan, J E, Wahlgren, D R, Grand, I, Patterson, T L, Webster, J S, Slater, J A, and Atkinson, J H, 'The contribution of job satisfaction to the transition from acute to chronic low-back pain' (1998) *Arch Phys Med Rehabil* 79, pp 366–74 (USA)

Examples of studies showing that other psychological stressors lead to back pain:

'Raising kids is a pain in the . . . ' *The Back Letter* (Dec 1994) 9(12), p 140

'Back pain under fire: Do police in a war zone suffer an increased risk of back problems?' *The Back Letter* (Oct 1996) 11(10), p 111

Lampe A, Stollner, W, Krismer, M, Rumpold, G, Kantner-Rumplmair, W, Ogon, M, and Rathner, G, 'The impact of stressful life events on exacerbation of chronic low-back pain' (1998) *J Psychosom Res* (May 1998) 44(5), pp 555–63

Chapter 10 – Developing and Maintaining a Healthy Organisation

Confederation of British Industry, 'Counting the costs: Absence and labour turnover survey' (2002)

Health & Safety Executive, 'Initiative evaluation report: Back to work' (2002) Contract Research Report CRR 441/2002

Kindler, H, *Conflict management: Managing disagreement constructively* (1988) Kogan Page (ISBN 185 0918120)

Useful organisations to contact:

Age Concern (charity concerned with the needs and aspirations of older people)
Website: www.ageconcern.org.uk

Andrea Adams Trust (charitable trust campaigning against workplace bullying)
Tel: 01273 704 900 (National Helpline)
Website: andreaadamstrust.org

Alcohol Anonymous (for people with an alcohol problem)
Tel: 0845 769 7555 (National Helpline)
Website: www.alcoholics-anonymous.org.uk

British Association for Counselling and Psychotherapy (BACP) (for list of counsellors and psychotherapists)
Tel: 0870 443 5252
Website: www.babcp.org.uk and www.counselling.co.uk

Cancer BACUP (providing up to date information, practical advice and support for cancer patients)
Tel: 020 7696 9003
Website: www. cancerbacup.org.uk

Citizens Advice Bureau (for free, confidential, impartial and independent advice)
(Addresses in phone book for local branches)
Website: www. nacab.org.uk

Cruse Bereavement Care (offers information and advice to anyone who has been affected by death)
Tel: 0870 167 1677 (Daytime helpline)
Website: crusebereavementcare.org.uk

International Stress Management Association UK (ISMA^{UK}) (for details of stress management professionals)
Tel: 07000 780 430
Website: www.isma.org.uk

MIND (Mental health charity)
Website: mind.org.uk (check website for local numbers)

Relate (branches throughout the country offer relationship counselling)
Tel: 01788 573 241 (and check local phone book)
Website: relate.org.uk

Samaritans (24-hour listening and befriending service for the depressed and suicidal)
Helpline: 0345 909 090
Website: www.samaritans.org.uk

UK Council for Psychotherapy (for a list of psychotherapists)
Tel: 0207 436 3002
Website: www.psychotherapy.org.uk

Victim Support (supporting victims of violence)
Tel: 0845 3030 900
Website: www.victimsupport.org

Westminster Pastoral Foundation (for counselling support)
Tel: 020 7361 4800
Website: www.wpt.org.uk

Chapter 11 – Personal Stress Management Strategies

Cousins, N, *Anatomy on an Illness as perceived by the patient: Reflections on healing and regeneration* (1981) Bantam Books

Covey, S R, *The 7 habits of highly effective people* (1990) Simon & Schuster (ISBN 0 684 85839 8)

European Agency for Safety & Health at Work, 'The state of occupational health in the European Union: Pilot Study' (September 2000) *OSH Monitoring* Issue 45

Table of Cases

Table of Statutes

Table of Statutory Instruments

Table of Statutory Instruments

Index

Q

A complete range of reference works for Health and Safety Professionals

❶ Order Details

(Qty)	Title	ISBN	Product Code	Unit Price	Total
	New for 2003				
	Workplace Accident Handbook	0 7545 2023 4	WAH	£60.00	
	Fire Safety Training Manual	0 7545 2183 4	FST	£60.00	
	Stress Management Handbook/Managing Stress in the Workplace	0 7545 1269 X	MSIW	£50.00	
	Risk Assessment Workbook - Retail	0 7545 1890 6	RAWR	£60.00	
	Risk Assessment Workbook - Education	0 7545 1891 4	RAWE	£60.00	
	Risk Assessment Workbook - Leisure	0 7545 1892 2	RAWL	£60.00	
	Risk Assessment Workbook - Health	0 7545 1887 6	RAWH	£60.00	
	Risk Assessment Workbook - Office	0 7545 1889 2	RAWO	£60.00	
	Risk Assessment Workbook - Manufacturing	0 7545 1888 4	RAWM	£60.00	
	Looseleafs				
	Health and Safety at Work*	0 7545 0815 3	HSWMW	£99.00	
	Environmental Law and Procedures Management*	0 7545 0811 0	ENLMW	£95.00	
	Handbooks				
	Guide to Managing Employee Health	0 7545 1886 8	MGEH02	£59.50	
	Managing Violence in the Workplace	0 7545 1967 8	MVW	£65.00	
	Corporate Killing: Manager's Guide to Legal Compliance	0 7545 1066 2	CKNL01	£50.00	
	Health and Safety at Work Handbook 15th Edition	0 7545 1751 9	HSW15	£70.00	
	Office Health and Safety 3rd edition	0 7545 1267 3	OHS3	£72.50	
	Practical Risk Assessment 4th edition	0 7545 2247 4	PRAH4	£50.00	
	Online				
	www.hsedirect.com			2 Users £444.15 one user £265.25	
				Total amount due	

❷ Confirm Delivery Details

E-mail ______________________

Name ______________________

Job Title ______________________

Company ______________________

Address ______________________

____

____

Postcode ____________ Tel ____________

❸ Make Payment Choice

LexisNexis Butterworths Tolley account, no. ______________________

or by credit card as follows: ☐ Mastercard ☐ Amex ☐ Visa

My credit card no. is ______________________

and its expiry date is ______________________

or with the attached cheque for £ ______________________
(made payable to **LexisNexis Butterworths Tolley**)

My VAT registration no. is ______________________

☐ Please supply the above product(s) with a 21 day money back guarantee (EU countries only)

Signature/Date ______________________

+We are pleased to be able offer you these titles on an annual subscription basis. Simply tick the appropriate box below.

☐ Please send me the above title(s). I understand that I will be sent all future editions automatically on publication, with a 10% discount. I am at liberty to cancel this arrangement at any time, or return the book if it does not meet with my requirements.

☐ Please send me the above title(s). I do not wish to participate in the subscription scheme, however, please advise me when a new edition becomes available.

❹ Return Your Order

Freepost
Paul Holliday Marketing Department Butterworths Tolley
FREEPOST 6983, London WC2A 1BR

Phone
Please phone Customer Services on: +44 (0)20 8662 2000

Fax
Please fax your order to: +44 (0)20 8662 2012

Online Bookshop
Visit our Online Catalogue and Bookshop at www.lexisnexis.co.uk to order securely online.

Bookshop
You can also order copies via our specialist bookshop which stocks a wide range of our titles. It is based at 35 Chancery Lane, London WC2A 1EL, nearest tubes are Chancery Lane and Temple.
www.lexisnexis.co.uk

☐ If you do NOT wish to be kept informed by mail of other LexisNexis products and services, please tick here.

☐ If you do NOT wish your mailing details to be passed on to companies approved by LexisNexis, to keep you Informed of their products and services, please tick here.

* Pay As You Go (PAYG)

I understand that I will continue to receive, until countermanded, future updating material issued in connection with this work and will be invoiced for these issues when I receive them.

21 Day Money Back Guarantee

The book(s)/product(s) mentioned on this order form, are available in EU countries only with 21 days money back guarantee. This allows you to try it/them out and assess its/their usefulness before buying. If you should decide not to purchase the book(s) /product(s), simply return it/them to us in good condition and your invoice will be cancelled and any payment made refunded. All orders placed with us are subject to our standard terms of trade, available on request and with the invoice. The book(s)/product(s) detailed on this order form is/are intended to be (a) general reference guide(s) to the law and cannot be (a) substitute(s) for legal advice.

www.lexisnexis.co.uk

Thank you for your order, prices are subject to change without notice.

LexisNexis™ UK

35 Chancery Lane London WC2A 1EL A division of Reed Elsevier (UK) Ltd
Registered office 25 Victoria Street London SW1H 0EX
Registered in England number 2746621 VAT Registered No.GB 730 8595 20